Particles on Surfaces 3

Detection, Adhesion, and Removal

Edited by
K. L. Mittal
IBM US Technical Education
Thornwood, New York

PLENUM PRESS • NEW YORK AND LONDON

AF770957

Library of Congress Cataloging-in-Publication Data

Symposium on Particles on Surfaces: Detection, Adhesion, and Removal (3rd : 1990 : San Diego, Calif.)
Particles on surfaces 3 : detection, adhesion, and removal / edited by K.L. Mittal.
p. cm.
"Proceedings of the Third Symposium on Particles on Surfaces: Detection, Adhesion, and Removal, held as part of the Twenty-first Annual Meeting of the Fine Particle Society, August 21-25, 1990, in San Diego, California"--T.p. verso.
Includes bibliographical references and index.
ISBN 0-306-44180-2
1. Particles--Congresses. 2. Surfaces (Technology)--Congresses. I. Mittal, K. L., 1945- . II. Fine Particle Society. III. Title.
TA418.78.S96 1990
620'.44--dc20 91-39707
CIP

Proceedings of the Third Symposium on Particles on Surfaces: Detection, Adhesion, and Removal, held as part of the Twenty-First Annual Meeting of the Fine Particle Society, August 21–25, 1990, in San Diego, California

ISBN 0-306-44180-2

© 1991 Plenum Press, New York
A Division of Plenum Publishing Corporation
233 Spring Street, New York, N.Y. 10013

All rights reserved

No part of this book may be reproduced, stored in a retrieval system, or transmitted in any form or by any means, electronic, mechanical, photocopying, microfilming, recording, or otherwise, without written permission from the Publisher

Printed in the United States of America

PREFACE

This volume chronicles the proceedings of the Third Symposium on Particles on Surfaces: Detection, Adhesion and Removal held as a part of the 21st Annual Meeting of the Fine Particle Society in San Diego, California, August 21-25, 1990. The first two symposia in this series were held in 1986 and 1988, respectively, and have been properly documented[1,2]. Like its antecedents the Third symposium was very well received, and the continuing success of these symposia reinforced our earlier belief that regular symposia on the topic of particles on surfaces were very much needed. Concomitantly, the fourth symposium in this series is planned in Las Vegas, July 13-17, 1992.

As pointed out in the Preface to the earlier two volumes[1,2], the topic of particles on surfaces is of tremendous interest and concern in a wide spectrum of technological areas. The objectives of the Third symposium were essentially the same as those of the earlier two and our intent here was to provide an update on the research and development activities in the world of particles on surfaces. Apropos, there has been a deliberate attempt every time to seek out new people to present their research results and we have been very successful in this mission.

The symposium program comprised 48 papers dealing with many and varied aspects of particles on surfaces. The subtopics ranged from quite fundamental to very applied. The interest level by the attendees was very high and many of them expressed that a good deal of up-to-date and technologically useful material was covered in this symposium.

Now turning to this volume, it contains a total of 24 papers divided into three parts as follows: Part I. Particle-Surface Interactions, Adhesion and General Papers; Part II. Particle Detection, Identification, Analysis and Characterization; and Part III. Particle Reduction and Removal. The topics covered include: relevance of surface energetics in departiculation; modelling particle accumulation on a surface; release of particles during spaceflight; identification and analysis of particles on different surfaces by a number of techniques, e.g., light scattering, infrared micro-profiling, and total reflection X-ray fluorescence spectrometry; particulate and defect reduction strategies for semiconductor devices; particle protection of semiconductor surfaces; and various techniques to remove particles from surfaces including ultrasonic, cryogenic, UV laser, and solvent drying.

It must be recorded here that, as in the past, all manuscripts were rigorously reviewed and all were revised (some quite extensively) before final acceptance and inclusion in this volume, and some could not pass muster. In other words, this book is not merely a collection of as-received manuscripts, rather it represents a set of papers which have been sifted by fellow researchers.

This and earlier two volumes[1,2] contain bountiful information on the various aspects of particles on surfaces, and I sincerely hope that anyone interested in this topic will find this latest volume a useful source to learn about recent R&D activities and future trends.

Acknowledgements: First, I am thankful to the Fine Particle Society for sponsoring this event. My thanks are due to the appropriate management of IBM Corporation for allowing me to organize this symposium and to edit this volume. My appreciation is extended to my wife, Usha, for her unflinching help in more ways than one. Peer reviews are sine qua non to maintain the high standard of publications, so my sincere thanks are extended to the reviewers for their many valuable comments. Last, but not least, the cooperation, sustained interest, and contribution of the authors is gratefuly acknowledged.

K.L. Mittal

IBM U.S. Technical Education
500 Columbus Avenue
Thornwood, NY 10594

RERERENCES

1. K.L. Mittal, Editor, Particles on Surfaces 1: Detection, Adhesion and Removal, Plenum Press, New York, 1988.

2. K.L. Mittal, Editor, Particles on Surfaces 2: Detection, Adhesion and Removal, Plenum Press, New York, 1989.

CONTENTS

PART I. PARTICLE-SURFACE INTERACTIONS, ADHESION AND GENERAL PAPERS

PART II. PARTICLE DETECTION, IDENTIFICATION, ANALYSIS AND CHARACTERIZATION

PART I. PARTICLE-SURFACE INTERACTIONS, ADHESION AND GENERAL PAPERS

RELEVANCE OF SURFACE ENERGETICS TO DEPARTICULATION OF DISK DRIVE SUBSTRATES

R. Nagarajan

IBM Corporation, MS 912/502
5600 Cottle Road
San Jose, CA 95193

A rational correlation has previously been established between the particulate removability of disk drive component materials in an aqueous ultrasonic system, and their Hamaker constants modified to reflect their non-elastic deformability. This work expands the scope of both the theory and experiments. A less-restrictive parameter, the surface energy, is used here to correlate with cleanability in a variety of cleaning systems, including water-based and solvent-based ultrasonics, ultrasonics in concert with low-pressure spray, centrifugation, and high pressure water jetting. The fundamental differences in the particle removal characteristics of these systems are highlighted. The influence of the extraction-based measurement procedure on estimated substrate cleanability is pointed out. It is concluded that for non-erodible materials, or for erodible materials cleaned with a non-erosive mechanism, cleanability decreases as surface energy increases. Two exceptions are noted: in ultrasonic cleaning, the tendency of the material to erode limits its cleanability, even if the substrate has very low surface energy; also, some state-of-the-art cleaning methods, such as high pressure water spray, display departiculation efficiencies that are virtually independent of surface energy. Implications of this study for future solvent-free cleaning of disk drive components are discussed throughout the paper.

INTRODUCTION

In a previous publication[1], we had introduced the concept of surface cleanliness measurement by extracting the substrate ultrasonically in a liquid, followed by quantitative determination of extracted particulate contamination in the liquid. A turbidimeter or a liquid particle counter was used for the latter part of the analysis. Two cleanability parameters were defined in that work. The first, the maximum cleaning potential, represents the cleanliness of a surface subjected to an exhaustive multistage ultrasonic cleaning procedure. The second, the initial cleanability, is a measure of what fraction of the cleaning action occurs in the very first stage of extraction. These two parameters, considered fundamental to surface cleaning, were then correlated to substrate/particle interfacial properties such as the Hamaker constant, as well as the bulk hardness of the substrate material. Good correlation was obtained for various materials when an appropriate ratio of these two terms was used as the correlating factor.

Particles on Surfaces 3, Edited by K.L. Mittal
Plenum Press, New York, 1991

While the importance of the Hamaker constant should come as no surprise in studying particle/substrate adhesion phenomena dominated by van der Waals attraction forces, the incorporation of the hardness factor into the model was an attempt to account for the deformability of the substrate and particle materials (which were taken to be the same). Several simplifying assumptions were made in order to arrive at the proportionality:

$$F \propto \frac{A^2}{H} \tag{1}$$

where F represents the adhesion force, A the Hamaker constant of the system, and H the bulk hardness of the solid material involved. To obtain this simple, yet rational, relationship, it was assumed that adhesion was predominantly due to intermolecular forces of attraction; that the spherical particle was non-deformable; that the adhesion process was not time-dependent; and that the elastic deformation term was negligible in comparison to the non-elastic component.

The follow-on work reported in this paper is an effort to refine the correlation model by eliminating some of the more restrictive assumptions. Two facts became apparent toward the conclusion of the previous work: Bulk hardness does not directly affect surface/particle adhesion phenomena, which are, unless large impaction velocities are involved, governed by interfacial interactions. Besides, the Hamaker constant is not easily measurable. Tabor[2] has carried out a thought experiment in which two bodies (a sphere and a flat) are taken from separation Z_0 to infinity. Calculation of the work done to create two new surfaces in place of the interface yields a very useful relationship linking the Hamaker constant to surface energies:

$$A = 12 \Pi Z_0^2 (\gamma_1 + \gamma_2 - \gamma_{12}) \tag{2}$$

where γ_1, γ_2, and γ_{12} are the surface energies of the sphere, the flat, and the interface, respectively; here, surface energy (or surface tension) of solids represents not only resistance to stretching due to Gibbs surface free energy, but also due to elastic forces within the solid. The analysis is restated here in terms of the more-easily determinable surface energy, and more recent data on surface cleaning are used to re-establish an even more useful correlation. Of course, the surface energy of solids is more difficult to measure than that of liquids[3] because of the ability of solids to support shear stresses, but among the advantages of using surface energy as the correlation parameter are that interfacial deformations can be accounted for in a fundamentally sounder way, and that surface roughness can easily be incorporated in the model as an additional influence. Sources of surface energy data for metals include the International Critical Tables[4], the Mechanical Engineer's Reference Book edited by Parrish[5], the Metals Handbook (Desk Edition) edited by Boyer and Gall[6], and other miscellaneous literature references. Surface energy of polymeric materials may be conveniently measured in a laboratory setting, as will be described later.

The previous investigation was limited to ultrasonic cleaning, and all cleanability parameters reported were obtained by the procedure of multiple ultrasonic extraction. This extended study encompasses various other cleaning and extraction methods for particulate contamination, including low pressure water spray, high pressure water spray, and centrifugation. These processes are influenced quite differently by the surface energy of the substrate, and this will be demonstrated by results presented here.

SURVEY OF RELEVANT LITERATURE

Tabor[2] has examined surface forces and interactions from the viewpoint of an experimental physicist. Among the interesting observations in that paper, the following are noteworthy: With simple pure metals, the contribution of van der Waals forces to the surface energy is only about one-tenth of that due to the metallic bond itself. In the case of a hard spherical particle in contact with a flat surface of the same material, their mutual attractive force, F, is given by:

$$F = 4 \Pi R_p \gamma \tag{3}$$

where R_p represents the radius of the sphere, and γ represents the surface energy of the sphere and the flat. In the case of deformable solids, 3 replaces 4 in this equation. Thus, in theory, a surface with lower surface energy is inherently more cleanable. In practice, however, this is not necessarily always the case. For example, particle removal from a surface coated with zinc stearate is generally easier to accomplish than from one coated with Teflon, even though their surface energies are comparable. A rougher surface is also theoretically more cleanable, as adhesion diminishes, according to Tabor, with increasing surface roughness. Again, this is not always encountered in practice, particularly when the roughness elements are such that the particle can nestle within the surface asperity; in this case the increased area available for intimate contact becomes the predominant factor. For instance, it is common practice in the paint industry to roughen the substrate before applying paint in order to obtain improved adhesion. The critical quantities here are the mean deviation of the asperity heights, σ, and the average radius, R, of the asperity tips. Adhesion of a particle to a substrate depends on the ratio of σ to the extension which an asperity junction can sustain before loss of adhesion. For an elastic solid of modulus E, and Poisson's ratio of 1/2, an adhesion parameter, θ, may be defined:

$$\theta \equiv \frac{E \sigma^{3/2} R^{1/2}}{2 R_p \gamma} \tag{4}$$

In general, for elastic solids, if θ exceeds a value of about 10, adhesion will be very small, and the surface should lend itself to precision cleaning.

Muller and Yushchenko[7] followed up Tabor's work with a more rigorous analysis of pressure distribution in and around the particle/substrate contact zone, particle shape in the vicinity of the contact, etc. Johnson et al.[8] have discussed the influence of surface energy on the contact between elastic solids, supporting their theory with experiments carried out using rubber and gelatin spheres. They concluded that the force of adhesion was independent of the contact radius, and, hence, of the elastic modulus. Chang et al.[9] have presented an improved model for the adhesion of contacting metallic rough surfaces. Their conclusion is that adhesion is quite large for clean surfaces even when they are relatively rough; however, roughness, in combination with surface contamination, causes a significant decrease in adhesion. Rabinowicz[10] suggests that the adhesional wear coefficient for sliding surfaces is influenced by the ratio of the interfacial energy of adhesion to the sum of the surface energies of the two contacting materials; Warren and Wert[11] have recently analyzed adhesive wear on the same basis. It is likely that the erosive wear of substrates is also affected by their surface energies, a topic dealt with later in this paper.

Ramharack and Nguyen[12] have measured contact angles with a Gaertner goniometer. They use the Zisman concept that by plotting the cosine of the contact angle versus the surface tension of a series of test liquids, the critical surface tension of the solid can be found by extrapolation to zero contact angle (i.e., complete wetting). For low-energy materials, e.g., polymers, surface energy can thus be determined from contact angle measurements. Mittal[13] has authored an excellent review paper dealing with the relationships between wettability, surface energy, and adhesion behavior of polymeric materials. While the use of contact angles as a means of determining surface energy is a simple measurement to make, it can be a complex one to interpret. Among the potential pitfalls to be borne in mind are the effects on the measurements of the saturation vapor pressure of the liquid on the surface, and of the solid-liquid interfacial energy.

Adhesion phenomena closely associated with industrial hygiene and air pollution have been studied by Corn[14]. Matsunaga[15] specifically addresses the measurement of surface contamination by organics using a contact angle measurement technique. He explicitly states that surface energy has a relation to surface contamination rates, and, by extension, to surface decontamination efficiency. The surface energy of most solids is much higher than that of liquids, and the lowering of free energy by contaminant chemicals is correspondingly much greater. In Matsunaga's experiments, surface energies of some metallic oxides were measured, and related to their surface (film-type) contamination. Thus, there is clearly a precedent for relating surface cleanliness, at least from a chemical contamination viewpoint, to measured contact angles. Here, a similar, but different, application of the same concept is attempted, where measured surface energies of substrates are related to their cleanability with respect to surface-adhered particles. The existence of such a relationship, intuitively obvious but not demonstrated at this time for sonic or spray cleaning of disk drive components, will enable the selection of materials and surface textures to be made on the basis of cleanability. The establishment and exploration of such a correlation constitutes the principal portion of the remainder of this paper.

RESULTS AND DISCUSSION

Aqueous Ultrasonic Cleaning

Surface energies for commonly-occurring types of materials in disk drives are presented in Table I. These were measured for representative parts using the contact angle approach at a consulting laboratory, and are used here to correlate with parameters governing particle removal by various mechanisms. For ultrasonic cleaning, two parameters were defined by Nagarajan[1]: the maximum cleaning potential (MCP), defined as:

$$\text{MCP} \equiv 1 - \frac{\text{Asymptotic turbidity/LPC count}}{\text{Initial turbidity/LPC count}} \tag{5}$$

and (initial) surface cleanability (SC), defined as:

$$\text{SC} \equiv \frac{\text{Initial turbidity/LPC} - \text{First stage turbidity/LPC}}{\text{Initial turbidity/LPC} - \text{Asymptotic turbidity/LPC}} \tag{6}$$

where turbidity and liquid particle count (LPC) are measures of ultrasonically extracted particulate contamination in the extraction medium, namely a deionized water-surfactant solution. The liquid particle counter used here can detect particles (and agglomerates) in the 5 to 100 μm size range with acceptable resolution. The cumulative LPC counts reported correspond to particles larger than 5 μm. The term 'asymptotic' refers to behavior observed when a non-erodible material is subjected to repeated ultrasonic extraction in a multi-stage environment where the medium is replaced with a fresh supply prior to each extraction. Less and less material is removed in every successive extraction, until the extracted count becomes indistinguishable from the 'noise' in the measurement, which happens to be particles generated by cavitation erosion of the substrate surface[16]. Clearly, an ultrasonic-based cleaner cannot clean the part any better than this state, and a part close to this level of cleanliness cannot be measured using ultrasonic

Table I. Surface Energies for Typical Disk Drive Component Materials

Material Type	Surface Energy (ergs/cm^2)
Silicone rubbers	24
Polycarbonate plastics	32
Epoxy coatings	39
Soft aluminum alloys	50
Hard aluminum alloys	65
Stainless steel alloys	87
Nickel-plated metals	106

extraction. This limiting condition, then, represents best-can-do ultrasonic cleaning. Initial turbidity/LPC counts represent the as-received state of cleanliness of the part, and the 'first stage' numbers correspond to one exposure to ultrasonic extraction. MCP and SC, then, respectively represent maximum-achievable and easily-achievable cleanliness levels for a substrate that is ultrasonically cleaned.

In Figures 1 and 2, respectively, these are plotted against surface energy. The trend may be interpreted to mean that as surface energy increases, there is an accompanying increase in ultimate cleanability of the surface when ultrasonic cleaning is employed. Stated differently, the high-energy surfaces used in disk drives, e.g., stainless-steels, nickel-plated metals, etc., appear to be more cleanable because they are harder, and therefore less erodible. Lower-energy surfaces, such as elastomers, plastics, and epoxy-coated metals, have their cleanliness degraded to varying extents under prolonged exposure to cavitation. Although this result, that low-energy surfaces are more difficult to clean than high-energy surfaces, may appear to contradict classical adhesion theory, it does not in reality. Surface erosion associated with ultrasonic scrubbing tends to dominate in the asymptotic regime discussed earlier, and leads to counter-intuitive particle removal characteristics. An ideal surface, then, for ultrasonic cleaning would be a low-energy material which is also resistant to cavitation erosion. The use of corona discharge to increase the surface energy of several elastomeric coatings has been investigated by Lawson[17]; the increase in surface energy, while detrimental from a particle-adhesion perspective, could help mitigate cavitation-erosion. Rose and Liston[18] have used a cold gas plasma for surface modification of polymers; use of electron beam irradiation for the same purpose has been studied by Kim et al.[19]. Pacansky and Waltman[20] have succeeded in lowering the surface energy of coatings via radiation curing. Oxidation, nitration, hydrolization, amination, sulfonation, cyclization, chlorination, flame treatment, special primers, and mechanical treatments have been tried as well. Oxygen containing groups have been found to be most effective at increasing surface energy. Ultrasonic cleaning will continue to be the popular method of choice for metal cleaning, especially in the wake of the decision to phase-out chlorinated solvents. Thus the development of coatings and treatments that promote particle removal, while simultaneously resisting cavitation-induced particle generation, will be of great importance in future.

It can be seen from Figure 2 that there is a non-monotonic relationship between surface energy and initial cleanability (ultrasonic) of surfaces. Plastics, rubbers and epoxy, which have surface energies below 50 ergs/cm^2, are characterized by very low van der Waals forces of adhesion; for these materials (the elastomers, especially), the reduced adhesion dominates over cavitation-induced erosion corresponding to limited ultrasonic exposure. Thus net cleanliness achieved, which may be defined as particles removed minus particles generated, is excellent with ultrasonic cleaning. It should be borne in mind that elastomeric surfaces can be very difficult to clean if they have a tackiness (or stickiness) to them, primarily arising from incomplete cure or inadequate cross-linking of

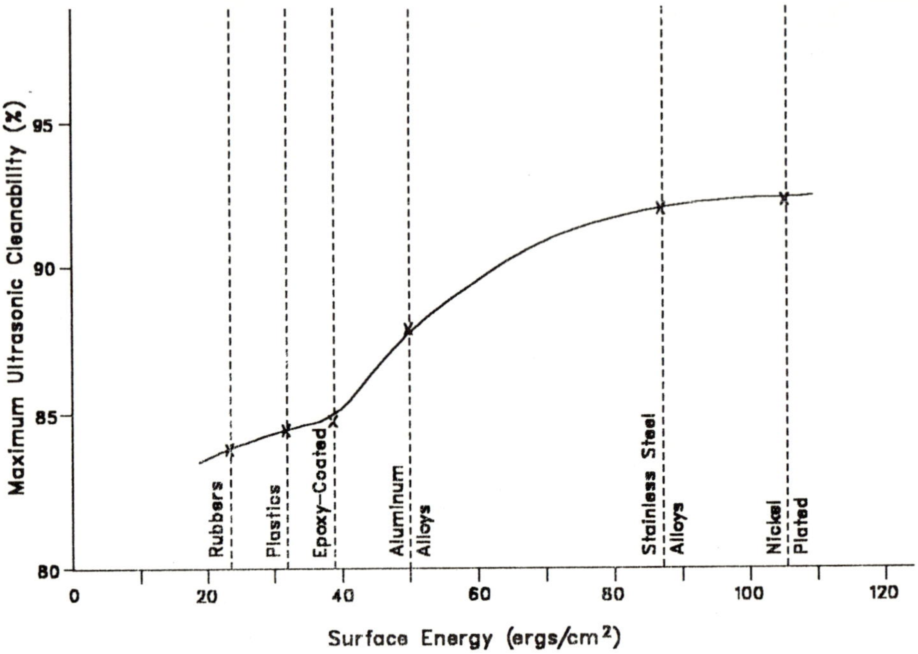

Figure 1. Effect of surface energy on the multiple ultrasonic cleaning limit for various materials.

monomers during polymerization. Such chemistry-dependent effects on adhesion are neglected in this analysis, and are not manifested in the data collected. The implicit assumption here, that all surfaces (metallic or polymeric) are fully chemically saturated, appears to be quite valid. With increasing surface energy, particles become more difficult to remove, while at the same time erosion still persists up to a surface energy of about 50 ergs/cm^2. Thus, certain epoxy coatings and soft aluminum alloys are subject to a double jeopardy, and constitute the minimum in the ultrasonic cleaning efficiency curve (Figure 2). Harder materials, such as stainless steel and nickel-plated substrates, do not erode appreciably, and therefore appear to have net cleaning efficiencies comparable to those for polymers. These trends are peculiar to ultrasonic cleaning, as will soon become apparent.

Freon Ultrasonic Cleaning

That the medium of cleaning, or extraction, does not influence the qualitative characteristics of these results is demonstrated in Figure 3. The current in-house cleaner at IBM San Jose for the 3390 disk drive program is a multi-stage Freon ultrasonic system (to be replaced by an aqueous ultrasonic/spray system in 1991). Parts cleanliness is tracked before and after in-house cleaning as part of a statistical process control strategy. This data-base is also extremely valuable for more fundamental studies, such as this one. The period of data-collection extends back more than a year for some parts. Data are collected in the form of turbidity or liquidborne particle count, depending on the material the part is made of, as well as its criticality from a functional standpoint. Percent cleaning efficiencies reported in Figure 3 were obtained by averaging cleanliness values for various parts before- and after-cleaning, grouping the parts by material type, and further averaging these numbers by material type. The shape of the surface energy-dependence is similar to that obtained as the initial cleaning curve (Figure 2), suggesting that the in-house cleaner, although it has several ultrasonic stages, is not

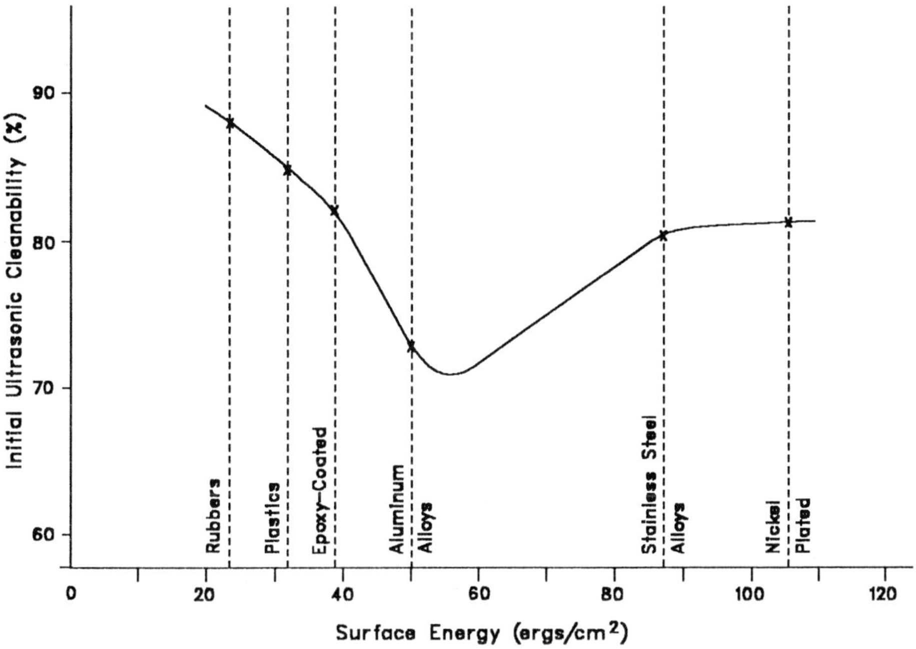

Figure 2. Effect of surface energy on the initial ultrasonic cleanability of various materials.

cleaning much better than a single-stage cleaner. Of course, Figure 3 also indicates that the qualitative relationship between ultrasonic cleaning efficiency and surface energy is independent of the medium in which the cleaning occurs. In quantitative terms, water has repeatedly been shown (e.g., by Ko[21]) to be more effective at removing particles from surfaces than solvents such as Freon.

Aqueous Ultrasonic/Spray Cleaning

The incorporation of a low-pressure spray wash/rinse following the ultrasonic stage in a cleaning system appears to improve its capability for particle removal by nearly an order of magnitude. There are two reasons for this: for erodible parts, the spray can rinse away debris generated by cavitation in the ultrasonic stage; in addition, for all parts, spray cleaning will remove larger particles than can be removed ultrasonically. Spray cleaning is intrinsically less sensitive to surface morphology than sonic cleaning[22], relying almost exclusively on fluid dynamics to impart a shear force to surface-adhered particles. This cleaning mechanism was studied earlier by Bhattacharya and Mittal[23]. The shear force still has to exceed the adhesion force in order for the particle to be dislodged from the surface, but energy is typically not delivered to the surface in a way that causes particle generation. The in-house cleaner for a future disk-drive program incorporates an ultrasonic stage followed by a spray wash/rinse, followed by hot-air dry. Similar cleaners will also replace current solvent-based in-house cleaners for other file programs within the next year. As part of system qualification and acceptance studies, several parts were run through this aqueous ultrasonic/spray cleaner, and cleaning efficiencies were estimated. These are shown for various material types as a function of surface energy in Figure 4. These results are more in line with expectations based on adhesion theory, i.e., the lower the surface energy, the greater the cleanability of the surface. The low-pressure (50 psi) spray apparently negates the perturbation in the observed trend due to erosion, and makes the data seem more rational.

It appears that this system is excellent for cleaning polymeric and light-metal parts. Harder substrates, which typically are machined and therefore are likely to be heavily contaminated with metallic chips and cutting fluids, are not cleaned as well by this system, but the cleaner does exhibit a lower performance limit of about 50%. This asymptotic behavior is, in many ways, a desirable feature for an in-house cleaner, which is expected to serve as a universal cleaner for all parts. It is not optimized for any single part, and clearly has its limitations in removing particles from machined metallic surfaces, but will not display much part-to-part variation in the case of metals and hard alloys. This would imply that parts such as large machined castings ought to be cleaned to the extent necessary (to the state-of-the-art, if the cleanliness of the part is deemed critical to the reliability of the product) at the supplier, and merely 'dusted-off' (to remove packaging debris, and other loose environmental contamination) in the in-house cleaner. The results also suggest that the presence of a hard, adherent coating on the substrate surface can prove to be a great 'equalizer'.

High Pressure Spray/Centrifugal Cleaning

Following a train-of-thought from the previous paragraph, a cleaner to be installed at a part supplier can be part-specific, i.e., designed to optimally clean one or two parts. Two systems, both incorporating high pressure spray, have been investigated thoroughly towards this end, and detailed results have been published elsewhere[22]. The first, centrifugal cleaning, is applicable to rotationally-symmetric substrates; the other, high pressure spray cleaning, is suited to large machined castings that can be held firmly in place with a fixture. In centrifugal cleaning, the part is spun about its center of gravity, creating coriolis and centrifugal accelerations. When the part is immersed in a liquid medium, these are converted to shearing forces on the contaminant particles. Low rpm wash, high rpm rinse, and higher rpm dry typically constitute the cleaning cycle in a spin-rinser/dryer, an equipment that has found extensive application in cleaning disk substrates, and is more recently being tested on other disk drive components[22]. Preliminary data, presented in Figure 5, indicate that the estimated cleaning efficiency of this

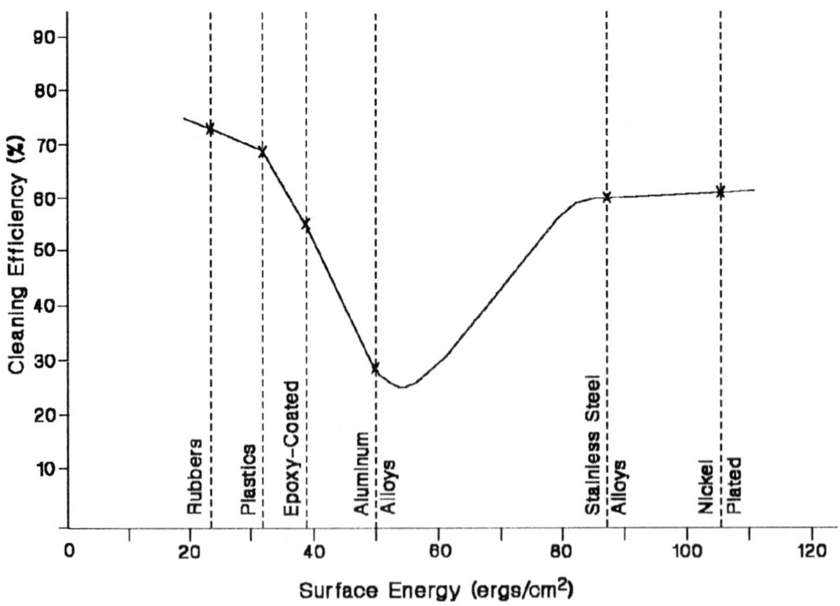

Figure 3. Effect of surface energy on the cleanability of various materials in a multistage Freon ultrasonic system.

method, based on ultrasonic extraction followed by liquidborne particle counting, is strongly coupled to surface energy, reaching a minimum of about 40% for soft aluminum alloys. No stainless steel or nickel-plated parts were tested as part of this study, and therefore the curve cannot be extrapolated beyond a surface energy of about 70 ergs/cm^2, corresponding to a relatively hard aluminum alloy. It was suspected in this case that the cleaning method was getting parts so clean that the method of measurement, involving ultrasonic extraction, was unduly influencing the final result. To verify this, the soft aluminum substrate was extracted with a low-pressure spray before and after centrifugal cleaning (using different samples from the same lot), and cleaning efficiency was calculated based on particle counts in the extract liquid. With spray extraction, the cleaning efficiency was determined to be close to 80%, instead of the 40% obtained by ultrasonic extraction. Thus the extraction procedure does significantly influence the measurement of the cleanability of a substrate, particularly if the surface happens to be abradable. Unfortunately, in the case of disk drive component cleaning and cleanliness measurement, surface topography is typically so complex that direct inspection becomes virtually impossible. High pressure sprays also exert shear forces on insoluble surface contaminants, with the spray pressure, volumetric flow rate, nozzle configuration and placement being the parameters that govern cleaning effectiveness[22]. Various non-polymeric substrates were high-pressure (> 2500 psi) water sprayed in the laboratory, and their respective cleanabilities were determined using both ultrasonic and low-pressure spray extraction. The results are shown in Figure 6. Data obtained by ultrasonic-extraction suggest an increase in cleaning efficiency with increasing surface energy of the part. This trend, reminiscent of the multiple ultrasonic cleaning trend reported in Figure 1, is actually indicative of the fact that high-pressure sprayed soft-metal (e.g., aluminum alloys) parts are so clean that they erode with a single application of ultrasonic energy, such as that employed in ultrasonic extraction as part of the cleanliness measurement procedure. Surface examination has revealed that high-pressure jet impingement does not, by itself, cause a metallic surface to degrade and shed particles. Again, to verify the suspicion that the extraction process itself was generating particles, and therefore

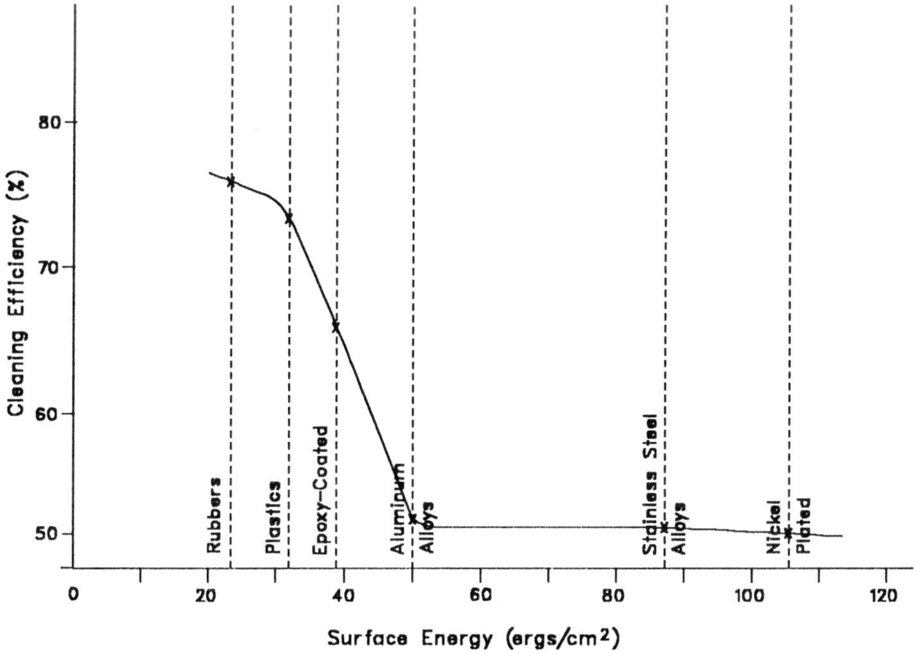

Figure 4. Effect of surface energy on the cleanability of various materials in an aqueous ultrasonic/spray system.

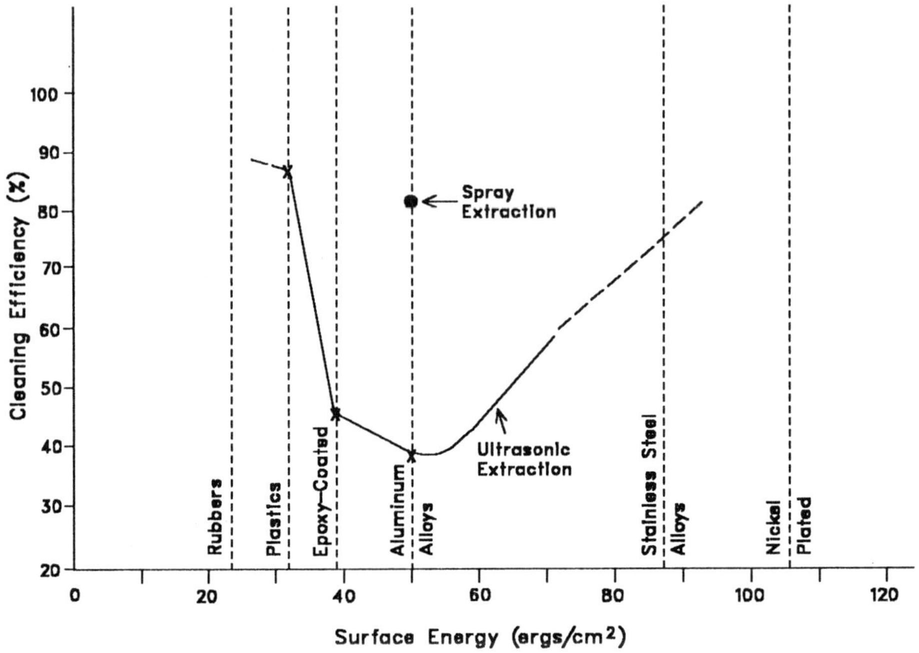

Figure 5. Effect of surface energy on the cleanability of various materials in a high-pressure spin-rinser/dryer.

skewing the data, all cleaned parts were extracted with a low-pressure spray, followed by an identical LPC measurement. The dashed line, which represents cleaning efficiency data obtained by spray extracting, reveals that high-pressure water exerts so powerful a force that particle adhesion forces to the substrate become insignificant in comparison. All surfaces are cleaned to nearly zero particulate contamination levels (> 5 μm) by high-pressure water, provided the surface lends itself to the technique. i.e., it must be fairly large, it cannot be polymeric, and it must be fixturable.

CONCLUSIONS AND FUTURE WORK

The surface energy of a substrate to be cleaned has been demonstrated to have a large bearing on its cleanability (with respect to particulate contamination) in a variety of cleaning systems. The relatively low surface-energy materials included in this analysis were typically polymers and soft-metals that tended to erode readily when exposed to sonic cavitation. This resulted in somewhat surprising dependences between cleanability and surface energy whenever ultrasonics was involved, either in cleaning, or as part of the extraction-measurement procedure. The intrusive effect of the ultrasonic extraction preceding particle counting or turbidity measurement in the extract liquid can be negated by employing low-pressure spray extraction instead. Surface cleanability, when estimated via spray rinsing of the substrate, correlated with surface energy in more recognizable, rational terms. Some state-of-the-art cleaning methods, such as high pressure water spray, yielded parts so uniformly clean that surface energy became a negligible factor; however, such methods cannot be applied to all substrates, and can be fairly expensive. Ultrasonic cleaning followed by low-pressure spray rinsing to remove any leftover or erosion debris also resulted in constant cleaning efficiencies across a surface energy range that encompassed most metallic and alloy materials. These results suggest that the latter system is an ideal general-purpose cleaner to clean a multitude of parts, whereas the former is best suited to specialized cleaning applications.

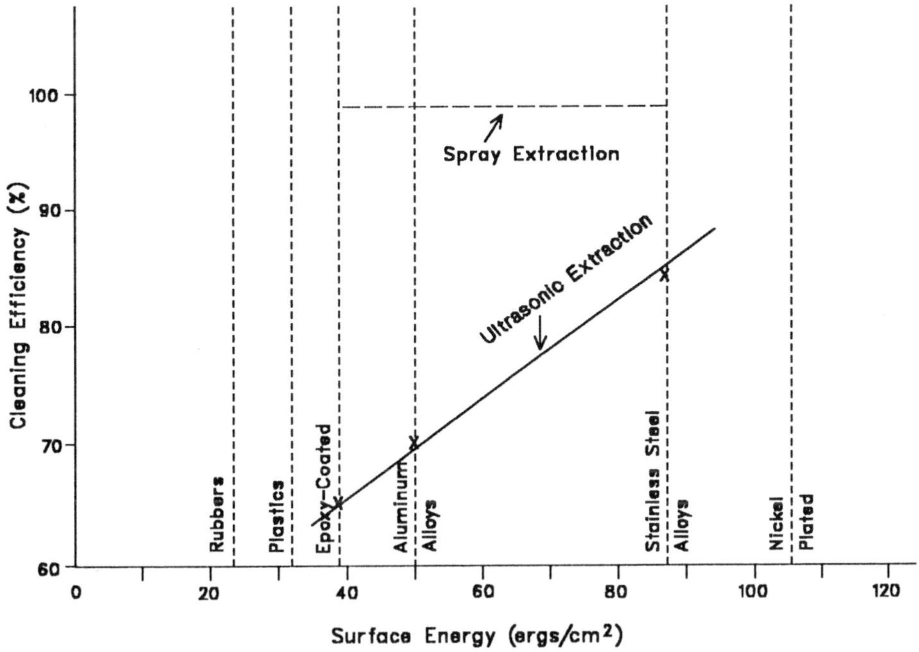

Figure 6. Effect of surface energy on the cleanability of various materials in a high-pressure water sprayer.

This work primarily used data obtained in a production environment. More controlled experiments are underway that are designed to establish fundamental relationships between surface finish/surface polish/surface coatings and surface energy (determined by means of contact angle measurements), and thereafter, between surface energy and cleanability. Flat test panels will be used for these tests so that cleanability may be estimated by direct surface inspection; this would eliminate all external effects associated with any energetic extraction. Substrates to be studied would include epoxy-coated panels where the epoxy paint would be cured under various (time/temperature) conditions, aluminum alloys prepared with different well-characterized roughness elements, stainless steel alloys of various compositions of interest, and other materials with surface characteristics commonly encountered in disk drives. The ultimate objective of this effort is to develop a design criterion for selection of materials based on their cleanability, estimated via their surface energy; and to develop a cleaning strategy for various substrates on the basis of the specific relationship between surface energy and surface particle removal using the method chosen. This systematic approach to parts design and manufacture will result in higher product yields, and greater field reliability of the assembled product.

ACKNOWLEDGMENTS

The author is grateful to Ed Gipstein and Ron Weaver for many helpful observations, and to Kash Mittal for suggesting this course of investigation. Marion Clark and Sylvia Flores provided exemplary assistance in data collection.

REFERENCES

1. R. Nagarajan, Surface properties and cleanability: A rational correlation, J. IES, 33, No. 1, 26-33 (January-February 1990).

2. D. Tabor, Surface forces and surface interactions, J. Colloid Interface Sci., 58, No. 1, 2-13 (1977).

3. E. U. Condon, and H. Odishaw, Editors, "Handbook of Physics," 2nd edition, pp. 5-94 to 5-95, McGraw-Hill, New York (1967).

4. E. W. Washburn, Editor, "International Critical Tables of Numerical Data, Physics, Chemistry, and Technology," McGraw-Hill, New York (1926).

5. A. Parrish, Editor, "Mechanical Engineer's Reference Book," CRC Press, Cleveland, OH (1973).

6. H. E. Boyer, and T. L. Gall, Editors, "Metals Handbook: Desk Edition," American Society of Metals, Metals Park, OH (1985).

7. V. M. Muller, and V. S. Yushchenko, Influence of surface forces on deformation and adhesion of elastic particles to a rigid substrate or to each other, Kolloidnyi Zhurnal, 52, No. 3, 500-510 (May-June 1980).

8. K. L. Johnson, K. Kendall and A. D. Roberts, Surface energy and the contact of elastic solids, Proc. Royal Soc. London A, 324, 301-313 (1971).

9. W. R. Chang, I. Etsion and D. B. Bogy, Adhesion model for metallic rough surfaces, Trans. ASME J. Tribology, 110, 50-56 (January 1988).

10. E. Rabinowicz, The dependence of the adhesive wear coefficient on the surface energy of adhesion, in "Proc. Int. Conf. on Wear of Materials," pp. 36-46, St. Louis, MO, publ. by ASME, New York (1977).

11. C. D. Warren and J. J. Wert, A model for adhesive wear, J. Adhesion Sci. Technol., 4, 177-196 (1990).

12. R. Ramharack and T. H. Nguyen, Fluoropolymers of very low surface energies, J. Polym. Sci.: Polym. Lett., 25, 93-98 (1987).

13. K. L. Mittal, Surface chemical criteria for maximum adhesion and their verification against the experimentally measured adhesive strength values, in "Adhesion Science and Technology," L. H. Lee, Editor, Part A, pp. 129-168, Plenum Press, New York (1975).

14. M. Corn, The adhesion of solid particles to solid surfaces, II. J. Air Pollution Control Assoc., 11, No. 12, 566-575 (1961).

15. T. Matsunaga, Relationship between surface energy and surface contamination, in "Surface Contamination: Genesis, Detection and Control," K. L. Mittal, Editor, Vol. 1, pp. 47-56, Plenum Press, NY (1979).

16. R. Nagarajan, Cavitation erosion in disk drive component cleaning, in preparation (1991).

17. D. F. Lawson, Corona discharge activation and reconstruction of elastomer surfaces, Rubber Chem. and Technol., 60, 102-110 (1986).

18. P. W. Rose, and E. Liston, Gas plasma technology and surface treatment of polymers prior to adhesive bonding, in PLASTICS 85, Proc. of the SPE 43rd Annual Technical Conf. and Exhibition, pp. 685-688 (1985).

19. H-C. Kim, J. H. Song, G. L. Wilkes, S. D. Smith and J. E. McGrath, Electron beam effects on polymers. II. Surface modification of bis-GMA substrates by functionalized siloxane oligomers, J. Appl. Polym. Sci., 38, 1515-1533 (1989).

20. J. Pacansky and R. J. Waltman, Low surface energy coatings on substrates prepared via radiation curing, Prog. Organic Coatings, 18, 79-87 (1990).

21. M. Ko, Comparison of Freon with water cleaning processes for disk drive parts, in "Particles on Surfaces 2: Detection, Adhesion and Removal," K. L. Mittal, Editor, pp. 283-296, Plenum Press, NY (1989).

22. R. Nagarajan, R. W. Welker and R. L. Weaver, Evaluation of aqueous cleaning techniques for disk drive parts, in "Proceedings of Microcontamination 90 Conference," Santa Clara, CA, pp. 113-131 (1990).

23. S. Bhattacharya, and K. L. Mittal, Mechanics of removing glass particulates from a solid surface, Surface Technol., 7, 413-425 (1978).

PARTICLE ADHESION TO SURFACE UNDER TURBULENT FLOW CONDITIONS

N. Vatistas

Dipartimento di Ingegneria Chimica
Chimica Industriale e Scienza dei Materiali
Università di Pisa
Via Diotisalvi 2
I-56100 Pisa, Italy

After a brief description of a proposed model of particle adhesion under turbulent flow conditions, the range of its validity is examined. The model is extended to include the experimentally observed rolling of the particles by turbulent bursts. A relationship between the sticking probability and the various parameters that affect the deposition process has been obtained.

INTRODUCTION

The particles in a fluid in turbulent flow are deposited on the wall according to the following mechanisms: the carrying of the particles to the surface, their adhesion to the surface and the re-entrainment of those particles which were only loosely adhered to the surface.

It is possible to study the first step (the carrying of the particles from the fluid to the wall surface) experimentally by coating the wall surface with an adhesive material, thus avoiding the more complex problem of adhesion and re-entrainment. Most theories of particle deposition are formulated on the assumption of perfect stickability; they are, therefore, theories of particle transport.[1,2] The non-perfect stickability has been taken into account by the introduction of the sticking probability, defined as the portion of particles reaching the wall which stay there.[3,4] The determination of the sticking probability is a complicated problem, due to the many parameters that affect it. The adhesion and re-entrainment steps are not independent and need to be studied together. They are affected by both the kinetics of adhesion and the flow dynamics in the wall region.

Experimental studies regarding the fine structure of turbulence in the wall region have revealed the presence of organized motion there. These studies make it possible to assume that the deposited particles are affected by the wall shear stress. The kinetics of adhesion depends both on the type of adhesion forces involved and the particles' mode of approach to the wall surface.

This paper elaborates on earlier attempts to obtain the sticking probability, by combining the adhesion and re-entrainment steps.[5,6] The new points considered are: a) the effect of loosely adherent particles on the adhesion process, and b) the rolling effect of these particles along the surface due to the burst phenomenon,[7] which were not previously taken into account.

Particles on Surfaces 3, Edited by K.L. Mittal
Plenum Press, New York, 1991

COMBINED ADHESION AND RE-ENTRAINMENT PROCESS

Descriptions of the fine structure of flow within the sublayer of the turbulent regime show that the fluid flow is not entirely laminar as in the conventional boundary layer theory.[8] A cyclic phenomenon (bursts) has since been observed involving fluid ejections from the sublayer and sweeps towards the solid-fluid interface from the outer flow region.[9] Cleaver and Yate[10] initially suggested that the effect of the bursts was the removal of some of the instantaneously adhered particles.

The introduction of the notion of adhesion time, arising from the non-instantaneous adhesion process on the one hand and the cyclic shear stress produced by the bursts on the other, allows us to combine the adhesion and re-entrainment processes in turbulent flow regimes.[6]

Regarding the fine structure of turbulence in the wall region, it is known that the mean period t_b between bursts is approximately given by:

$$t_b = 100\frac{\nu}{u^{*2}} \tag{1}$$

During this time only a small portion p of a given area of the wall surface is affected by the bursts, resulting in the partial removal η of loosely adherent particles, i.e. those remaining on the wall surface for less than the adhesion time (t_a). The latter derives from a balance between the forces drawing particles to the wall surface (increasing with time) and those periodically removing them from it owing to the bursts.

To determine the relations deriving from the combined adhesion and removal mechanism, let us consider the behaviour of n_{tb} particles carried to the unit area of the wall surface over a period of time t_b. The number of particles remaining on the wall surface after t_b is the fraction $\gamma = 1 - \eta p$ of n_{tb}, while the product γ^2 gives the fraction after $2t_b$, etc. For the particles remaining on the wall surface the adhesion process is completed after the adhesion time $t_a = t_a^+ t_b$. The portions remaining on the wall surface vs time are

$$f(t) = \begin{cases} 1 & \text{at } t = 0 \\ \gamma & \text{at } t = t_b \\ \gamma^2 & \text{at } t = 2t_b \\ \gamma^3 & \text{at } t = 3t_b \\ \cdots & \cdots\cdot \\ \cdots & \cdots\cdot \\ \gamma^{t_a^+ - 1} & \text{at } t = (t_a^+ - 1)t_b \\ \gamma^{t_a^+} & \text{at } t = t_a^+ t_b \\ \gamma^{t_a^+} & \text{at } t = (t_a^+ + 1)t_b \\ \gamma^{t_a^+} & \text{at } t = (t_a^+ + 2)t_b \\ \cdots & \cdots\cdot \end{cases}$$

The particles surviving the period of adhesion time t_a adhere to the wall surface and are not removed by the effect of the bursts. The relative fraction value is

$$S = \gamma^{t_a^+} \tag{2}$$

which also corresponds to Parkins' notion of *sticking probability*,[3] defined as the ratio of the number of particles adhering to the surface to the number originally arriving there.

The total portion of the particles arriving at the surface but adhering only loosely to it is

$$1 + \gamma + \gamma^2 + \gamma^3 + \cdots + \gamma^{t_a^+ - 1} = \frac{1 - \gamma^{t_a^+}}{1 - \gamma} \tag{3}$$

High values of loosely adherent (unstable) particles should affect the validity of equation (2). A modified equation of sticking probability is here put forward to include the effect of loosely adherent (unstable) particles.

PORTION OF WALL SURFACE OCCUPIED BY UNSTABLE PARTICLES

The portion of the wall surface occupied by loosely adherent particles can be calculated if the form and dimension of the deposited particles are known. Assuming that the particles are spherical, the portion of wall surface occupied by loosely adherent particles is

$$x_u = \frac{1-\gamma^{t_a^+}}{1-\gamma}\frac{\pi d_d^2}{4} n_{tb} \tag{4}$$

where n_{tb} is obtained by means of the transfer rate V_t

$$CV_t = \frac{\rho_p \pi d_p^3}{6}\frac{n_{tb}}{t_b} \tag{5}$$

which gives

$$n_{tb} = t_b \frac{6}{\pi d_p^3}\frac{CV_t}{\rho_p} \tag{6}$$

By substitutings n_{tb} in equation (4) with the value derived from equation (6) we obtain the portion of the wall surface occupied by unstable particles

$$x_u = \frac{1-\gamma^{t_a^+}}{1-\gamma}\frac{t_b}{t_t} \tag{7}$$

where t_t represents the transport time

$$t_t = \frac{2}{3}\frac{C}{\rho_p}\frac{d_p}{V_t}$$

Equation (7) relates the portion of the wall surface occupied by unstable particles (x_u) to the dimensionless adhesion time (t_a^+), as shown in Fig. 1 for some values of the time ratio t_b/t_t.

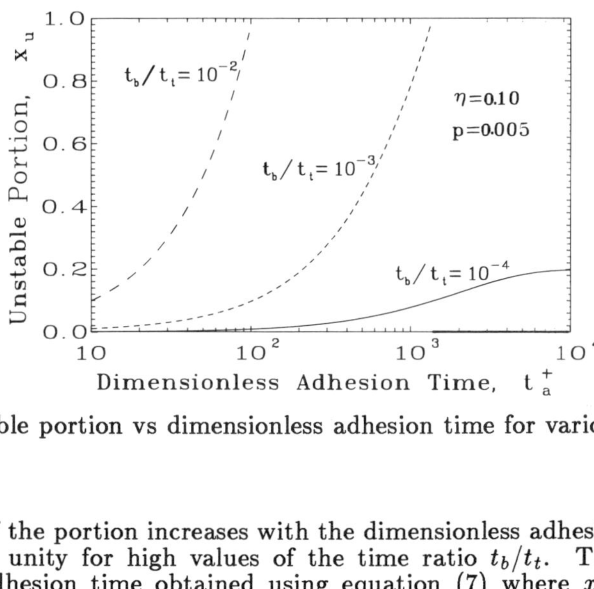

Figure 1. Unstable portion vs dimensionless adhesion time for various values of time ratio.

The value of the portion increases with the dimensionless adhesion time and may be greater than unity for high values of the time ratio t_b/t_t. The values for the dimensionless adhesion time obtained using equation (7) where $x_u = 1$ are to be considered as limiting values since when $x_u = 1$ no portion of the wall surface is any

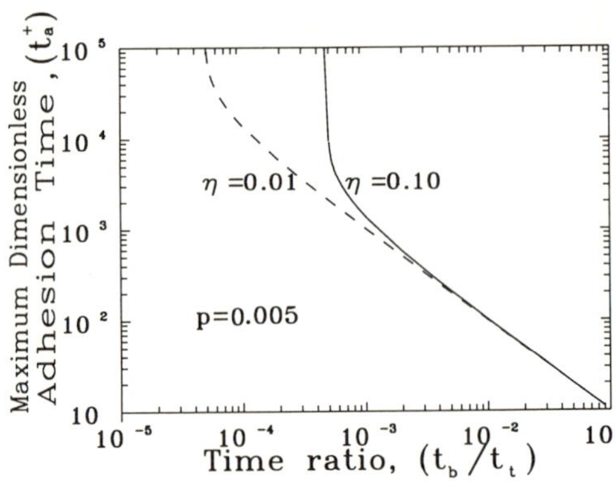

Figure 2. Maximum dimensionless adhesion time vs time ratio for two different values of removal efficiency of bursts.

longer occupied by fully adherent particles. The limiting (maximum) values t_a^+ vs t_b/t_t are shown in Fig. 2 for some values of the removal efficiency.

EFFECT OF UNSTABLE PARTICLES ON THE STICKING PROBABILITY

The particles that have fully adhered to the wall surface constitute the new stable wall surface, while those that have only loosely adhered to it constitute the new unstable surface.

The combined adhesion and removal mechanism proposed above assumes that the particles are carried to a stable wall surface. Hence it is probably not valid for the unstable portion of the wall surface, for which the deposition mechanism is likely to be intermediate between the following two extreme cases:

a) the proposed mechanism also occurs in the unstable wall regions

b) the proposed mechanism does not occur in the unstable regions.

If the former situation prevails this will not affect the sticking probability derived using equation (2). If the latter prevails, the sticking probability becomes

$$S = \gamma^{t_a^+}(1 - x_u) \tag{8}$$

Using the value for x_u derived from equation (7) we obtain

$$S = \gamma^{t_a^+}\left[1 - \frac{1-\gamma^{t_a^+}}{1-\gamma}\frac{t_b}{t_t}\right] \tag{9}$$

The results obtained by means of equation (9) for two different values of the removal efficiency are shown in Figs. 3 and 4. In the deposition process studied here, the time ratio t_b/t_t is related to the mass-transfer phase while the other terms are related to the adhesion/removal phase.

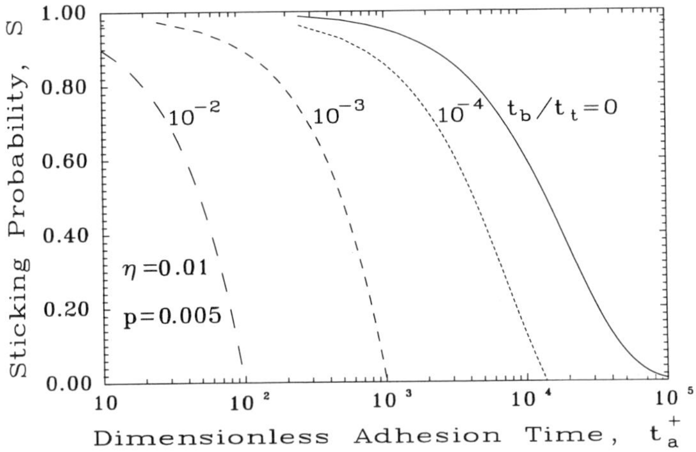

Figure 3. Sticking probability vs dimensionless adhesion time, for low removal efficiency and various values of time ratio.

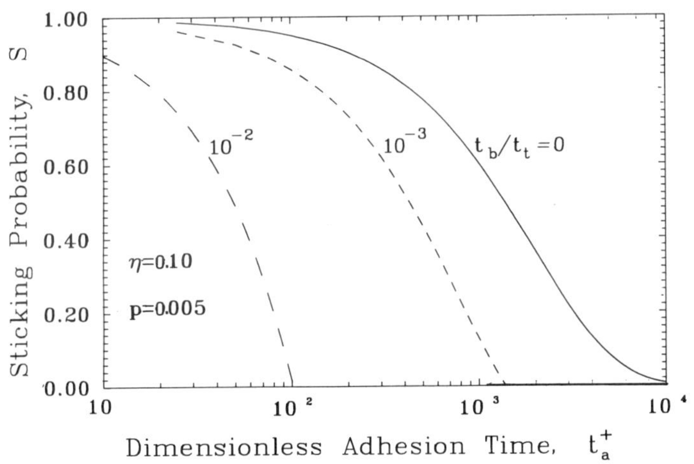

Figure 4. Sticking probability vs dimensionless adhesion time for high removal efficiency and various values of time ratio.

On comparing the results shown in Figs. 3 and 4 it is easy to note that if the mass-transfer term is high the sticking probability depends on the mass-transfer and not on the adhesion/removal phase; whereas if it is low then the whole process depends largely on the adhesion/removal phase.

COMBINED ADHESION, RE-ENTRAINMENT, AND ROLLING PROCESS

Recent experimental research[7] regarding the effect of turbulent bursts on deposited particles has shown that the bursts remove only a portion of the particles they disturb, while another portion rolls along the surface. In both cases the adhesion process is interrupted, the difference being that in the former the particles return to the fluid, while in the latter the adhesion process begins all over again at the particles' new points of arrival. The combined adhesion and removal mechanism proposed above has been extended to include both removal and rolling.

It is to be noted that a particle that has only loosely adhered to the wall surface and has then been subject to the rolling effect has the same age (degree of adhesiveness acquired over time) as a newly arrived particle. This means that we can now specify that not all particles that arrive at the same time and remain on the wall surface have the same age.

During a period of time t_b a small portion p of the wall surface is affected by the bursts, resulting in the partial removal η and rolling κ of the particles occupying that area. Hence a portion $\xi = \kappa p$ of the loosely adherent particles is subject to the rolling effect, another ηp to removal, while the rest $\gamma = 1 - (\eta + \kappa)p$ are in no way affected by the bursts.

To determine the relations deriving from the combined adhesion, removal and rolling mechanism, let us once more consider the behaviour of n_{tb} particles carried to the unit area of the wall surface over a period of time t_b.

The quantity $\gamma = [1 - (\eta + \kappa)p]$ is the portion of particles subject to aging, while the quantity $\xi = (\kappa p)$ is the portion of particles subject to rejuvenation. Hence, while initially $(t = 0)$ all the n_{tb} particles are equally ageless , by the end of the given period $(t = t_b)$ the respective ages (α) of the various portions are

$$f(\alpha) = \begin{cases} \xi & \text{age } \alpha = 0 \\ \gamma & \text{age } \alpha = t_b \end{cases}$$

while their sum is

$$\gamma + \xi \tag{10}$$

By the end of a second period $(t = 2t_b)$ the respective ages (α) of the various portions are

$$f(\alpha) = \begin{cases} \xi(\gamma + \xi) & \text{age } \alpha = 0 \\ \gamma\xi & \text{age } \alpha = t_b \\ \gamma^2 & \text{age } \alpha = 2t_b \end{cases}$$

the sum of all portions is

$$(\gamma + \xi)^2 \tag{11}$$

By inference, by the end of the adhesion time period $(t = t_a^+ t_b)$ the respective ages (α) of the various portions are

$$f(\alpha) = \begin{cases} \xi(\gamma + \xi)^{t_a^+ - 1} & \text{age } \alpha = 0 \\ \gamma\xi(\gamma + \xi)^{t_a^+ - 2} & \text{age } \alpha = 1t_b \\ \gamma^2\xi(\gamma + \xi)^{t_a^+ - 3} & \text{age } \alpha = 2t_b \\ \gamma^3\xi(\gamma + \xi)^{t_a^+ - 4} & \text{age } \alpha = 3t_b \\ \cdots\cdot & \cdots\cdot \\ \cdots\cdot & \cdots\cdot \\ \gamma^{t_a^+ - 2}\xi(\gamma + \xi) & \text{age } \alpha = (t_a^+ - 2)t_b \\ \gamma^{t_a^+ - 1}\xi & \text{age } \alpha = (t_a^+ - 1)t_b \\ \gamma^{t_a^+} & \text{age } \alpha = t_a^+ t_b \end{cases}$$

while the portion of unstable particles is

$$(\gamma + \xi)^{t_a^+} - \gamma^{t_a^+} \tag{12}$$

By the end of a further period $[t = (t_a^+ + 1)t_b]$ the respective ages (α) of the various portions are

$$f(\alpha) = \begin{cases} \xi[(\gamma+\xi)^{t_a^+} - \gamma^{t_a^+}] & \text{age } \alpha = 0 \\ \gamma\xi(\gamma+\xi)^{t_a^+-1} & \text{age } \alpha = 1t_b \\ \gamma^2\xi(\gamma+\xi)^{t_a^+-2} & \text{age } \alpha = 2t_b \\ \gamma^3\xi(\gamma+\xi)^{t_a^+-3} & \text{age } \alpha = 3t_b \\ \cdots\cdot & \cdots\cdot \\ \cdots\cdot & \cdots\cdot \\ \gamma^{t_a^+-2}\xi(\gamma+\xi)^2 & \text{age } \alpha = (t_a^+-2)t_b \\ \gamma^{t_a^+-1}\xi(\gamma+\xi) & \text{age } \alpha = (t_a^+-1)t_b \\ \gamma^{t_a^+}\xi & \text{age } \alpha = t_a^+ t_b \\ \gamma^{t_a^+} & \text{age } \alpha = (t_a^+ + 1)t_b \end{cases}$$

while the portion of unstable particles is

$$(\gamma+\xi)^{t_a^++1} - (2\xi+\gamma)\gamma^a \tag{13}$$

By the end of a still further period $[t = (t_a^+ + 2)t_b]$ the respective ages (α) of the various portions are

$$f(\alpha) = \begin{cases} \xi[(\gamma+\xi)^{t_a^++1} - (2\xi+\gamma)\gamma^{t_a^+}] & \text{age } \alpha = 0 \\ \gamma\xi[(\gamma+\xi)^{t_a^+} - \gamma^{t_a^+}] & \text{age } \alpha = 1t_b \\ \gamma^2\xi(\gamma+\xi)^{t_a^+-1} & \text{age } \alpha = 2t_b \\ \gamma^3\xi(\gamma+\xi)^{t_a^+-2} & \text{age } \alpha = 3t_b \\ \cdots\cdot & \cdots\cdot \\ \cdots\cdot & \cdots\cdot \\ \gamma^{t_a^+-2}\xi(\gamma+\xi)^3 & \text{age } \alpha = (t_a^+-2)t_b \\ \gamma^{t_a^+-1}\xi(\gamma+\xi)^2 & \text{age } \alpha = (t_a^+-1)t_b \\ \gamma^{t_a^+}\xi(\gamma+\xi) & \text{age } \alpha = t_a^+ t_b \\ \gamma^{t_a^+}\xi & \text{age } \alpha = (t_a^+ + 1)t_b \\ \gamma^{t_a^+} & \text{age } \alpha = (t_a^+ + 2)t_b \end{cases}$$

while the portion of unstable particles is

$$(\gamma+\xi)^{t_a^++1} - (\gamma+\xi)(\gamma+3\xi)\gamma^{t_a^+} \tag{14}$$

By inference, by the end of an (a+n)th period $[t = (t_a^+ + n)t_b]$ the portion of unstable particles is

$$(\gamma+\xi)^{t_a^++n} - (\gamma+\xi)^{n-1}\big[\gamma + (n+1)\xi\big]\gamma^{t_a^+} \tag{15}$$

The minimum value of this expression must be zero, meaning that n may not be greater than the limiting value N, derived from

$$N = \left[\left(1+\frac{\xi}{\gamma}\right)^{t_a^+} - 1\right]\left(1+\frac{\gamma}{\xi}\right) \tag{16}$$

N vs t_a^+ for various values of k is shown in Fig 5. The effect of η on the above equation is negligible.

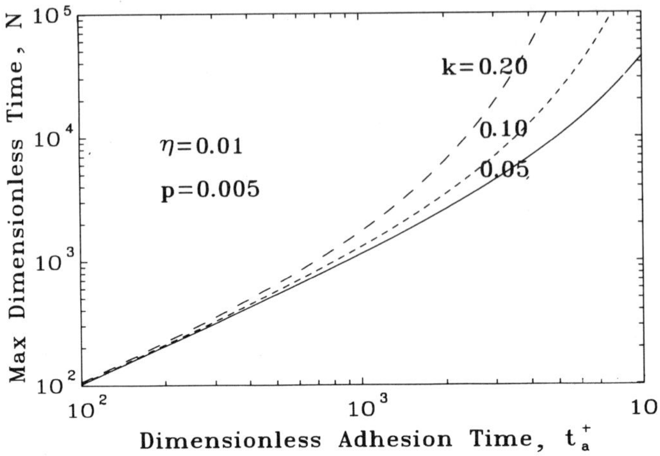

Figure 5. Maximum dimensionless time vs dimensionless adhesion time for various values of rolling efficiency of bursts.

By the end of adhesion time ($t = t_a^+ t_b$) the particles have become stable. The portion of stabilized particles is derived by means of equations (10)-(16) according to the following law: the portion of particles stabilized after the period $t = (t_a^+ + n)t_b$ is given by multiplying $\gamma^{t_a^+}$ by the value of the unstable portion at $t = (n-1)t_b$.

$$
f(t) = \begin{cases}
\gamma^{t_a^+} & \text{at } t = t_a^+ t_b \\
\gamma^{t_a^+}\xi & \text{at } t = (t_a^+ + 1)t_b \\
\gamma^{t_a^+}\xi(\gamma+\xi) & \text{at } t = (t_a^+ + 2)t_b \\
\gamma^{t_a^+}\xi(\gamma+\xi)^2 & \text{at } t = (t_a^+ + 3)t_b \\
\cdots\cdots & \cdots\cdots \\
\cdots\cdots & \cdots\cdots \\
\gamma^{t_a^+}\xi(\gamma+\xi)^{t_a^+ - 1} & \text{at } t = 2t_a^+ t_b \\
\gamma^{t_a^+}\xi[(\gamma+\xi)^{t_a^+} - \gamma^{t_a^+}] & \text{at } t = (2t_a^+ + 1)t_b \\
\gamma^{t_a^+}\xi[(\gamma+\xi)^{t_a^+ + 1} - (\gamma+2\xi)\gamma^{t_a^+}] & \text{at } t = (2t_a^+ + 2)t_b \\
\gamma^{t_a^+}\xi[(\gamma+\xi)^{t_a^+ + 2} - (\gamma+\xi)(\gamma+3\xi)\gamma^{t_a^+}] & \text{at } t = (2t_a^+ + 3)t_b \\
\cdots\cdots & \cdots\cdots \\
\cdots\cdots & \cdots\cdots \\
\gamma^{t_a^+}\xi[(\gamma+\xi)^{t_a^+ + (N-1)} - (\gamma+\xi)^{N-2}(\gamma+N\xi)\gamma^{t_a^+}] & \text{at } t = (2t_a^+ + N)t_b
\end{cases}
$$

The sticking probability is the sum of the portions stabilized

$$
\begin{aligned}
S = {} & \gamma^{t_a^+} + \gamma^{t_a^+}\xi\left[1 + (\gamma+\xi) + (\gamma+\xi)^2 + \cdots + (\gamma+\xi)^{t_a^+ + N - 1}\right] - \\
& - \gamma^{2t_a^+}\xi\left[1 + (\gamma+2\xi) + (\gamma+\xi)(\gamma+3\xi) + (\gamma+\xi)^2(\gamma+4\xi) + \cdots \right. \\
& \left. \cdots + (\gamma+\xi)^{N-2}(\gamma+N\xi)\right]
\end{aligned} \tag{17}
$$

The above equation may be written in a more convenient form, thus

$$
\begin{aligned}
S = {} & \gamma^{t_a^+} + \gamma^{t_a^+}\xi\left[1 + (\gamma+\xi) + (\gamma+\xi)^2 + \cdots + (\gamma+\xi)^{t_a^+ + N - 1}\right] - \\
& - \gamma^{2t_a^+}\xi\left[1 + (\gamma+\xi) + (\gamma+\xi)^2 + \cdots + (\gamma+\xi)^{N-1}\right] - \\
& - \gamma^{2t_a^+}\xi\left[1 + 2(\gamma+\xi) + 3(\gamma+\xi)^2 + \cdots + (N-1)(\gamma+\xi)^{N-2}\right]
\end{aligned} \tag{18}
$$

or

$$S = \gamma^{t_a^+} + \gamma^{t_a^+} \xi \sum_{i=0}^{t_a^+ + N-1} (\gamma+\xi)^i - \gamma^{2t_a^+} \xi \sum_{i=0}^{N-1} (\gamma+\xi)^i - (\gamma^{t_a^+} \xi)^2 \sum_{i=0}^{N-2} (i+1)(\gamma+\xi)^i \quad (19)$$

While there should be no difficulty in calculating the respective sums of the initial geometric progressions, the sum of the final combined arithmetic and geometric progression is[11]

$$\sum_{i=0}^{N-1} (1+i)(\gamma+\xi)^i = \frac{1 - N(\gamma+\xi)^N}{1-(\gamma+\xi)} + \frac{(\gamma+\xi)\left[1-(\gamma+\xi)^{N-1}\right]}{\left[1-(\gamma+\xi)\right]^2} \quad (20)$$

Finally, combination of equations (19) and (20) gives the sticking probability.

$$S = \gamma^{t_a^+} \Bigg\{ 1 + \frac{\xi}{1-(\gamma+\xi)} \Bigg\{ 1 - \gamma^{t_a^+} \frac{1-\gamma}{1-(\gamma+\xi)} - (\gamma+\xi)^N \left[(\gamma+\xi)^{t_a^+} - \gamma^{t_a^+}\right] + $$
$$+ \gamma^{t_a^+} \xi (\gamma+\xi)^{N-1} \left[N + \frac{\gamma+\xi}{1-(\gamma+\xi)} \right] \Bigg\} \Bigg\} \quad (21)$$

The sticking probability vs dimensionless adhesion time is shown in Figs. 6 and 7.

Increased rolling efficiency reduces the sticking probability, while a comparison of Figs. 6 and 7 shows that the rolling effect is more pronounced the lower the removal efficiency.

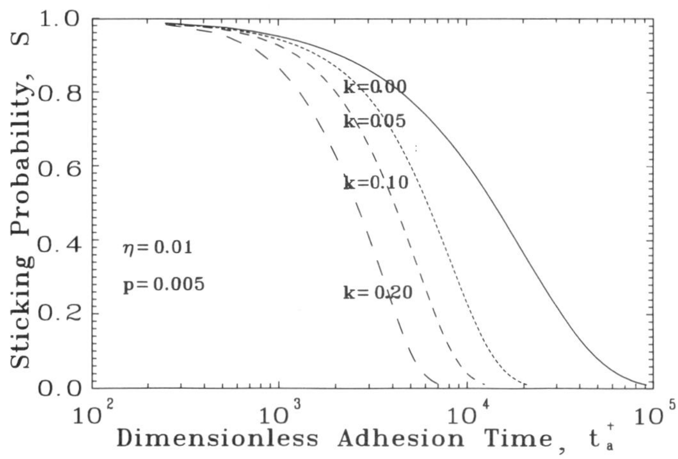

Figure 6. Sticking probability vs dimensionless adhesion time for low removal efficiency and various values of rolling efficiency of bursts.

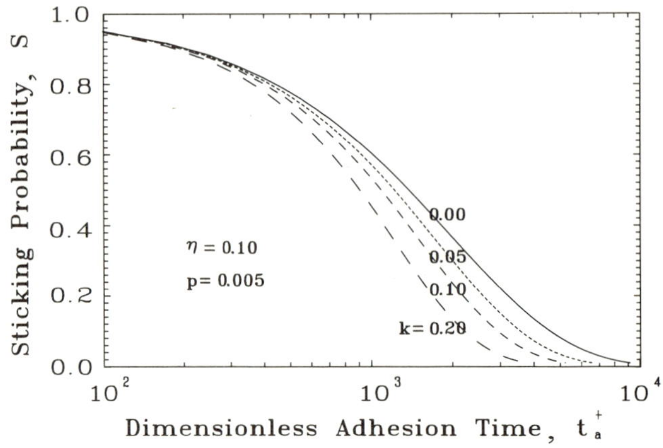

Figure 7. Sticking probability vs dimensionless adhesion time for high removal efficiency and various values of rolling efficiency of bursts.

Equation (21) is valid if we assume that the portion of unstable particles on the wall surface is small.

CONCLUSIONS

An attempt to include the influence of the adhesion on the particle removal phase was proposed in a previous paper, but without taking the mass-transer effect into account. The proposed mechanism has here been completed by including this effect. Dimensionless parameters affecting the whole mechanism have been introduced, their limiting values found and their effect on the sticking probability determined.

The mechanism has also been extended to include the experimentally observed rolling of the particles on the the wall surface due to turbulent bursts. In spite of the complexity of the derived mechanism, analytical solutions for the sticking probability have been obtained, but without for the moment considering the mass-transfer effect.

ACKNOWLEDGEMENT

The research described here was carried out with financial assistance from the Italian C.N.R.

GLOSSARY OF SYMBOLS

C	bulk particle concentration, kg/m^3
d_p	particle diameter, m
N	maximum dimensionless time, equation (18)
n_d	number of deposited particles/unit area, particles/m^2
n_{db}	number of particles deposited between bursts/unit area, particles/m^2
n_{tb}	number of particles transported between bursts/unit area, particles/m^2
n_u	number of unstable particles/unit area of surface, particles/m^2
p	portion of wall surface affected by bursts during t_b
S	sticking probability, dimensionless
t	time, s
t_a	adhesion time, s
t_a^+	dimensionless adhesion time, $\left[= \dfrac{t_a u^{*2}}{100\nu}\right]$

t_b mean period between bursts, s, $\left[= 100\frac{\nu}{u^{*2}}\right]$

t_t mass transport time, s, $\left[= \frac{2}{3}\frac{C}{\rho_p}\frac{d_p}{V_t}\right]$

u^* wall friction velocity, m/s, $\left[= \left(\frac{\tau}{\rho}\right)^{1/2}\right]$

V_t transfer rate/unit surface area, m/s

x_u portion of wall surface occupied by unstable particles, dimensionless

Greek letters

α age of particle, s

γ portion subject to aging, $[1 - (\eta + \kappa)p]$

η removal efficiency of bursts

κ rolling efficiency of bursts

ν kinetic viscosity, m^2/s

ξ portion subject to rejuvenating, (κp)

ρ_p density of deposit, kg/m^3

τ_w wall shear stress, N/m^2

REFERENCES

1. P. G. Papavergos and A.B. Hedley, Chem. Eng. Res. Des., 62, 275 (1984)
2. N. Epstein, in "Advances in Fouling Science and Technology." L. F. Melo, T. R. Bott and C. A. Bernardo, editors, p. 143, NATO ASI Series, Kluwer Academic Publishers, Dordrecht, 1988
3. W. E. Parkins, in "Proc. Tripartite Conf. on Transport of Materials in Pressurizzed-Water System," Report AECL-1265, Paper 9, Atomic Energy of Canada Ltd., 1961
4. S. K. Beal, Nucl. Sci. & Eng., 40, 1 (1970)
5. N. Vatistas, in "Advances in Fouling Science and Technology." L. F. Melo, T. R. Bott and C. A. Bernardo, editors, p. 165, NATO ASI Series, Kluwer Academic Publishers, Dordrecht, 1988
6. N. Vatistas, Chem. Eng. Sci. 44, 1603 (1989)
7. B. P. K. Yung, H. Merry and T. R. Bott, Chem. Eng. Sci., 44, 873 (1989)
8. B. J. Cantwell, A. Rev. Fluid Mech. 13, 457 (1981)
9. H. Schlichting, "Boundary-Layer Theory," McGraw-Hill, New York, 1955
10. J. W. Cleaver and B. Yates, Chem. Eng. Sci. 31, 147 (1976)
11. J. H. Perry, "Chemical Engineer's Handbook," p.2.14, Sixth Edition, McGraw-Hill, New York, 1984

MODELLING PARTICLE ACCUMULATION ON A FILTER SURFACE

P. Schmitz and D. Houi

Institut de Mécanique des Fluides, URA-CNRS 0005
Avenue Camille Soula, 31400 Toulouse, France

A statistical model is proposed to model the accumulation of micrometer-sized particles on surfaces. The particle displacement and adhesion rules, defined here, can be used to simulate the formation of aggregates with very different structures. Porous or non-porous surfaces may be considered. The model is applied to analyse particle deposition at the filter surface during a cross-flow microfiltration operation.

INTRODUCTION

The main phenomenon always occurring in all the filtration processes is the decrease of filtrate flux with time. This dramatic loss of efficiency, resulting in a considerable increase of energy consumption, is due to the fouling of the filtering medium. Indeed, the solid particles and the macromolecules, in suspension in the fluid to be treated, can create deposits on the surface or inside the filter medium. The formation of such aggregates is related to complex physico-chemical mechanisms.

If we focus our attention on the solid particle aggregation, the main relevant questions concern the nature of the phenomena and the conditions leading to the retention of the particles on or throughout a porous medium, the modification of the material structure due to the deposition and its effect on the flow pattern.

The main phenomena to be understood and modelled are:
- The deposition of particles on the collector surface,
- The nature of deposit morphology and its evolution,
- The reentrainment conditions of the particles.

In this paper we present a statistical model which simulates the solid particle accumulation at the filter surface in cross-flow microfiltration.

Particles on Surfaces 3, Edited by K.L. Mittal
Plenum Press, New York, 1991

AGGREGATION IN CROSS-FLOW MICROFILTRATION

Surface particle aggregates, generally experienced in cross-flow microfiltration operations, are commonly believed to be one of the main factors responsible for decreasing filtrate flux with time. This disastrous fouling phenomenon needs to be analysed and modelled, by understanding first the behaviour of a suspended particle in the proximity of the filter surface in cross-flow filtration conditions.

A preliminary study, at a macroscopic scale, has been conducted to determine the hydrodynamic field, without particles, in a porous tube which simulates an industrial hollow fiber used for drinking water production . This idealized approach has provided adequate boundary conditions to investigate the local flow field near the filter surface. This 2D hydrodynamic analysis, at a microscopic scale, has shown two flow configurations, leading or not to the possibility of particle entrainment by suspension tangential flow [1]. These results have been used to confirm the deposition of some particles on the filter when they are submitted to Stokes entrainment force and to van der Waals attractive force. Individual particle trajectories near the porous surface have characterized the impact of these two classical effects on the particle behaviour.

The purpose of this study was to predict and furthermore to visualize the accumulation of micrometer-sized particles on a porous surface, for a dilute suspension under cross-flow filtration hydrodynamic conditions.

The final objective of this work will be to predetermine the structure of the fouling surface aggregates, especially the time dependent macroscopic physical properties : permeability and thickness, in order to better model the mass transfer laws at the filter surface.

Aggregation Model

This approach is two-dimensional. All the particles are assumed to be micrometer-sized, spherical and of the same diameter. We also assume that they are at low concentration in the fluid in order to consider that they enter individually the filtration region.

The simulation of the deposition mechanism consists in releasing particles one by one from a randomly selected initial location upstream the porous wall, following a linear trajectory towards the filter surface (Figure 1). The surface is assumed to be porous with uniform porosity, ϵ.

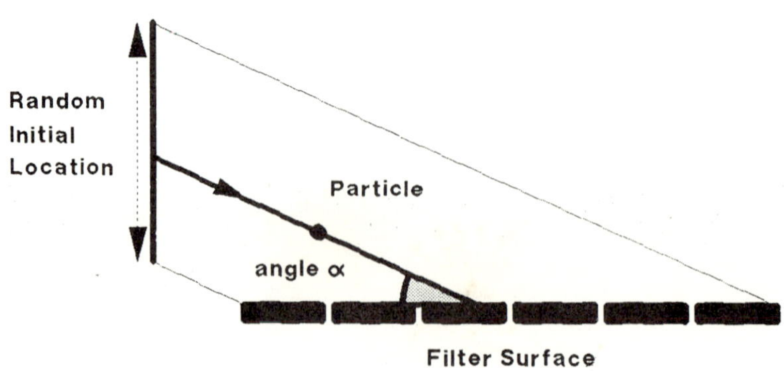

Figure 1. Principle of the Statistical Simulator.

Specific moving and sticking rules are proposed to define the particle behaviour, from its initial starting point to its aggregation. These empirical rules are used to model simply the motion of a suspended particle when it makes contact with the surface or with another particle already aggregated. We believe that this method describes the complex physical reality where numerous phenomena such as hydrodynamical conditions, van der Waals and double layer forces, concentration, density and shape effects, are involved.

Motion Rules: Two parameters are employed to take into account the hydrodynamic conditions at the proximity of the surface.

The initial linear displacement is controlled by the incidence angle α and the next one by the number of reentrainments N_r. This specific parameter determines how many times the particle is allowed to recover its first incident direction of angle α instead of taking a vertical trajectory after contact and rotation around a previously aggregated particle (see Figure 2).

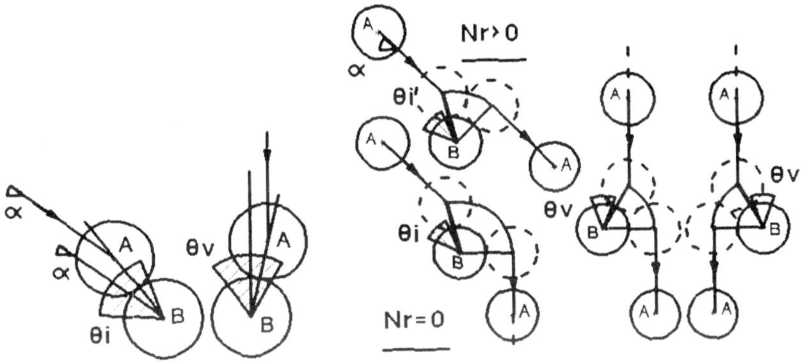

Figure 2. Motion and Sticking Rules.

The incidence angle α characterizes the relative magnitude of the radial suction flow over the tangential entrainment flow in the vicinity of the filter surface.

The number of reentrainments N_r quantifies the interference of the flow at the interface and even inside the growing aggregate.

Sticking Rules: Two other parameters are defined to specify the particle capture mechanisms. We assume here that the balance of forces acting on particle, when it makes contact, determines if the particle would stick or not.

A moving particle of center A is immediately captured when it makes contact with the porous surface. When it touches a previously aggregated particle of center B, it is also captured provided that the centerline (AB) is inside the circular sector defined by the sticking angle θ as shown in Figure 2. Then we define two parameters θ_i and θ_v which characterize adhesion in incident and vertical directions, respectively.

The incident sticking angle θ_i specifies the relative importance of flow entrainment force over particle-particle adhesion force.

The vertical sticking angle θ_v is used to take into account the complex balance of forces occurring inside the aggregate.

Results

Particle motion and sticking rules, such as those defined here, can be used to simulate the accumulation of particles on surfaces, with very different structures, allowing the study of very different physical problems. Here we are specially interested in the particle accumulation on a filter surface during a cross-flow microfiltration operation.

For the moment, the numerical values given to the parameters of this statistical model are empirical but they will be estimated later by analysing light microscope video recordings of aggregation mechanisms occurring during a cross-flow microfiltration process.

At first, it is necessary to investigate the whole range of aggregates we can simulate by varying the most important model parameters in order to determine their influence on the morphology of the aggregates[2].

As an example, we present here two aggregates obtained from our statistical simulator, for specific values of parameters. These simulations have been performed for a filter surface considered as a nucleopore membrane of 15 % porosity and a pore diameter of 1 µm. The size of solid particles is 0.2 µm, five times smaller than the pore size, allowing suspended particles to pass through the pores. The deposits have been limited to 2000 accumulated particles.

Contact-capture type: The first case, presented in Figure 3, models a filter fouling due to a suspension filtration where particle-particle and particle-wall adhesion forces are much greater than hydrodynamic entrainment forces. The assumptions considered here are of the contact-capture type i.e., a suspended particle hitting the filter surface or a particle of the aggregate is immediately stopped and captured. Thus the number of reentrainments N_r has been chosen equal to 0 and θ_i about 60°. The structure of the aggregate is dendritic and its porosity has a mean value about 51 %.

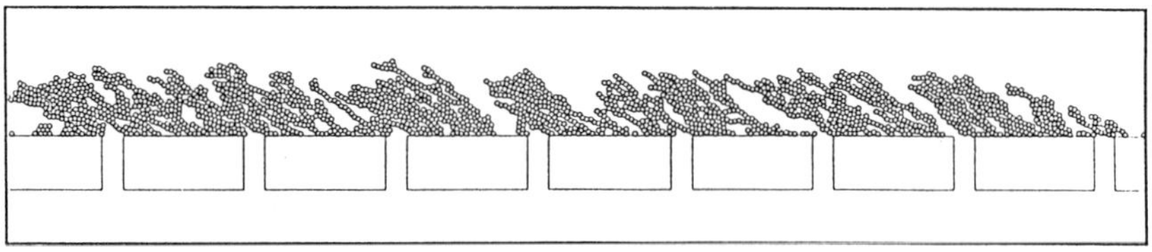

Figure 3. Statistical Simulation,
$\alpha = 20°$, $\theta_i = 60°$, $\theta_v = 1°$, $N_r = 0$.

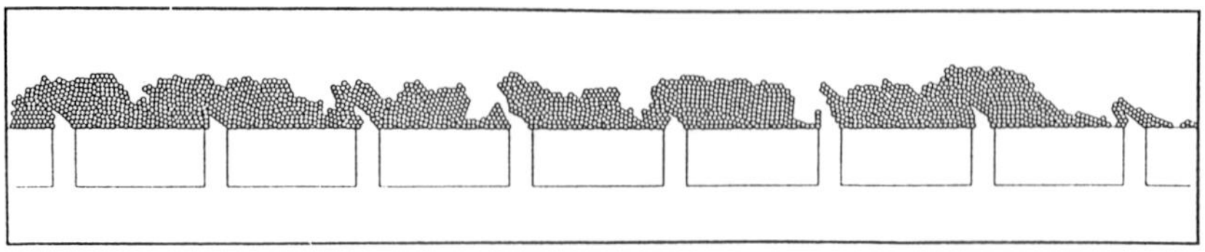

Figure 4. Statistical Simulation,
$\alpha = 20°$, $\theta_i = 1°$, $\theta_v = 1°$, $N_r = 3$.

Reentrainment type: The second simulation, shown in Figure 4, characterizes a very compact accumulation which has been already observed experimentally [3] during a well-mud filtration on a flat mineral membrane (see Figure 5). In this case the flow has a high impact on particle reorganization, at the interface and even inside the aggregate. The filter surface is very attractive and prevents aggregated particles from being carried into the flow again. Physically, it means that tangential shear flow erodes the deposit and especially reduces the vertical growth before complete space filling. N_r has been taken equal to 3, and θ_i and θ_v have very small values. The mean porosity is about 29 % and the mean thickness is obviously lower than the one calculated in the first case.

Of course, we have verified that the same type of structure cannot be obtained with a different set of empirical constants, corresponding to different physical situations[2].

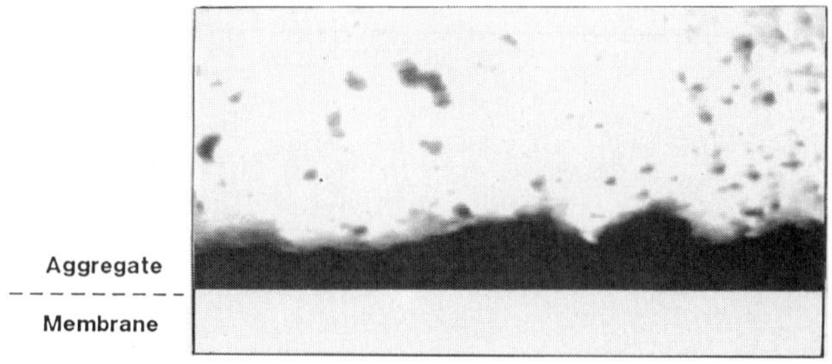

Figure 5. Aggregate Formation during the Dilute Well-Mud filtration.

Heterogeneity

The Root Mean Square is employed to characterize the mean thickness T_d of the simulated particle accumulations on surfaces. It is defined as follows :

$$T_d = \left(\frac{1}{N_p} \sum_{1}^{N_p} y_i^2 \right)^{\frac{1}{2}},$$

where y_i is the distance between the center of particle i and the flat surface. Some authors [4, 5] have demonstrated that the degree of heterogeneity of the aggregate can be deducted from the following relation :

$$T_d = B.N_p^{\lambda},$$

in which the power λ can be expressed as a function of D_e and D_f, the euclidian and fractal dimensions, respectively, as follows :

$$\lambda = \frac{1}{1 - D_e + D_f}$$

As shown in Figure 6, the Root Mean Square T_d is found to be proportional to the number of aggregated particles N_p, leading to λ equal to 1. It is important to notice that statistical aggregates are perfectly homogeneous. It proves that capture mechanisms used in the model do not induce heterogeneity effects.

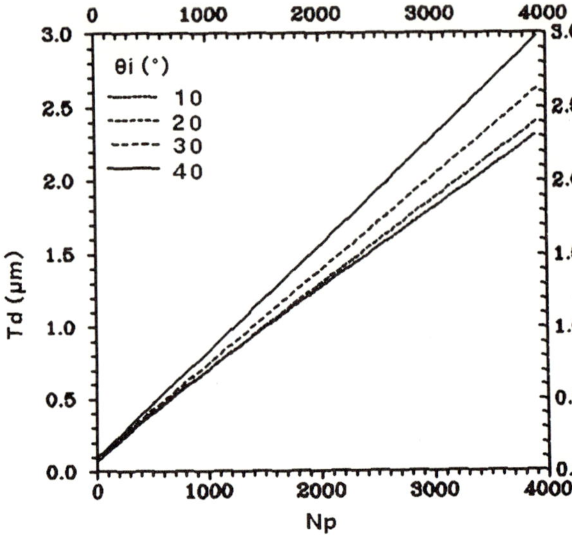

Figure 6. Variation of the Mean Thickness as a Function of the Number of Aggregated Particles.

CONCLUSION

A statistical model describing the formation of a deposit consisting of monodispersed spherical particles has been developed. Simulations are found to be in good agreement with experimental aggregates carried out under similar hydrodynamical and physico-chemical conditions.

The results are interpreted by defining two macroscopic mean parameters: porosity and thickness. Further 3D simulations, taking into account concentration and flow values, will provide the mass transfer law at the interface between porous medium and fluid when analysing time-dependent thickness and permeability parameters.

ACKNOWLEDGEMENTS

This work was supported by the "Société Lyonnaise des Eaux-Dumez" and the French "Centre National de la Recherche Scientifique"

REFERENCES

1. P. Schmitz, C. Gouverneur, D. Houi and M. Madianos, Vth World Filtration Congress Proceedings, **1**, Nice, France (1990).
2. P. Schmitz, Thèse de Doctorat de l'I.N.P.T., Toulouse, France (1990).
3. A. Ritter and D. Houi, 5th Institut Français du Pétrole Research Conf. on Expl./Prod., Arles, France (1990).
4. D. Bensimon, Phys. Rev. Lett., **51**, 1394 (1983).
5. P. Meakin, J. Colloid Interface Sci., **105**, 240 (1985).

PARTICLES IN ULSI GRADE CHEMICALS AND THEIR ADHESION TO SILICON SURFACES

Mitsushi Itano, Ichiro Kawanabe, Frederick W. Kern, Jr.,
Masayuki Miyashita, Reed W. Rosenberg and Tadahiro Ohmi

Department of Electronics
Faculty of Engineering
Tohoku University
Sendai 980, Japan

Ryosuke Fukushima and Shigeyuki Akiyama
Horiba Ltd.

Despite the advancement of dry processing in ULSI manufacturing, wet chemical cleaning remains the only reliable method for controlling particulate contamination on semiconductor wafer surfaces.

It is necessary to accurately measure the number of particles in all liquid chemicals and understand the mechanisms of particulate adhesion so that this form of contamination may be minimized on silicon surfaces. We believe that the particle concentration and zeta potential, and the condition of the wafer surface are all factors which determine the number of particles which deposit on surfaces immersed in liquids.

A new laser particle counter, the Horiba PLCA 520, with specifications for detecting particles to 0.2 μm in chemicals was used in this study. This particle counter is unique in its ability to use either pressure or suction fluid sampling. Thus, the question of whether particle counts are inflated through the inclusion of bubbles during suction sampling may be addressed. No appreciable difference between suction and pressure sampling was found, with two exceptions: nitric acid and ammonium hydroxide. Calibration tests, as well as pressure/suction sampling characterizations of common ULSI-grade processing chemicals are presented.

We have demonstrated that particles adhere to bare silicon wafers in proportion to their zeta potential and concentration in liquids. Furthermore, it is shown that the number of particles deposited in acidic and alkaline solutions are quite different, in keeping with published data on zeta potentials as a function of solution pH.

Particles on Surfaces 3, Edited by K.L. Mittal
Plenum Press, New York, 1991

INTRODUCTION

Wet chemical process technologies will continue to be important in ULSI fabrication of semiconductor devices. With the increase in pattern densities for ULSI devices, wet chemical processes are needed to remove increasingly numerous, small particles in order to obtain higher device yields and enhance circuit reliability. In order to minimize particulate contamination of wafer surfaces, it is necessary to understand the mechanisms of particle deposition and removal. Factors such as the number of particles in solution, the zeta potential of those particles and the potential of the wafer surface are all important factors in the adhesion of, and the force necessary to remove particles.

Particle Counts in Chemicals

The ability to measure particles in liquid chemicals accurately and safely is discussed in this paper. The use of a laser particle counter continues to be the most widely-used method for determining particle concentrations in liquids[1]. Normally these devices use suction sampling (i.e. drawing the liquid through the counter using a downstream pump) to minimize system-generated contamination. It is widely believed that the counting at subambient pressures is responsible for overcounts due to the presence of bubbles, either from vaporized chemicals or gasses dissolved in the chemicals. The counter used in these experiments allowed both pressure and suction sampling of liquids; thus this issue can be addressed without the usual experimental uncertainty arising from the use of different counters. We found no statistically significant difference for pressure or suction sampling with the exception of nitric acid (HNO_3) and ammonium hydroxide (NH_4OH). In the case of NH_4OH, we speculate that the higher suction counts were due to the evolution of NH_3 from the solution. In the case of HNO_3, however, we have determined the cause to be degradation of the chemical by ultraviolet light.

Particulate Deposition Mechanisms

Many researchers have attempted to correlate the number of particles found on a wafer surface to particulate concentrations in chemicals and DI water[2-6]. The works of Niida[5] in Japan and Riley and Carbonell[6] in the United States were the first to relate deposition mechanisms on silicon wafers specifically to colloid chemistry and physics.

Much work has been done in the field of colloid chemistry to understand stability and coalescence mechanisms in lyophobic suspensions. Concepts such as the electrical double layer and zeta potential have proven quite useful in describing these phenomena and predicting the behavior of colloidal suspensions with changes in ionic strength and pH[5,7]. Charge imbalances result when otherwise electrically neutral particles enter aqueous suspension. A tightly bonded shell of ions (the Stern layer) forms on the surface of the particle. This charge imbalance is counteracted by a more diffuse outer shell of oppositely-charged ions. It is this shell, and the number and charge of ions incorporated in it which gives rise to the zeta-potential of the particle. In solutions of low ionic strength, the outer shell is quite large as a greater volume of fluid must be incorporated in order to accommodate a sufficient number of ions to neutralize the charge imbalance of the Stern layer. Electrical repulsion of associated shells keep adjacent particles separated, preventing coalescence. As the ionic strength is increased, the shells shrink, as less volume is needed to achieve the required number of ions to balance the charges. Thus, particles can get much closer to each other before electrical repulsion occurs. In some cases, van der Waals attractive forces between particles can overcome this repulsive force and the particles coalesce.

We propose thinking of semiconductor wafers as very large particles. Ionic double layers exist at the wafer-liquid interfaces and behave exactly as they do in colloidal suspensions.

EXPERIMENTAL

Particle Counter Calibration

Any particle counter must be able to provide accurate measurements of the concentration of particles in various fluids to its published performance limits and over a wide range of particle concentrations. With this in mind, we set out to test the Horiba PLCA 520 counter with a series of recovery tests using polystyrene latex (PSL) spheres ranging in size from 0.344 to 2.904 μm and concentrations from 4.3 to 900 particles per milliliter. We were unable to obtain 0.2 μm particles, the published detection limit of the counter, but as we will show later, the functionality of the parameters measured with size is sufficiently small to allow us to extrapolate the data obtained at 0.344 μm to its detection limit.

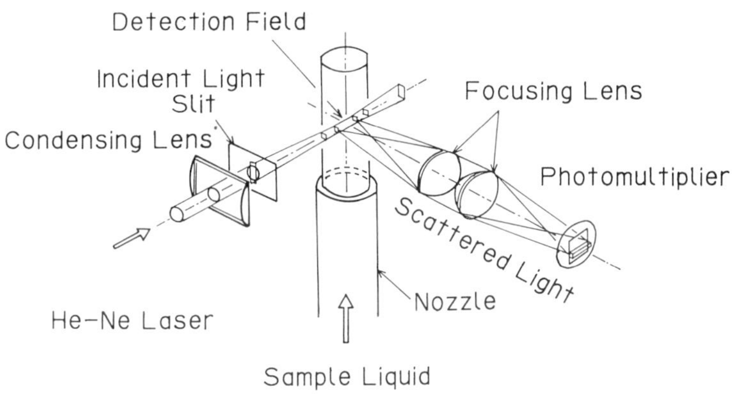

Figure 1. Schematic representation of the sample cell and light path ofthe Horiba PLCA 520 particle counter

Figure 1 shows the measurement principle used in the Horiba particle counter. A helium-neon (He - Ne, 632.8 nm) laser is focused by a condensing lens and slit through the detection field (a rectangular prism with its major axis perpendicular to the direction of fluid flow). Light scattered by the particles at an angle of 90° to both the fluid flow and laser light path is collected by a photomultiplier. The intensity of the scattered light is measured and interpreted as a particle of a given size. Six channels of resolution, corresponding to particles equal to or greater than 0.2 μm, 0.5 μm, 1 μm, 2 μm 5 μm and 10 μm, are available.

The particle counter is designed for either suction or pressure sampling. Figures 2a and 2b show the flow diagrams for suction and pressure sampling systems, respectively. In the suction sampling system, chemical is drawn from the bottle through the sensing unit by a syringe pump. The required flowrate for the system is set at 10 ml/min. Due to the design of the sample cell, where the light beam intercepts 10% of the cross-sectional area available for flow, this means that the effective sample rate is 1 ml/min.

In the pressure sampling system, the bottle of chemical to be analyzed is placed in a pressure vessel. Filtered nitrogen is used to pressurize the system to approximately 1 kg_f/cm^2, forcing the chemical from the bottle through the counter. The flowrate is controlled by a metering valve, located after the counting cell.

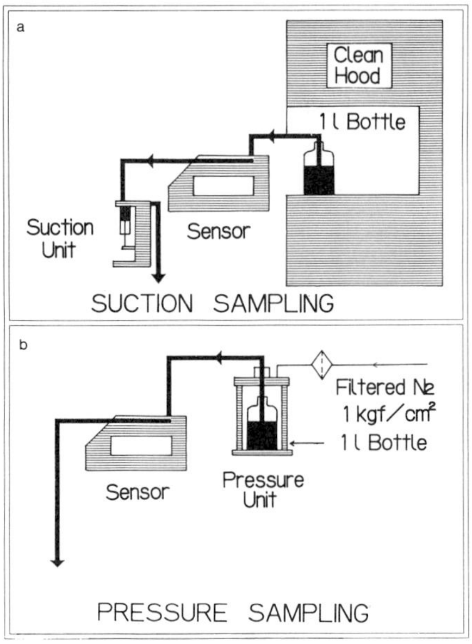

Figure 2. Figure 2a depicts the scheme used for suction sampling of chemicals. A bottle of the chemical to be tested is placed in a clean hood. A fixed volume of chemical is drawn through the sensor by the supplied suction unit.

Figure 2b shows the scheme used for pressure sampling. The bottle of chemical is placed in the sealed pressure unit. Filtered nitrogen at a pressure of 1 kg_f/cm^2 is supplied to the pressure unit, forcing the chemical through the sensing unit at a constant rate of flow.

Figure 3 shows the correlation between particle counts obtained by the manufacturer of the Horiba PLCA 520 with SEM measured values for 0.2 - 0.5 μm particles in various chemicals. The results are in very close agreement.

Our first test was to compute the counting efficiency (the ratio of the number of particles detected by the counter and the known particle concentration of the sample) as a function of particle size. 304 datum points were collected for this measurement. The counting efficiency for each point was computed, and linear regression was then used to curve-fit the data. We found a slope of 0.017 and an intercept of 0.972 for these data (with a standard error for the slope and intercept of 0.023 and 0.043, respectively), indicating that the counting efficiency is both very close to 1 and is quite invariant with particle size. The results of curve fitting, average counts and 95% confidence interval for each size of particles tested are summarized in Figure 4. A counting efficiency very close to unity, combined with the fact that we observed no fall-off of that efficiency, even for very small-sized particles, gives us confidence that we may extrapolate our findings (5.6% of the range investigated) from 0.344 μm to the published detection limit of 0.2 μm, and conclude that the counter would exhibit similar performance at this level. The projections for performance at this level have been included on all applicable graphical results. Quite a bit of scatter is visible for the counts with 0.988 and 2.904 μm particles. We believe this is not due to the counter, rather in the extremely low particle concentrations (4.3 P/ml and 5.5 P/ml for 0.988 and 2.904 μm particles, respectively) used for some of these standard solutions, as described in the next paragraph. Next, we assessed the ability of the counter to be unaffected by the particle concentration of the chemical being measured. The results of this test are summarized in Figure 5 which shows the average counting efficiency (for

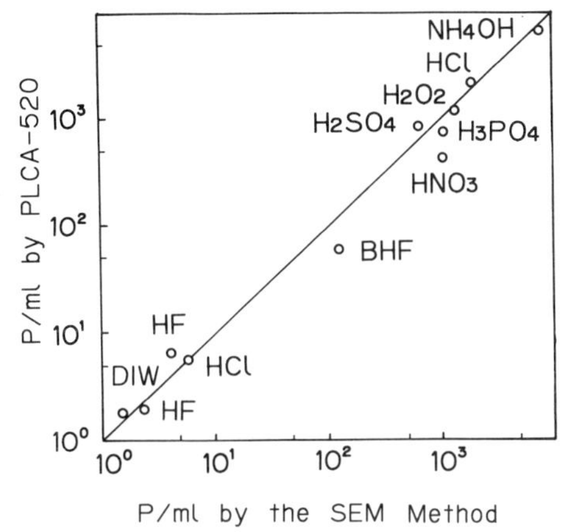

Figure 3 Correlation of the particle counts obtained using the Horiba PLCA 520 and those using the SEM method for 0.2 - 0.5 μm particles in various chemicals.

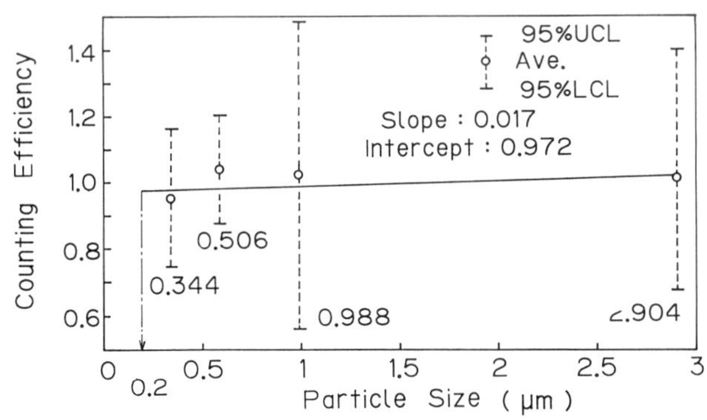

Figure 4. Counting efficiency as a function of particle size determined using 0.344, 0.506, 0.988 and 2.904 μm PSL spheres at various concentrations in DI water.

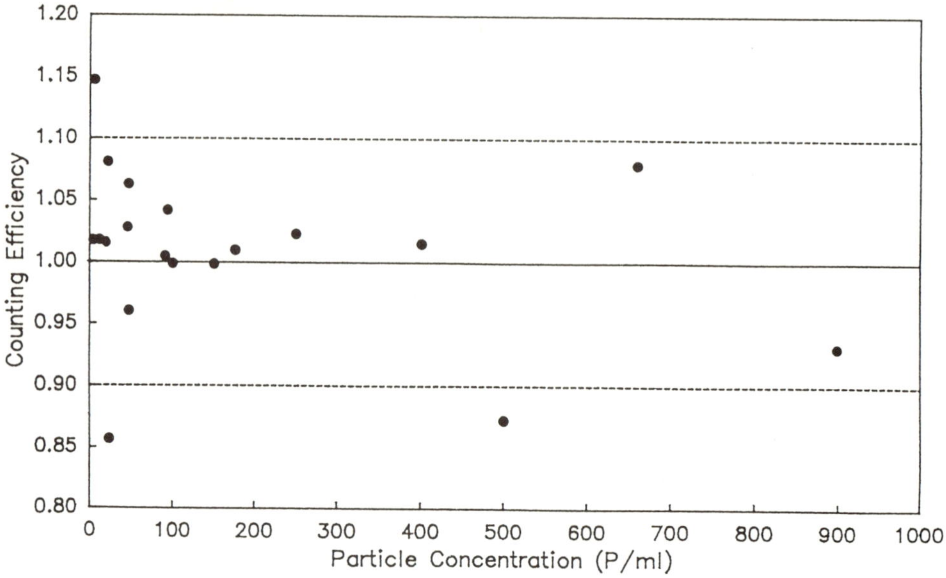

Figure 5. Counting efficiency as a function of particle concentration in DI water, for all particle sizes tested (0.344 - 2.904 μm).

all particle sizes) vs. the concentration of the particles being counted. While there is quite a bit of scatter, it is safe to conclude that, for all but the most dilute concentrations, efficiencies are well within the ±10% band. At low concentrations, it is expected that the background counts of, and perhaps microbubbles in, the water being used to formulate the standards was quite likely the source of the scatter. All standard solutions were formulated from stock suspensions of monodisperse PSL spheres. Background counts in the DI water used for dilution could add to the actual number of particles present and hence could be responsible for counting efficiencies which are significantly greater than the theoretical 1.0.

It is perhaps most informative to examine the overall ability of the counter to provide an accurate count. To estimate this parameter, the function for counting efficiency vs. particle size was used to compute a standard deviation for the overall data by the following formula:

$$\sigma = \sqrt{\frac{\Sigma\,(\eta - \langle\eta(s)\rangle)^2}{(n-1)}}$$

η = Experimental counting efficiency

$\langle\eta(s)\rangle$ = Least squares counting efficiency as a function of particle size

n = Number of data points

These results were then used to compute Z-scores for counting efficiencies of 0.75, 0.80, 0.90, 1.10, 1.20 and 1.25 (±25%, ±20% and ±10% of the actual value η = 1.00). Standard statistical techniques to estimate the probability of obtaining particle counts within this range were then used. The results of these probability computations are shown in Figure 6. In general, we would estimate that the counter has a precision of about ±25%, as we found about a 75% probability that counts would be in this range relative to the actual sample counts. For example, if a chemical with a particle concentration of 200 P/ml were sampled 100 times, 75 of those samples could be expected to contain from 150 to 250 particles per milliliter. Once again, precision is better if higher particle concentration samples are to be measured (generally above 100 P/ml). If low concentration samples (presumably those of greatest interest to the semiconductor industry) are measured, precision may be improved by longer sampling intervals and/or multiple sampling.

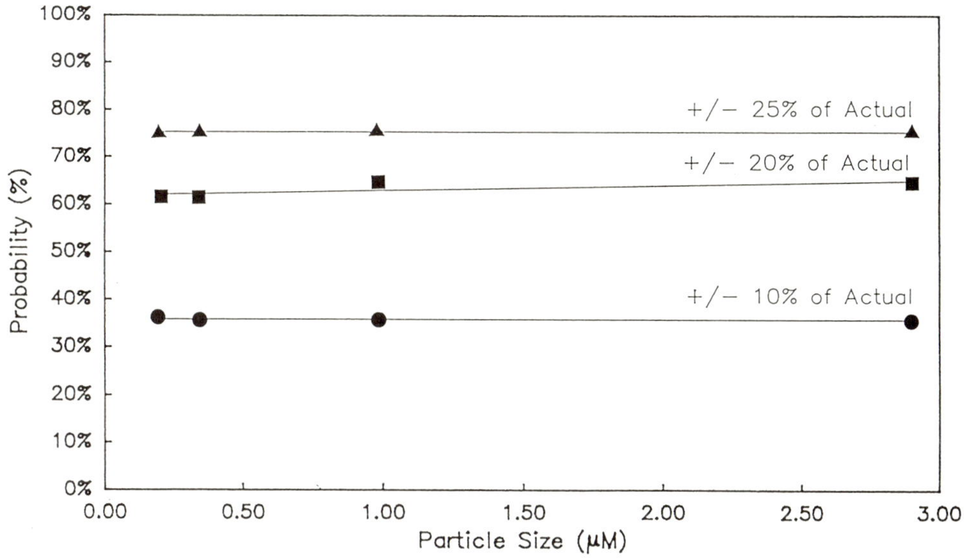

Figure 6. Probability that the measurement obtained will be within specified tolerance of the actual particle concentration.

Particle Counts in Chemicals

For pressure and suction sampling, five 1ℓ bottles each of various ULSI-grade chemicals were sampled. The first bottle of each was tested using suction sampling. Bottles 2, 3 and 4 were sampled using both pressure and suction. The remaining bottle was sampled using only pressure. Suction sampling data were also used to evaluate particle stratification in chemicals. In each "suction only" bottle, the following sampling scheme was employed:

(1)	First 200 ml		Discard
(2)	Bottom Sample	(Sample #1)	(10 × 10 ml)
(3)	Next 100 ml		Discard
(4)	Center Sample	(Sample #2)	(10 × 10 ml)
(5)	Next 100 ml		Discard
(6)	Upper Sample	(Sample #3)	(10 × 10 ml)
(7)	Surface Measurement	(Sample #4)	(5 × 10 ml)

No stratification, gross variation in counts between samples 1, 2 and 3, was found (Figure 7); however, considerably higher counts were found in sample #4, indicating the possibility of particle concentration at the surface of chemicals.

The results are depicted, graphically (for particles ≥0.2 μm in size) in Figure 8, and listed in Table I. The average particle count, upper 95% confidence limit and lower 95% limit or testing limit is shown for each bottle sampled. In the case of bottles which were sampled using both pressure and suction, the combined results (pressure + suction), as well as individual results for each sampling technique, are shown. From this figure, it is quite evident that difference in results obtained from pressure and suction sampling was less than the random differences between counts from the same bottle using the same sampling technique.

A word should be said regarding the data reported. The upper 95% confidence limit is the statistic which should be used when reporting particle concentration in chemicals. Averages are of little value, as particle counts can reasonably be expected to exceed these values, and the chemical user cannot tell by what extent actual particle counts may exceed this value. The upper 95% confidence limit puts a reasonable maximum bound on this parameter and has been used to characterize other similar measurements, such as airborne particle counts used in cleanroom certification[8], so its significance should be familiar to practitioners in the field of contamination control.

Where applicable, the lower 95% confidence limit is also shown in Figure 8. Given a wide variation in readings, and/or an average count close to zero, it is mathematically possible for this value to be less than zero (a condition which makes no sense in the real world). For this reason, the detection limit of the test, defined as 1/total sample volume, is substituted for such cases. In particle counting, it is mathematically impossible to obtain a guarantee of zero particles by other than destructive means (counting the whole bottle of chemical). Any string of zeroes in a counting test can only guarantee that the lower limit is less than one particle in the total volume of fluid run through the counter.

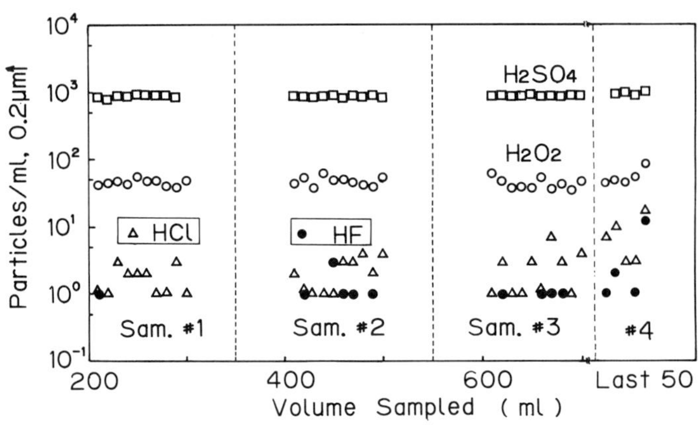

Figure 7. Particle counts in various chemicals as a function of the volume sampled.

Ammonium Hydroxide

Thirty percent ammonium hydroxide, in equilibrium with a partial pressure of ammonia gas of 0.75 atmosphere at 70°F[9] is a chemical which, as expected, gives very high particle counts when sampled using suction. Figure 9 shows the results of suction and pressure sampling of this chemical. Even using pressure to sample, quite a bit of scatter is evident. Initially, counts rise quickly and then slowly return to a steady-state level. We believe this is due to redissolution of the pressurizing gas. To test this, we pressurized one bottle at the pressure to be used for sampling (1 kg_f/cm^2) 30 minutes prior to taking readings. The results of this test are shown by the data points denoted with an asterisk (*) in Figure 9. The mean particle concentrations with an upper 95% confidence level are shown by the dashed and solid lines, respectively, in the pressure sampling region of the figure. They should be compared to similar lines in the suction sampling region to demonstrate the improvements inherent with this technique.

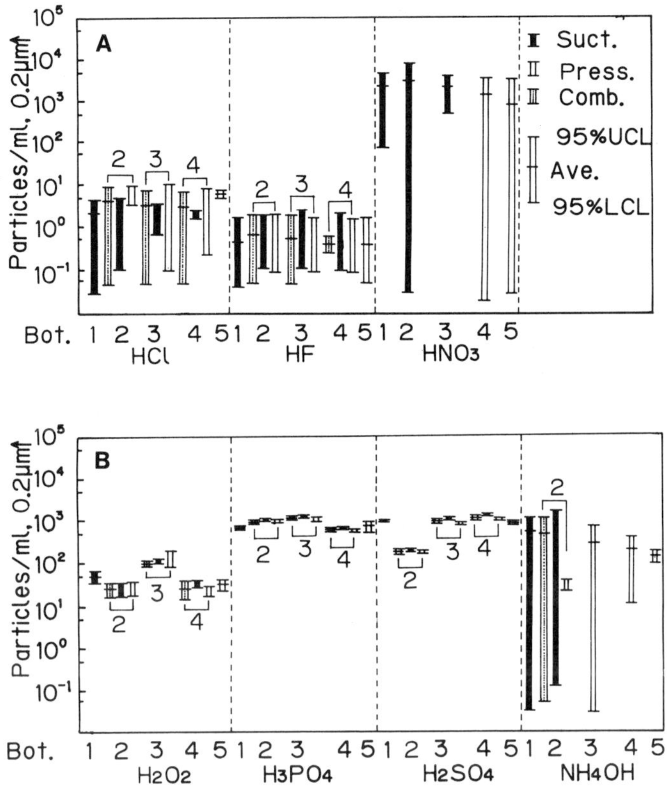

Figure 8. The results of suction and pressure particle counting in ULSI-grade semiconductor chemicals. The average and 95-th percentile are shown for concentrations of particles ≥ 0.2 μm in size

Table I. Upper 95% confidence limit for particle concentrations in ULSI-Grade Chemicals

	Particle Size	
	≥ 0.2 μm	≥ 0.5 μm
HCl	4.7-10.0	1.2-5.0
HF	0.59-2.0	1-2
HNO_3	3,300-7,800	600-700
H_2O_2	37-107	8-29
H_3PO_4	632-1123	78-221
H_2SO_4	950-1,200	17-860
NH_4OH	390-1,800	64-67

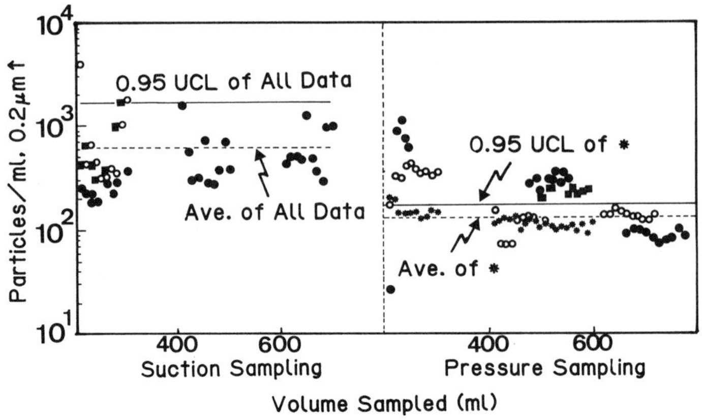

Figure 9. Suction and pressure sampling of ammonium hydroxide. Different symbols depict the concentration vs. sampling volume (or time at pressure) for various bottles measured. The best results were obtained for the data depicted by asterisks (*), where the chemical was pre-pressurized 30 minutes before sampling.

Nitric Acid

Nitric acid exhibited a more interesting problem. Particle counts under suction and pressure sampling obeyed an exponential increase with time. We suspect that this could be due to the uv-catalyzed decomposition of HNO_3 to NO_2.

$$4HNO_3 \xrightarrow{h\nu} 4NO_2\uparrow + O_2\uparrow + 2H_2O$$

This effect is shown in Figure 10. A fairly strong exponential dependence of particle counts with time is seen for suction sampling. Pressure sampling yields a much less-defined relationship but its variation with time is still evident. The time dependence was eliminated by shielding HNO_3 from fluorescent light during subsequent counting, as demonstrated by the "dark sampling" data.

PARTICLE ADHESION MECHANISMS

An attempt was made to correlate the number of particles found on wafers to the particle counts in processing chemicals. While some correlation was observed, there were many cases of clean wafers coming out of relatively dirty chemicals; thus confirming what many other researchers in the field have already found: It is rather difficult to predict on-wafer particle counts just from particle concentrations in liquids.

The phenomena of colloid chemistry could be used to explain particle adhesion to wafer surfaces. In very clean water, a protective ionic barrier can form on the wafer surface. As it would be of the same charge (usually negative) as the double layer surrounding suspended particles, they would be repelled from the wafer surface. If the ionic strength of the solution were increased, the wafer ionic barrier and particle double layers would shrink, until the particles would be attracted to the wafer surface. Figures 11 and 12 show the results of two tests we conducted.

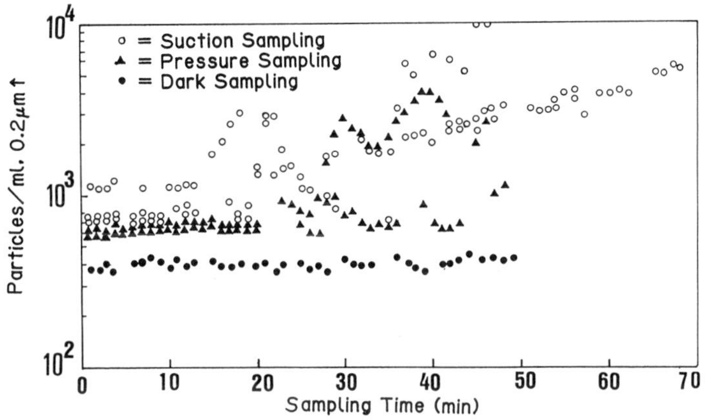

Figure 10. Particle concentration measurements as a function of time for various sampling techniques (pressure, suction and light-excluded) for nitric acid.

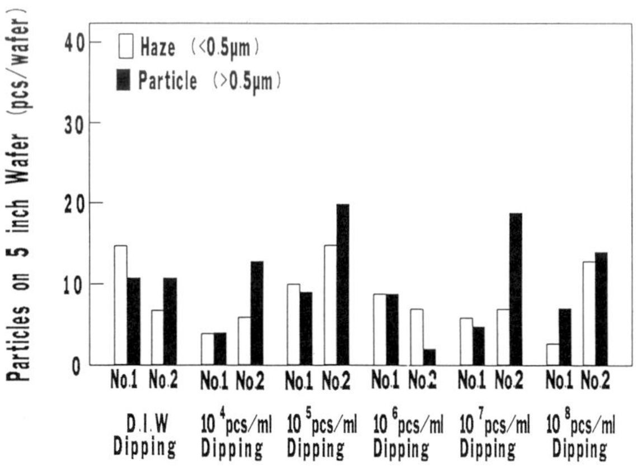

Figure 11. Adhesion of PSL spheres (0.506 μm) to the wafer surface as a function of their concentration in DI water

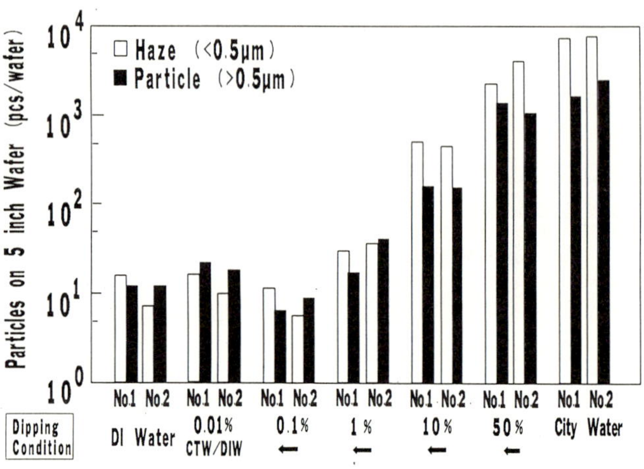

Figure 12. Adhesion of PSL spheres (0.506 μm) to a wafer surface from a mixture of tap (city) water (volume concentration shown) and DI water.

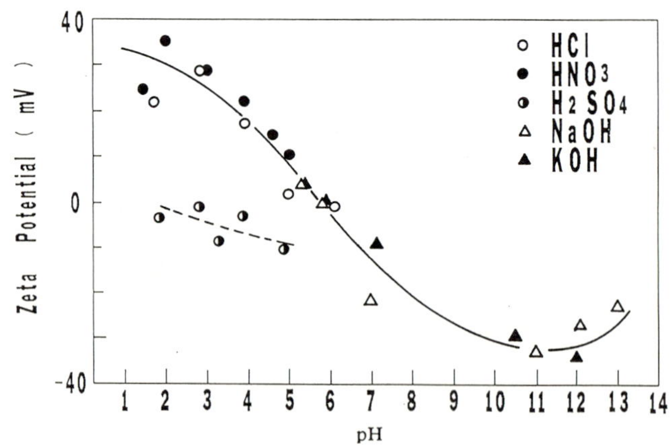

Figure 13. The influence of pH on the zeta potential of Fe_2O_3. (from Niida, et. al.[4])

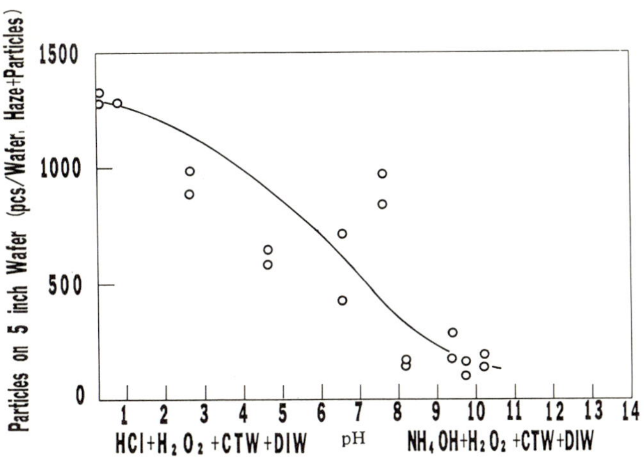

Figure 14. Number of particles on a 5" wafer as a function of seed solution pH. Acidity/basicity was controlled by the addition of HCl and NH_4OH to DI water.

Wafers were first cleaned using a solution of 0.5% wt. hydrofluoric acid in water. The wafers were then soaked for 10 minutes in a solution seeded with 0.506 μm PSL spheres, then rinsed and dried. In the first experiments (Fig. 11) the seeding solution was formulated from deionized water. In figure 12, the seeding solutions were made from increasingly concentrated mixes of tap (city) water with DI water.

Figure 11 shows that in the very low ionic strength (resistivity = 18.2 MΩ·cm) ultrapure water available at the Mini Super Cleanroom of Tohoku University[10], the number of particles deposited on the wafer surface were independent of the particulate concentration in the seeding solution (10^4 - 10^8 P/ml).

Figure 12 shows the effect of increasing ionic strength (adding city water). As predicted from colloid science, once the double layer (and boundary layer at the wafer surface) collapses, as a result of increased ionic strength, the particles will coalesce (adhere to the wafer surface). A further proof of the application of classical colloid theory to this situation is shown in Figures 13 and 14. Figure 13 shows the well-known effect of pH on zeta-potential[5]. Figure 14 shows the curve we obtained for particles on wafer surfaces as a result of modifying the pH of the seeding solution using hydrochloric acid (HCl) and ammonium hydroxide (NH_4OH). It can be seen that they match almost exactly.

The effect of ionic strength on adhesion mechanisms should be contrasted to the work of Riley and Carbonell[6]. In their study, they worked only with low ionic-strength seeding solutions. As a result they found that particles could easily be removed in subsequent clean rinses. During the final rinse step, in cases of low ionic strength (Fig. 11), our results were a confirmation of theirs.' However, when the ionic strength of the seeding bath was increased, we found it impossible to remove particles by rinsing. Only after drying did they find it impossible to remove particles once they had deposited on the wafer surface. We believe this to be due to a thin film of water remaining when wafers are transported between chemical baths. However thin, it is still sufficiently thick for electrical layers to exist; thus particles remain suspended and do not form electrostatic bonds to the wafer surface until this water film is removed in the drying step. In the case of high ionic strength solutions, where the diffuse layer surrounding both particles and the wafer surface is thin, particle deposition, by van der Waals attraction, can take place while liquid films are still present.

CONCLUSIONS

We have confirmed that optical particle counters can give reliable and accurate readings in the range of 0.2 μm and that suction sampling can be used when obtaining particle counts in most of the chemicals commonly used by the semiconductor industry. Chemicals with high viscosity (e.g. H_2SO_4 and H_3PO_4), which were thought to give artificially high counts under suction sampling, could be accurately measured using *either* suction or pressure sampling with identical results. Only ammonia, a chemical with a very high vapor pressure, required pressure sampling. It is therefore suggested that suction sampling be used first with all chemicals, and if counts for those with elevated vapor pressures appear high, pressure sampling should then be tried. Nitric Acid was a special case, undergoing a gas-producing, light-catalyzed reaction. Based upon our findings, one producer of this material has started bottling his nitric acid in double shell bottles (light absorbing outer shell with an unpigmented liner).

A more interesting set of conclusions may be drawn from our work regarding the adhesion mechanisms. In general, we have found that the behavior of semiconductor wafer surfaces can be described by the well-documented principles of colloid science. The importance of ultra-clean processing[11,12], i.e. the use of ultra-pure DI water and low particulate chemicals, is further emphasized by these principles. First, particulate contamination levels may be related to the ionic strength of process fluids. This means that ultrapure (low ionic strength) water is essential to obtain low particle counts on wafers as well as is necessary for rinsing of soluble species from wafer surfaces. Second, we have shown that particles which adhere to wafers exposed to high ionic-strength solutions cannot be removed by subsequent pH-neutral rinses. The necessity of maintaining first (rough) rinses as well as final rinses at low levels of ionic strength and particulate concentration is obvious. Finally, we should remember that many semiconductor processes make use of high ionic strength acidic solutions, and colloid science teaches that these are the conditions for maximum particle adhesion; thus the particulate concentrations in these chemicals must be controlled as low as is possible.

ACKNOWLEDGEMENT

This research has been carried out in the Mini Super Clean Room of the Department of Electronics, Faculty of Engineering, and in the Super Clean Room of the Laboratory for Microelectronics, Research Institute of Electrical Communication at Tohoku University.

REFERENCES

1. K. L. Mittal, editor "Particles in Gasses and Liquids 1: Detection, Characterization and Control", Plenum Press, New York (1989)
2. L. D. Michaels. V. B. Menon, R. P. Donovan and D. S. Ensor, Particle Deposition at the Solid-Liquid Interface, Proceedings, 34th Annual Technical Meeting of the Institute of Environmental Sciences, 438-442 (1988)
3. H. Mishima, T. Yasui, T. Mizuniwa, M. Abe and T. Ohmi, Particle-Free Wafer Cleaning and Drying Technology, IEEE Trans. on Semiconductor Manufacturing, 2 No. 3, 69-75 (August, 1989)
4. A. Saito, K. Ohta, M. Watanabe and H. Oka, Particle Deposition Mechanism onto Silicon Wafer, Extended Abstract of the 21st Conference on Solid State Devices and Materials, (409-412), Tokyo, Japan (August, 1989).
5. T. Niida, Controlling for Particle Adhesion in Liquids - a Basic Way of Thinking, Chemical Engineering Institute of Japan, Kansai Branch, Osaka, Japan 14-25, (September, 1989)
6. D. J. Riley and R. G. Carbonell. The Deposition of Liquid-Based Contaminants onto Silicon Surfaces Proceedings, 36th Annual Technical Meeting of the Institute of Environmental Sciences, 224-228 (1990)

7. A. W. Adamson, "Physical Chemistry of Surfaces, 4th Edition", John Wiley & Sons, New York (1982)
8. D. W. Cooper, Statistical Analysis Relating to Recent Federal Standard 209 (Cleanrooms) Revisions, J. Environ. Sci. 31 No. 5, 48-52 (1988)
9. Robert H. Perry and C.H. Chilton, "Chemical Engineers' Handbook", 5th ed., pp. 3-68, McGraw Hill Book Company, New York (1973)
10. Y. Kasama, Y. Yagi, T. Imaoka, M. Kawakami and T. Ohmi, Advanced DI Water System with Low Dissolved Oxygen for ULSI Processing, Proceedings, 36th Annual Technical Meeting of the Institute of Environmental Sciences, 344-349 (1990)
11. T. Ohmi, Future Trends and Applications of Ultra-clean Technologies, Technical Digest, International Electron Device Meeting, Washington, D.C. (1989)
12. T. Ohmi, Ultra-clean Technology: ULSI Processing's Crucial Factor, Microcontamination, 6, No. 10, pp. 49-58 (October, 1988)

EVALUATION OF POLYMERIC MATERIALS USED IN THE MANUFACTURE OF DISK HANDLING CASSETTES

G. J. Stone and A. Fukushima

IBM Corporation
Storage Systems Products Division
5600 Cottle Road
San Jose, CA 95193

A test matrix was developed to evaluate materials used to construct containers intended for use with parts susceptible to contamination. Types of materials tested include polypropylene (glass-filled), ULTEM (polyetherimide), polycarbonate, ABS (acrylonitrile-butadiene-styrene), PBT (polybutyleneterephthalate), PEEK (polyether-ether-ketone), and black antistatic polycarbonate. The ULTEM and polycarbonate proved to be the most satisfactory for handling magnetic recording disks during various manufacturing processes.

INTRODUCTION

Manufacture of high density disk storage files involves a significant amount of disk handling between finishing the disk and its final mounting inside the safe file environment. It is important that the disks remain free of contamination, particulate as well as organic vapors, throughout the process. Thus, during manufacturing the disks must be handled and transported in special disk handling cassettes. The cassettes must not shed debris or transfer material onto the disks during the operation.

To decide on a suitable material, a test matrix was developed so that each type of material could be equally evaluated in terms of abrasion resistance, abrasion generated particles, extractable organics, and inorganic additives (such as fillers and colorants). A test matrix or simulative test is the only positive way to evaluate polymeric materials for a specific use such as this.[1] Pressure-velocity limits and wear factors can be determined and materials can be ranked according to their ability to resist wear. But as the authors in this first reference point out: "There probably are as many plastic wear tests as there are people performing plastic wear tests." The test described in this article is different in that we are most concerned with loose debris generated by wear and with the material that might transfer to the disks by adhesive wear.

The types of materials chosen for testing were limited by their moldability and rigidity. They included blue glass-filled polypropylene (manufactured by Fluoroware), ULTEM – three colors (Fluoroware), polycarbonate – four colors (Empak), blue ABS (Empak), PBT – two colors (Empak), brown PEEK

Particles on Surfaces 3, Edited by K.L. Mittal
Plenum Press, New York, 1991

(Empak), and black antistatic polycarbonate (Empak). The different colored materials are necessary for tracking parts. Colored materials could contain various fillers and/or dyes.

It was understood from the beginning that glass-filled polypropylene might not be a suitable material because it is composed of 33% glass-filler which could shed. However, glass-filled polypropylene was included in this study because similar cassettes made from this material had been used for other less sensitive operations. The addition of glass-filled polypropylene into the sample set proved to be very useful for comparisons. For disk handling purposes it is the worst case material.

EXPERIMENTAL PLAN

A procedure was developed to test each material for abrasion resistance, abrasion generated particles, extractable organics and/or outgassing, and inorganic additives (such as fillers and colorants). Although ASTM Test Method D 1044-78 (Resistance of Transparent Plastics to Surface Abrasion) and ASTM D3702-78 (Wear Rate of Self-Lubricating Materials in Thrust Washer Configuration) provide some information on the different materials, these were not adequate for this evaluation due to the complexity of the wear process.[2, 3] Shore D hardness measurements were made for each material to see if hardness and abrasion resistance might be related (see Table I).

All of the cassette types or samples of materials were cleaned prior to testing by using a water-based high-pressure spray cleaner (manufactured by ATCOR). Each cassette was loaded with five unfinished 130-mm disks. The disks were bare aluminum substrates with a very fine polish. The loaded cassettes were placed into cleaned polypropylene containers and placed onto a Barnstead/Thermolyne, Maxi-Mix 3 orbital laboratory shaker. The shaker has a digital meter which indicates the rate at which the stage is shaking. At 450 RPMs the disks could be heard clicking against some of the harder cassette materials such as the ULTEM. The shaker was allowed to operate for three days so that each cassette was subjected to 1.4 million shakes. There was overall agreement throughout manufacturing that this was a reasonable simulation of extensive line use where disks are loaded daily in and out of cassettes, loaded cassettes are dipped into cleaning baths, and moved about by hand or on carts.

Following the shaking operation each disk was examined under intense light to determine qualitatively whether cassette material had transferred to the disks. A few examples of such transfer are presented in optical photographs, see Figures 1, 2, and 3. It may have been possible to predict which materials would transfer debris based upon the hardness of the material. Table I includes the measured hardness of each material together with a good correlation by using the Shore "D" tester. The harder ULTEM transferred the most, then polycarbonate, followed by the softer PBT and PEEK which had the least amount of debris. The act of abrasion requires movement but the softer materials may conform and act as a damper. Because of this particular type of movement the softer polymers may have maintained a rubbery state between polymer and disk surfaces whereas the harder polymers allowed some disk movement. With disk movement, wear particles could have been generated by physically chipping the polymer surface or as a result of frictional heating. Frictional heating does result in wear for many polymers.[4]

The next step involved removing abrasion generated particles from the cassettes following the shake test. The cassettes were washed in clean polypropylene containers to remove the particles. Controls using empty polypropylene containers came out reasonably clean (see Figure 4). Figure 4 is a low magnification scanning electron microscope (SEM) view of the debris collected onto the surface of a 3-micrometer (pore size) Nuclepore polyester membrane. The photograph attempts to show the edge of the filtration area on the membrane; this will become clearer with other examples. Each cassette was then extracted by using 2.5 L of filtered water with two drops of concentrated Tergitol® added. The solutions and cassettes were agitated with ultrasonic action for two minutes.

Table I. Summary of Results and Observations for Disk Handling Cassette Materials Following a 1.4 Million Shake Test

Sample ID	Organic Analysis Observations	% Ash	# of Disks with Visible Debris (5 disks/ cassette)	Hardness Shore "D" Scale	SEM Result	SEM Observations of Filtered Solutions
ULTEM (natural)		< 0.01 ash observed	3 disks, very light	87		Light amount of debris, few chunks
ULTEM (black)		< 0.01 ash observed	present on all disks	87		Clean, few large organic pieces
ULTEM (red)		0.42% - white ash Cd, Ba/S	not tested	87	Suspect	Cd and Ba/S particles
Blue ABS	Suspect (1)	1.7% - white ash Si/Al/Ti/K	3 disks, very light	75	Suspect (3)	Very clean
Polypropylene/glass filled (blue)		33.0% - glass Si/Ca/Al, Ti	1 disk, very light	75	Fail (4)	Heavy deposit glass rods and spheres
Polycarbonate (gray)		0.7% - brown ash Ti/Ca	3 disks, very light	82		Light amount, fibers metal flakes
Polycarbonate (tan)		0.4% - brown ash Ti/Sb/Fe	1 disk, small spots	83		Clean, fibers few organic chunks
Polycarbonate (blue)		0.5% - blue ash Ti/Co/Al/Cr	3 disks, small spots	84		Light amount, fibers metal flakes
Polycarbonate (green)	Suspect (2)	not tested	not tested	82	Suspect	Used, can't be cleaned
PEEK (brown)		0.03% - yellow ash	2 disks, very light	82		Clean, some dust or dirt
PBT (blue)		0.93% - yellow ash Si/Al/Ti/ S/Ca	Clean	76	Suspect (5)	Significant, fibers, calcium
PBT (white)		0.01% - white ash	2 disks, spots, flakes	75		Light amount, some dust and dirt
Antistatic (black) polycarbonate		0.04% - no ash	not tested	80	Fail (6)	Heavy, carbon pieces, Fe flakes

Notes:

(1) Outgassing

(2) High extractables from surface contamination, samples could not be cleaned

(3) Significant amount of inorganic material in ash

(4) Glass filler is continuously removed by ultrasonic action

(5) Significant amount of inorganic material in ash and surface contamination

(6) Marks on paper and heavy deposit left on membrane

Forty milliliters was removed in order to perform a turbidity measurement and the remainder was filtered through a 3.0-μm Nuclepore polyester membrane. Turbidity values were recorded only to see if they could be useful for future studies. The contamination levels or trends observed on the membranes by using SEM correlated very well with the turbidity values. All of the membranes were then examined using a scanning electron microscope fitted with an Energy dispersive X-ray spectrometer (SEM/EDX). Observations about the types and relative amounts of contamination that were removed are summarized in Table I; some examples are presented in Figures 5 through 14. Figures 15, 16, and 17 present contrasting photographs of the excessive amount of debris extracted from the used glass-filled polypropylene cassettes mentioned earlier. Figure 16 is a high-magnification image showing the glass (containing elements Si, Ca, and Al) rods and spheres used as fillers in this material.

Testing for Fillers and Colorants

A test for the percentage of inorganic matter present in the cassette material was accomplished by ashing 4-gram samples, placed in crucibles, in a muffle furnace operated at a temperature of 900°C. The remaining ash was transferred to carbon stubs for examination by SEM/EDX. A few examples of the different types of ashes are shown in Figures 18, 19, and 20.

ORGANIC ANALYSIS

Method

The cassettes were received in opaque polyethylene bags after having been cleaned in an ATCOR Cleaner. They were handled with nylon gloves and forceps to avoid contamination by human contact. Each cassette was thoroughly rinsed with 100 mL of 1,1,2-trichloro-1,2,2-trifluoroethane (Freon), Pacific Pac distilled in glass, followed by 50 mL of Baker Resi-analyzed methanol. A known amount of dioctyl phthalate-D4 was added to each extract as an internal standard. The extracts were allowed to dry at room temperature in a fume hood. Each residue was then reconstituted with chloroform, EM Science distilled in glass. These particular solvents were chosen for their virtual lack of organic residue.

The extracts were analyzed by direct probe introduction into a Hewlett-Packard 5987A mass spectrometer. Each sample was heated from ambient to 800°C at 75°C/minute, and electron impact mass spectra were recorded at 2-second intervals throughout the heating process. The results are shown in Table II.

Comments

The total organic residue found was (except for the green polycarbonate) less than 100μg per cassette. Although there is no established specification, considering that the area of actual disk surface that would come into contact with the cassettes is very small, and the time of exposure is very short, such small amounts of organic residue should pose no problem.

One disturbing factor was observed in this study. The green polycarbonate cassette had been used at one other plant: it was cleaned but we found at least twice as much residue on it as on any other cassette. We have no other details on the history of this particular cassette, but it leaves open the question of how durable and/or clean these cassettes will be after a period of use on the line.

Table II. Summary of Organic Analysis.
Low numbers indicate organic extractables are at levels that should not be of concern to the operation.

Cassette	m/e 57	59	89	129	149	185
Blue Polycarbonate	11.20		8.49	.86	4.37	
Grey Polycarbonate	21.84		11.43	1.26	3.39	
Green Polycarbonate	46.71		18.28	17.81	6.19	2.56
Tan Polycarbonate	22.70		7.93	2.15	4.21	5.03
Red Ultem	7.75		6.73	1.31	4.28	
Black Ultem	13.47		11.66	2.18	6.20	
Blue ABS	12.18		14.39	23.00	3.15	
Black ABS	4.35	3.32	3.68		1.70	

Notes: Major Identifying Peak

m/e 57 = Hydrocarbons, Oils
59 = Fatty Amide (Antistatics)
89 = Tergitol (Surfactant in Atcor Wash)
129 = Adipate (Plasticizer)
149 = Phthalate (Plasticizer)
185 = Sebacate (Plasticizer)

RESULTS

The results are summarized in Table I. The red ULTEM material was suspect at a very early stage because cadmium, barium, and sulfur particles were removed from the surface by ultrasonic action. The presence of these elements in the ash suggests that compounds containing them were added to the ULTEM possibly to help produce the color. Due to their presence it is recommended that the red ULTEM cassettes be used only in line-operations where the disks are carefully washed afterwards.

Figure 1. Optical photograph (10x) of debris transfer from black ULTEM cassette to disk.

Figure 2. Optical photograph (30x) of debris transfer from white PBT cassette to disk.

Figure 3. Optical photograph (30x) of debris transfer from blue polycarbonate cassette to disk.

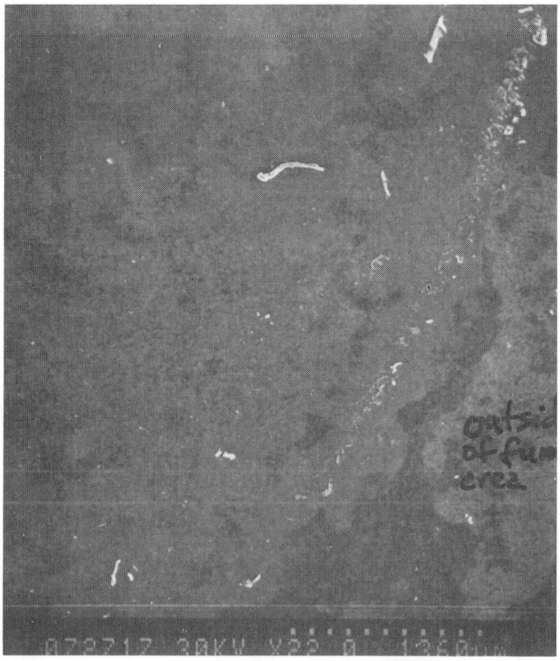

Figure 4. SEM photograph (22x) of debris collected onto membrane from control blank.

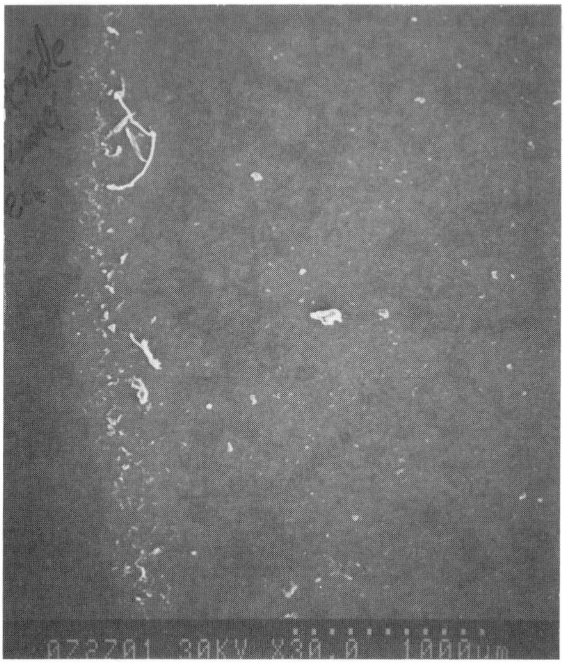

Figure 5. SEM photograph (30x) of debris collected onto membrane from natural or ULTEM cassette.

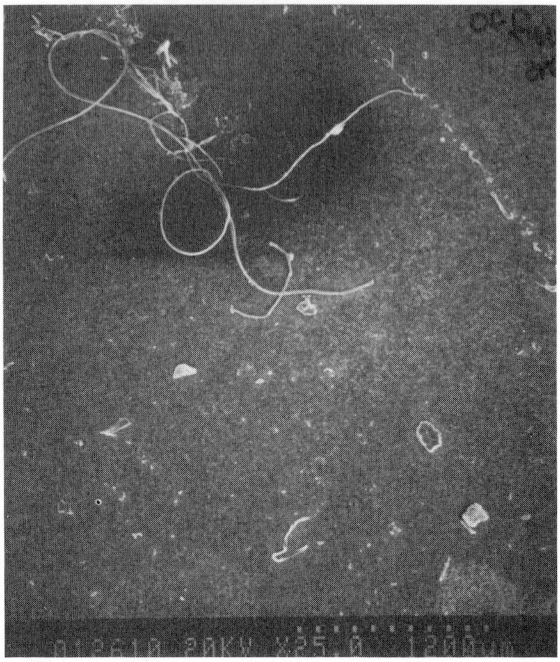

Figure 6. SEM photograph (25x) of debris collected onto membrane from gray polycarbonate cassette.

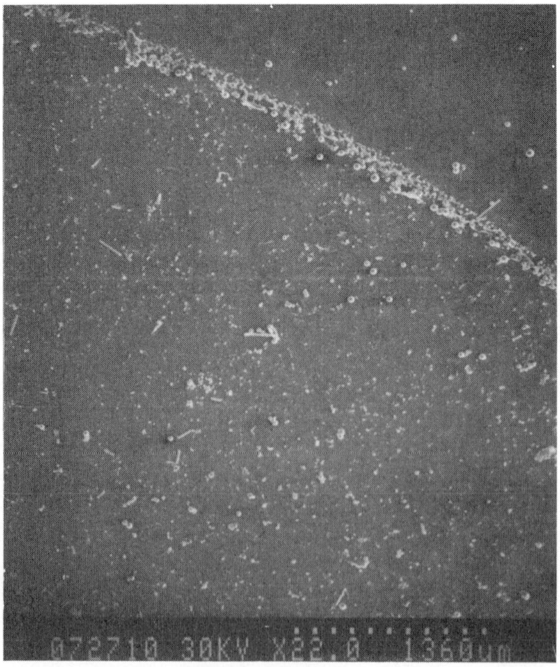

Figure 7. SEM photograph (22x) of debris collected onto membrane from new unused glass filled blue polypropylene cassette.

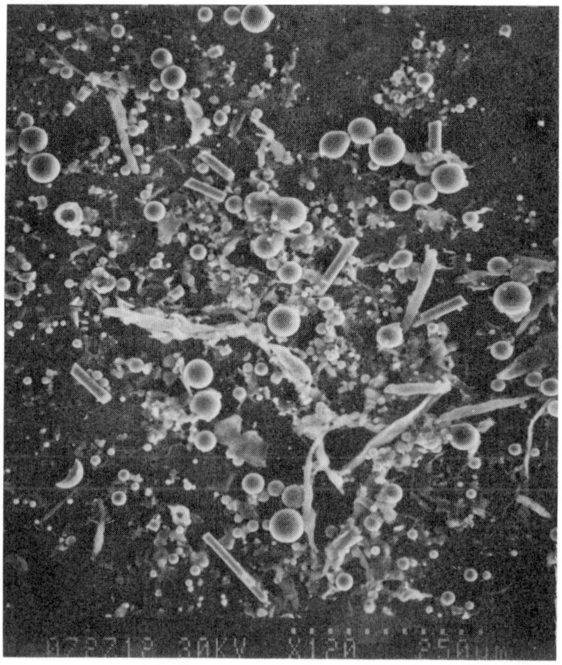

Figure 8. SEM photograph (120x) of debris collected onto membrane, same sample as Figure 7.

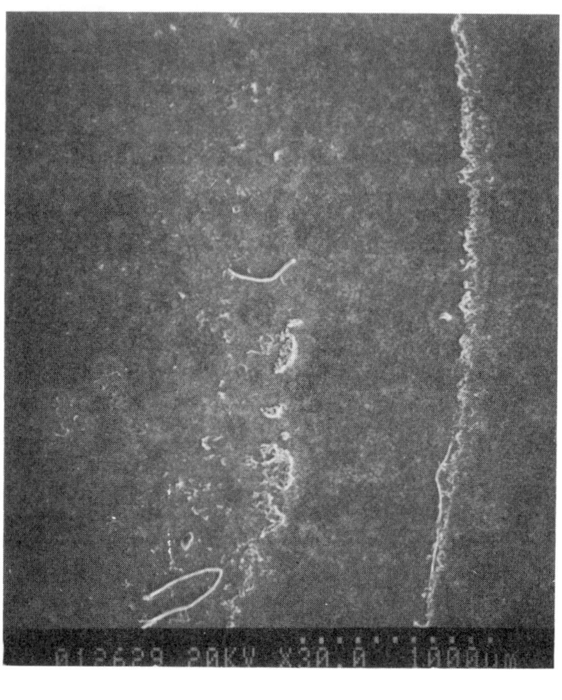

Figure 9. SEM photograph (30x) of debris collected onto membrane from white PBT cassette.

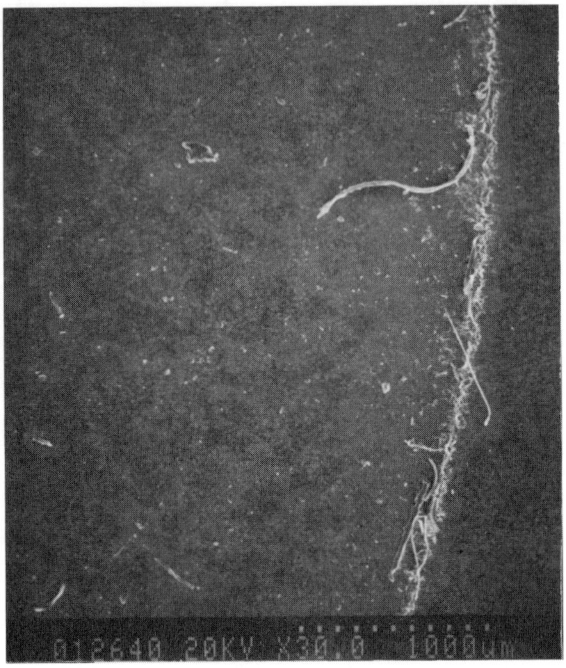

Figure 10. SEM photograph (30x) of debris collected onto membrane from blue PBT cassette.

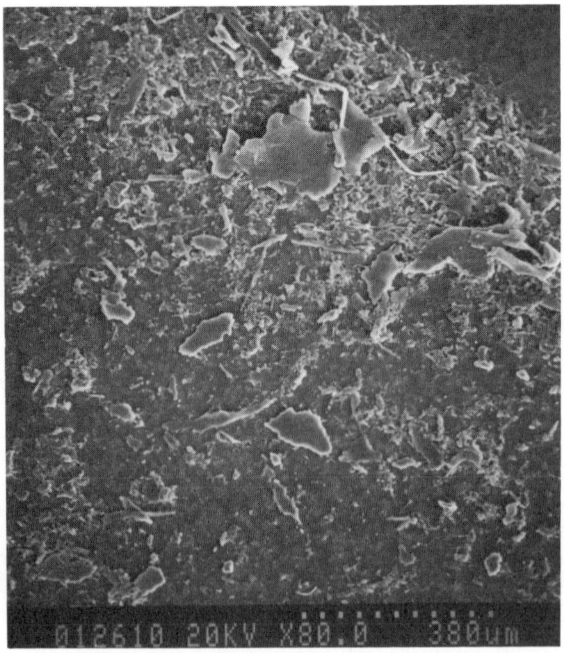

Figure 11. SEM photograph (80x) of debris collected onto membrane from anti-static black polycarbonate cassette.

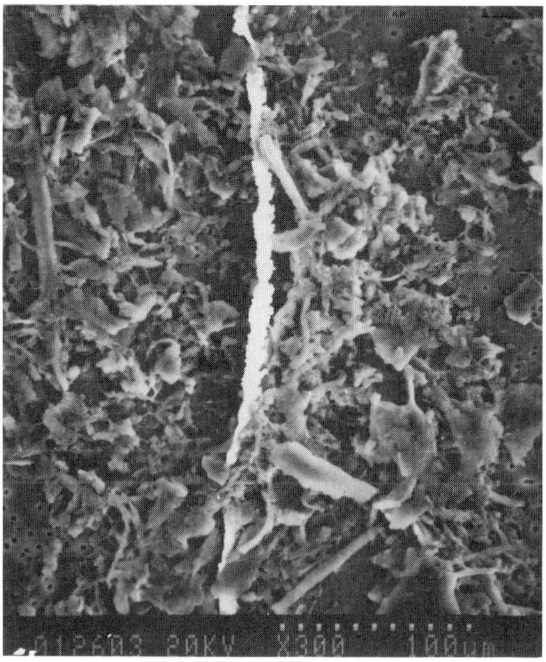

Figure 12. SEM photograph (300x) of debris collected onto membrane, same sample as Figure 11.

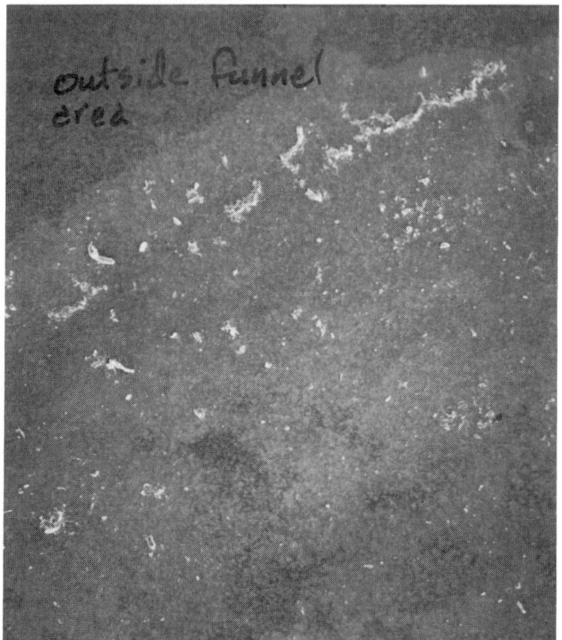

Figure 13. SEM photograph (30x) of debris collected onto membrane from tan PEEK cassette.

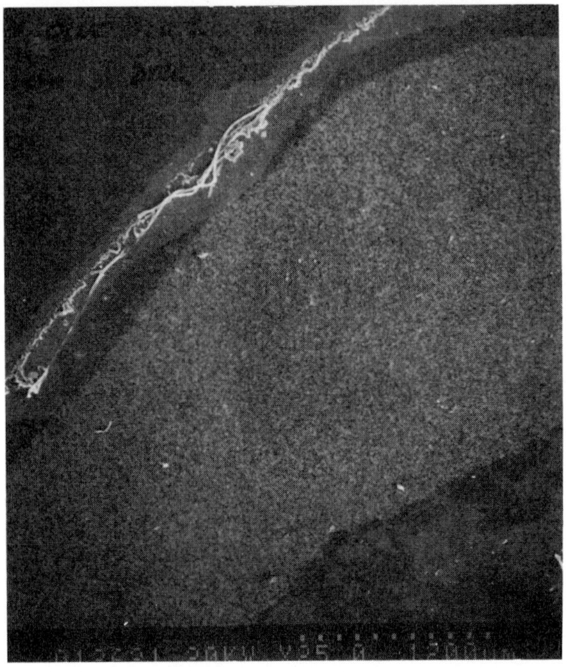

Figure 14. SEM photograph (25x) of debris collected onto membrane from tan polycarbonate cassette.

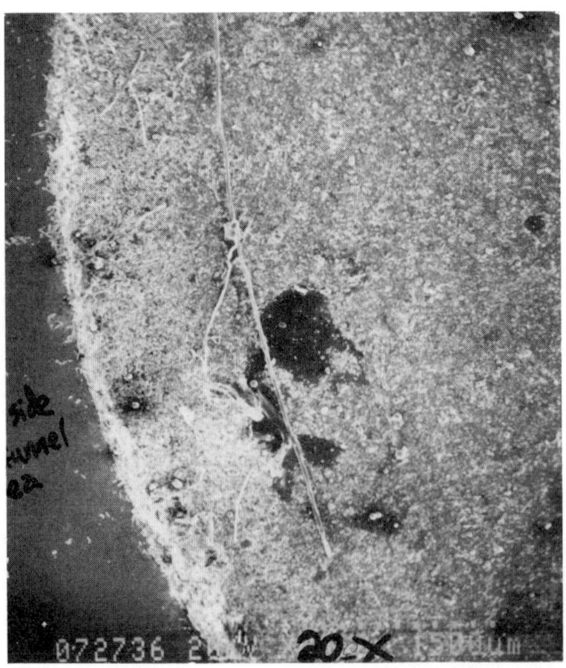

Figure 15. SEM photograph (20x) of debris collected onto membrane from used (ATCOR cleaned) glass filled blue polycarbonate cassette.

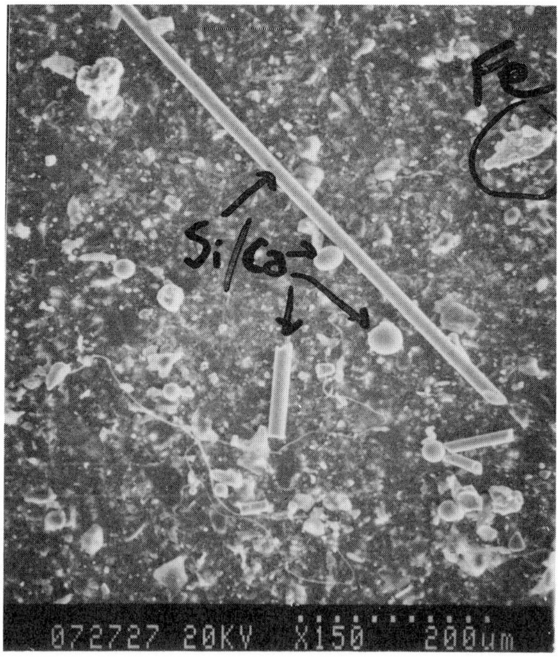

Figure 16. SEM photograph (150x) of debris collected onto membrane (note large glass rods some greater than 0.5 mm in length), same sample as Figure 15.

Figure 17. SEM photograph (220x) split image to demonstrate glass spheres present in debris collected onto membrane, same sample as Figure 15.

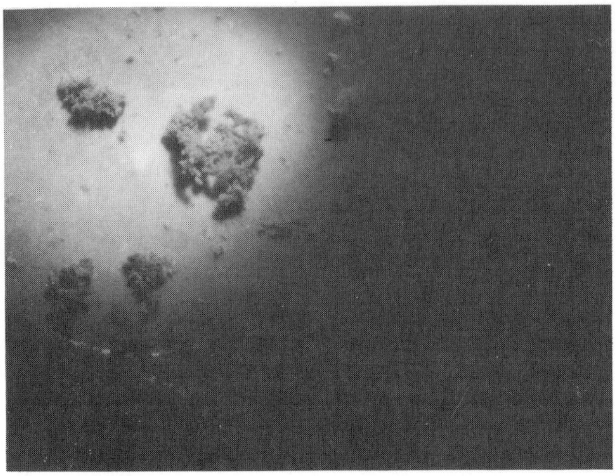

Figure 18. Optical photograph (20x) of ash remaining in the bottom of crucible following burn-off in muffle furnace, antistatic black polycarbonate.

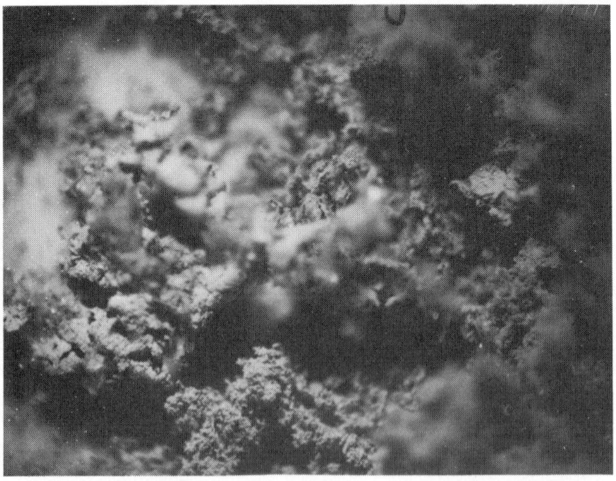

Figure 19. Optical photograph (200x) of ash remaining in the bottom of crucible, blue polycarbonate.

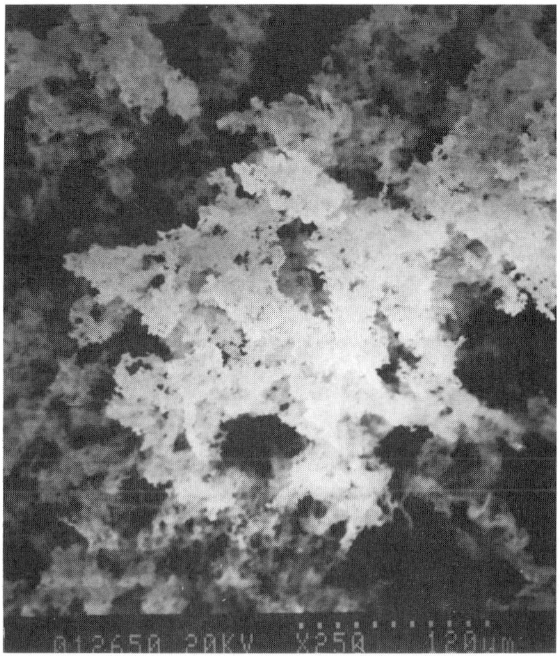

Figure 20. SEM photograph (250x) of ash remaining from tan polycarbonate cassette material following burn-off in muffle furnace.

Other materials considered suspicious or unacceptable include blue ABS, blue PBT, green polycarbonate, and black antistatic polycarbonate. The problems with the green polycarbonate have already been discussed in the organic section. The blue ABS is not only suspect for the outgassing portion (see Table II, m/e 129 possible plasticizer present) but was also found to contain almost 2% inorganic material. More work needs to be done to determine whether or not 2% is a problem for disk handling cassettes. The blue PBT is suspect because of the ultrasonically extracted particles rich in silicon and calcium. The black antistatic polycarbonate material failed because of the high amount of extractable particulates, see Figures 11 and 12. The material was doubtful at an early stage when it was observed that the cassette could be used to mark on paper which indicated the polymer would transfer material to disks.

The abrasion and wear tests indicate that the harder ULTEM polymers, depending on how they are used, may generate debris over time. The results show that the polycarbonates are the best selection for the disk line. The polycarbonates transfer very little material to the disks, release small numbers of particles (see Figure 5) and organics, and the amount of inorganic particulates is less than one percent. Interestingly, the polycarbonate cassettes cost about one half that of ULTEM.

The authors recommend that after a material has been selected it should be analyzed again after the cassette has been subjected to use on the line. To illustrate the importance of this recommendation, compare the photographs in Figures 7 and 15. Figure 15 shows the amount of extracted material from a used glass-filled polypropylene cassette: however it is important to remember that the cassette had been cleaned in an ATCOR cleaner before the extraction. Figure 7 shows the particles extracted from a new, never before used glass-filled polypropylene cassette that was also cleaned before the extraction. There is clearly more contamination on the membrane from the used cassette. The difference is due to the fact that after some use the filler (silicates) is more exposed on the surfaces and thus it is easier to shed particles.

CONCLUSIONS

The results of this study show that a simulative test should be developed and performed when selecting polymeric materials to be used with other materials that could cause abrasion. In this study the parts are magnetic recording disks which must be kept free of all forms of contamination such as abrasion generated particles, extractable organics, and inorganic additives.

All of the specifications that are available for different materials such as abrasion loss, coefficient of friction, density, are useful and help to narrow down the field when selecting materials. However, due to the complexity and the large number of different wear mechanisms, simulative tests such as the one described in this paper are necessary to observe how the materials will actually behave.

Finally, no simulative test can exactly match the manufacturing line environment; thus, the polymeric parts in question must be examined again after actual use on the line.

REFERENCES

1. T. Grove and K. Budinski, Predicting polymer serviceability for wear applications, in "Wear Tests for Plastics: Selection and Use," R. G. Bayer, Editor, STP 701, p. 18, ASTM, Philadelphia, PA, 1978.
2. ASTM, D1044 - 78.
3. J. K. Lancaster, Tribology, 6, No. 6, 219 (Dec. 1973).
4. K. Tanaka and Y. Uchiyama, Friction, wear and surface melting of crystalline polymers, in "Polymer Science and Technology," Volume 5B, L.-H. Lee, Editor, p. 499, Plenum Press. NY, 1975.

THE RELEASE OF PARTICLES DURING SPACEFLIGHT

J. B. Barengoltz

Jet Propulsion Laboratory
California Institute of Technology
Pasadena, California 91109

Particles have been observed by spacecraft sensors, sometimes in a manner that threatened mission objectives. A selective history of such missions is presented to demonstrate that particles in the vicinity of spacecraft are common and that the particles originate with the flight system. Some basic experimental and analytical work to understand particle adhesion and release mechanisms and to predict the release of particles is reviewed.

INTRODUCTION

Particles released from the surfaces of spacecraft have apparently been detected since very early in the exploration of space. None of the early "sightings" were accomplished by sensors intended to detect particles. Rather, particles were inferred from anomalous loss-of-lock events by the star trackers of unmanned interplanetary spacecraft. In 1964, acting on a suggestion by Boundy that micrometeoroid impacts could release contaminant particles from Mariner 4 (Mars 1964), Spencer satisfactorily explained the observed Canopus tracker anomaly rate.[1] The advent of manned spaceflight produced the first direct observation of particles by an appropriate sensor, the human eye. John Glenn's report of "firefly swarms" is probably the earliest unambiguous detection of particles in the vicinity of a spacecraft. However, these particles likely arose from a different mechanism than in the unmanned occurrences, namely frozen liquid waste droplets. A solar coronagraph on Skylab provided some quantitative data on particles unrelated to vehicle activities, such as thruster firings and waste dumps.[2,3,4] More recently, particles have been observed by Shuttle astronauts. On Space Transportation System (STS)-3, videotapes of the cargo bay cameras document some of the particle environment.

The Shuttle program represents the first attempts to specifically measure the particle environment in the vicinity of a spacecraft or vehicle. This effort was part of a comprehensive agenda to characterize all of the contaminant environments associated with the STS for the benefit of flown payloads. The Camera/Photometer instrument[5] of the Induced Environment Contamination Monitor (IECM) made measurements on STS-2, 3, 4 and 9 (Space Lab 1). Other instruments, such as the Particle

Particles on Surfaces 3, Edited by K.L. Mittal
Plenum Press, New York, 1991

Analysis Cameras for Shuttle (PACS) have been flown.[6] More instruments have been proposed, such as a new coronagraph.[7]

From its modest beginnings, the study of particles in the vicinity of spacecraft and vehicles has become rather important. Early manifestations caused usually minor difficulties at the spacecraft system level, rather than interference with scientific instruments. However, occasionally serious system problems arose, as in the case of Mariner Venus/Mercury 1973 (MVM'73) or the Tracking and Data Relay Satellite System (TDRSS)-B. The particles were of no scientific interest, only a problem to be avoided. Typical predictions for typical instruments indicated unobservable levels of degradation, and none was observed. Even when an effect might have occurred, as for the Infrared Astronomical Satellite (IRAS)[8], changes were made to render the detection system insensitive to particles in the field of view. There is an element of irony that only an instrument that is sensitive to particles (in the sense of being adversely affected) has a chance of observing them. With the expectation of improved scientific instruments that may be sensitive to this environment, the development and design of instruments to detect and measure particles is finally at hand.

UNMANNED SPACECRAFT

Many early unmanned planetary spacecraft flown by the Jet Propulsion Laboratory (JPL) have experienced occasional loss-of-lock of Canopus (the reference star) by the star tracker. Mariner 4 experienced about one such event per day, for example.[1] These anomalies are understood to be a result of sunlight scattered into the sensor by a particle (or group of particles) in its field-of-view. The particle must scatter in a sufficient intensity and for a sufficient duration to cause the spacecraft to go into a roll search (looking for Canopus). Spencer performed an analysis based on micrometeoroid impacts on the solar panels releasing contaminant particles from the surface.[1] His early analysis crudely treated particle adhesion and surface response to impact. The meteoroid model in use then is also outdated. Nevertheless, all the features of a modern analysis are present, including forces to affect the trajectory of a released particle: solar radiation pressure, spacecraft electrostatic field, and spacecraft gravitational field.

For subsequent missions, greater particulate cleanliness control was attempted to minimize the frequency of this particle interference with the star tracker. Contamination control was also practiced for the proper performance of scientific instruments and as a planetary quarantine requirement (especially for Mars missions). In general, some improvement in bright particle interference was achieved for Mariner 7 (Mars 1969) and Mariner 9 (Mars Orbiter 1971). However, this interference was usually considered a nuisance rather than a hazard by flight operators. The events typically only caused an undesirable consumption of attitude control propellant (during the search to reacquire the reference star).

A notable exception is Mariner Venus/Mercury 1973. This spacecraft had very relaxed contamination control, probably because the planetary quarantine requirements, which historically drove contamination control, were trivial for Venus and Mercury. As a result of a larger (expected) level of contaminant particles and a mission that approached the Sun, the frequency of bright particle events became a serious matter for MVM'73. The star tracker's searches had caused the consumption of a significant fraction of the attitude control gas on-board by the end of the Venus fly-by. The margin for the Mercury phase of the mission was minimal.

The project elected to go to a manual attitude control mode for the rest of the mission. With the automatic roll search inhibited, the marginal gas supply lasted for more Mercury encounters than originally planned, on this very successful mission. For the present discussion, MVM'73 demonstrated that system surface particulate contamination and its release during flight could cause a system failure.

The next interplanetary mission was the two Viking spacecraft launched to Mars in 1975. These spacecraft met stringent planetary quarantine requirements; in fact the Lander was partially sterilized. The Orbiter also was relatively clean because of the chance of an accident at Mars, as well as the needs of the celestial sensors and the scientific experiments. The difference in cleanliness requirements led to some interesting modeling developments for Viking, as discussed in the section on Analysis below. As reported by Hoffman and Koukol[9] in an excellent review of the contamination control for both Viking and Voyager, the Viking spacecraft had fewer bright particle occurrences than previous Mariner spacecraft. The events were generally correlated to spacecraft activities (pyrotechnic device firings and science platform slewing). Thus, mechanisms other than micrometeoroid impacts were relevant. The frequency decreased from about 1 particle per day early in the mission to about 1 per 100 days after the Orbiter had been in Mars orbit for more than a year. The rate reduction reflected the eventual depletion of the particle supply; i.e., the generation of new particles through material degradation did not occur. Also, the lower solar illumination near Mars was a factor. A smaller particle (8 μm diameter)

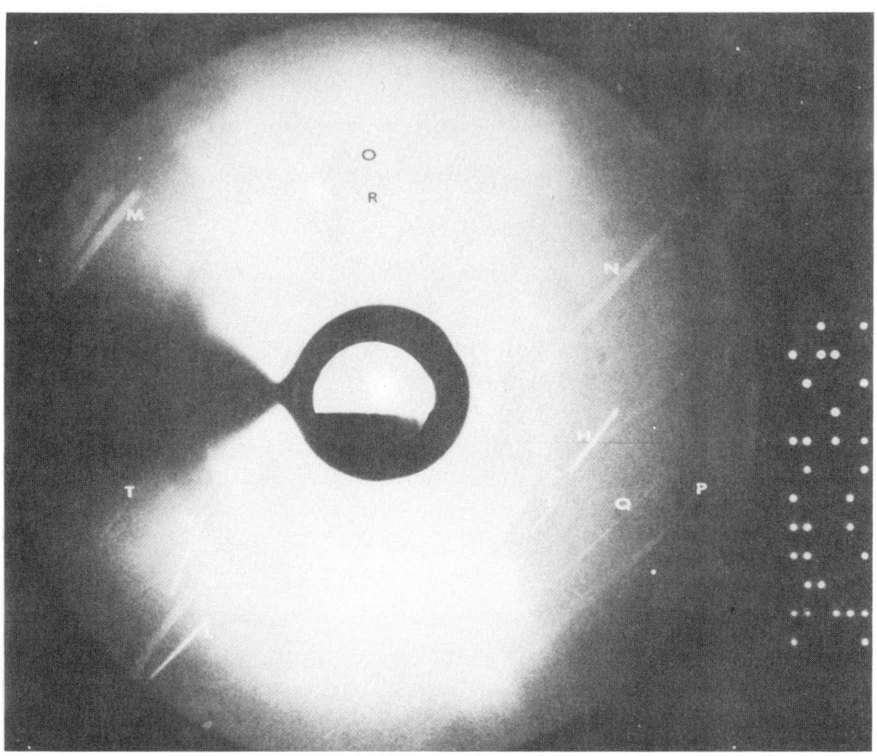

Figure 1. Particle tracks on a nine second exposure flight photograph by the Skylab S052 white light coronagraph (Ref. 3).

is sufficiently large to be observed by the star tracker near Earth, while a 15 μm particle (presumably fewer in number) is the minimum near Mars.

The early results for Voyager, two spacecraft launched in 1977 to the outer planets, were similar. A slightly lower typical frequency was observed, but a larger number of events occurred during propulsion burns.[9] Of course, the decrease of illumination with heliocentric distance would significantly reduce the occurrences beyond the orbit of Mars.

A more recent problem due to particles in the field-of-view of celestial sensors occurred with the first successful TDRSS satellite. Evidently particles either from the STS Orbiter, from the Inertial Upper Stage (IUS) - satellite stack, or from the deployment procedure caused difficulties for the IUS celestial attitude sensors. This is conjecture by this author and others; little real information has been published.

MANNED VEHICLES

After qualitative particle sightings by Gemini and Apollo astronauts, the sightings on Skylab included a photographic record (stills and videotape). Figure 1 is a typical photograph that exhibits

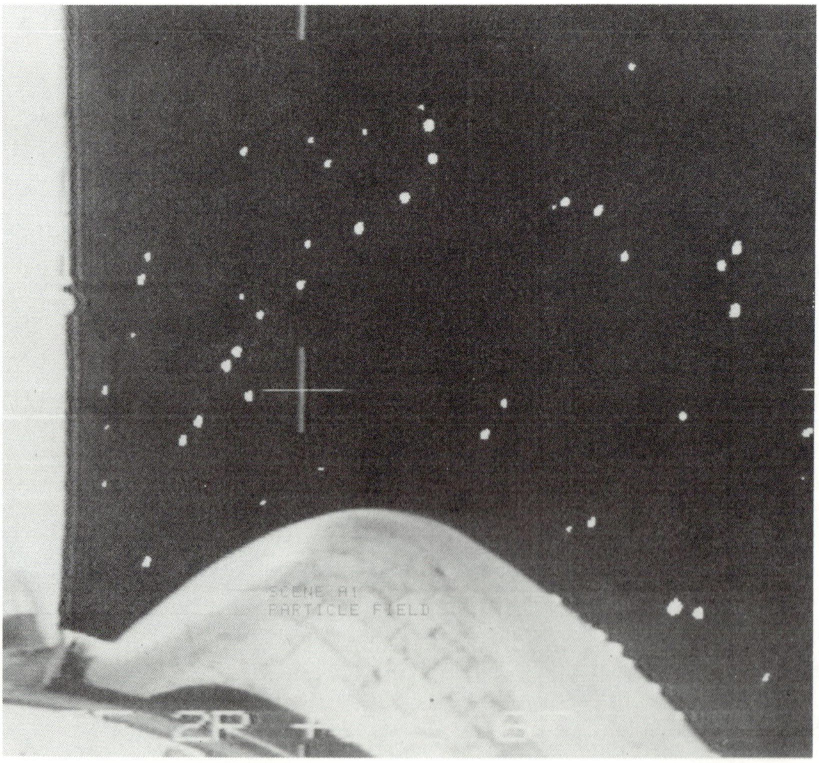

Figure 2. Particles in one frame of a videotape taken by the STS-3 astronauts using the Shuttle cargo bay television camera. This view is looking aft from the cargo bay forward bulkhead and slightly to port of the tail.

particles.[3] Schuerman, et al.[2,3,4,7], using this record from a solar coronagraph, were able to bound the size of the observed particles as less than 200 μm. Particles down to the detection limit of 5 μm were seen. The differential size distribution in this range was approximated as inversely proportional to the 1.3 power of the size.[4,7] The typical speed of the particles was 1 ms^{-1}. During quiescent periods, the number density of particles within 200 m of the coronagraph, the detection limit, varied from approximately 10^{-9} to 2×10^{-8} cm^{-3} (from 5 to 100 particles per field-of-view).[3]

As might be expected, Shuttle astronauts have also observed particles. Lousma and Fullerton videotaped particles with the Orbiter cargo bay television cameras (STS-3) during the third day. A frame of that videotape is shown in Figure 2. This data set was analyzed at JPL by Kuykendall and this author to determine the particle trajectories and to estimate speed and size. Figure 3 is an example of a composite that shows the trajectories. Every particle was moving away from the Orbiter. With no photometry, the speed and size could only be estimated as though the particles were near the aft end of the Orbiter. The effect of atmospheric drag as a size determinant was also considered. With the distance assumption, the typical size of these particles was 1 to 4 cm, and their speed was about 1 ms^{-1}. If the particles were actually closer to the camera, then they were smaller and slower.

Both the IECM Camera/Photometer[5] and the PACS[6] instruments were

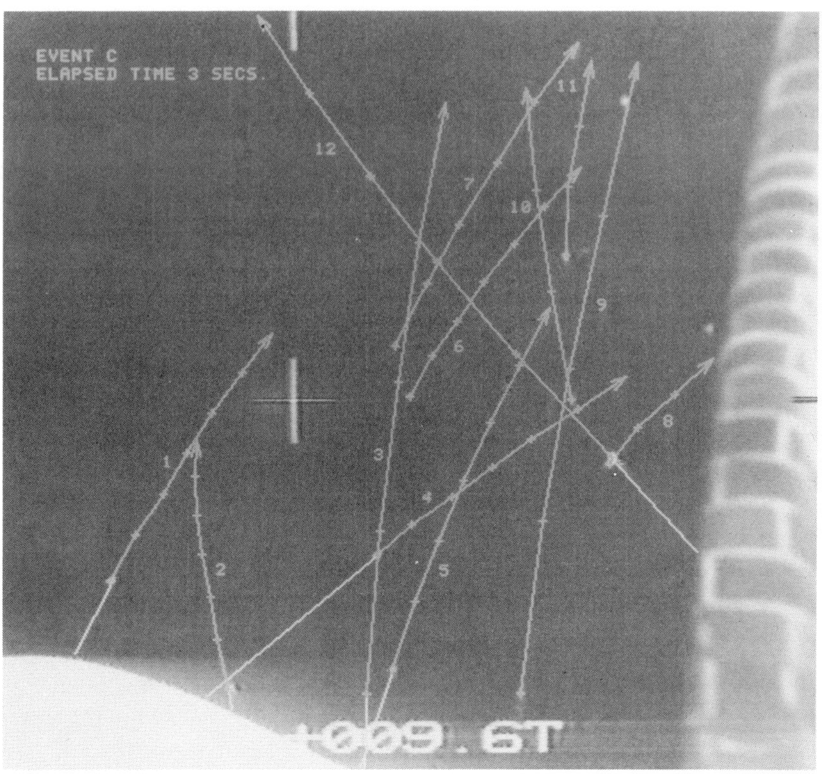

Figure 3. A composite of many frames by Kuykendall which shows the trajectories of the particles. The frames are similar to Figure 2 except view is slightly to starboard of directly aft.

designed to measure the particulate environment. Both attempted to resolve the apparent size-distance ambiguity by employing a stereo pair. The IECM instrument also used absolute photometry for this purpose. A typical particle event[10] is shown in Figure 4. Clifton and Owens[5] interpreted the Camera/Photometer data for the various flights as: particle mean speed 1.2 ms^{-1}, with a similar median; mean particle radius 53 μm during quiescent periods; and mean particle radius during water dumps 102 μm. They derived size and speed distributions. Many of the particles were smaller than 40 μm; the detection limit was 10 μm. They also provided an estimate of background brightness due to unresolved particles and molecules of less than 4×10^{-3} erg cm^{-2} s^{-1} sr^{-1} (9×10^{-13} mean solar brightness).

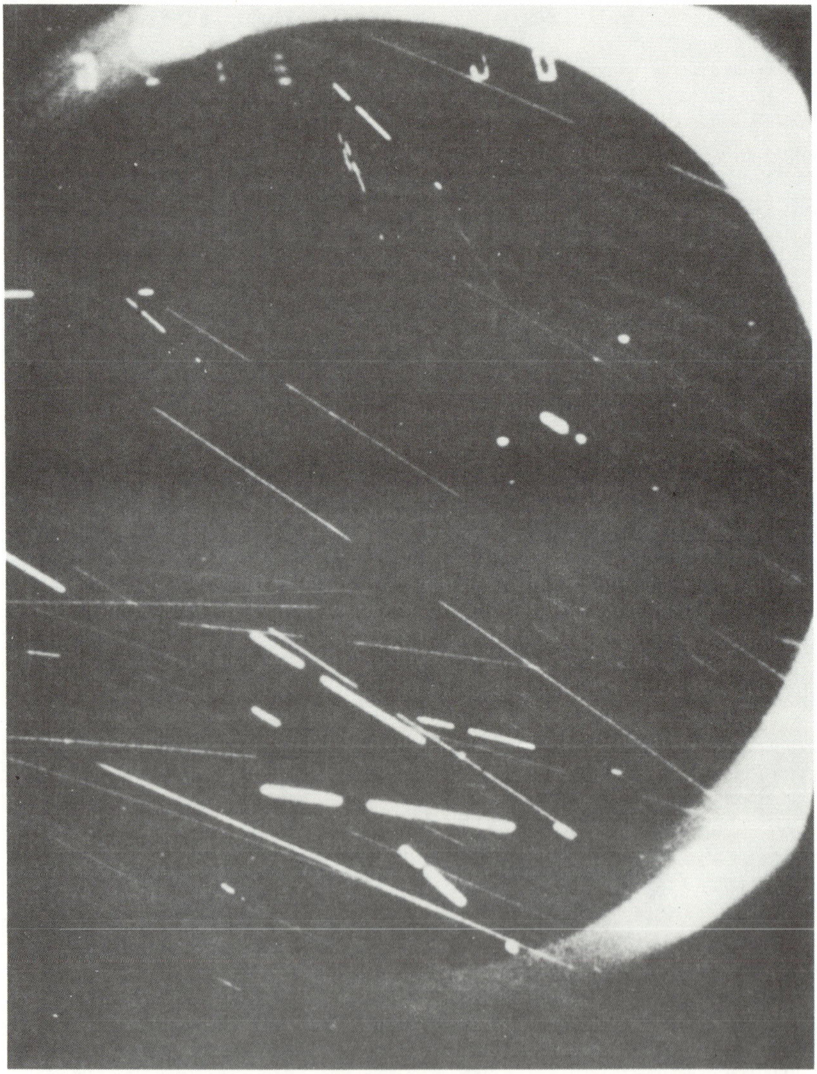

Figure 4. Particle tracks on a two second exposure flight photograph by the IECM Camera/Photometer. This photograph is typical of a particle generating event and correlates with the timing of an STS Orbiter Reaction Control System firing (Ref. 10).

The PACS was equipped with a strobe flash to detect particles even when the light direction from the Sun was not suitable for viewing them. Unfortunately, only one camera functioned properly during its STS-61C flight. Due to the malfunction, particle speed and size could only be treated in analogy to the analysis of STS-3. However, PACS' ability to see particles when present proved of extreme importance. While the Camera/Photometer showed quiescent periods late in the mission where no particles were observed[5], Green et al.[6], interpreted the PACS data to show some particles under all conditions. They also stated that thermally-induced surface motion (as at local sunrise) generates particles. This observation may be a release mechanism at work. Both instruments showed a general decrease in particles (or sightings) with the passage of time. The temporary increase observed during vehicle activity then decayed. Clifton and Owens[5] provided a characteristic clearing time of 5 minutes for a water dump.

MODELS OF RELEASED PARTICLES

The first real effort to model the release of contaminant particles from spacecraft external surfaces after the early work[1] for Mariner 4 was undertaken in the planetary quarantine program. The original interest was the contamination of another planet by the sloughing of such particles (presumably carrying terrestrial microbes). Then, for thermal design reasons, the decision was made to fly Viking with the bioshield partially removed. The bioshield was to protect the Lander from contaminants from the unsterile Orbiter, so only the half of the "clam shell" farthest from the Orbiter was to be jettisoned after launch and deployment. Thus began a comprehensive research activity to estimate the transfer of particles from the Orbiter to the Lander during Viking's trip to Mars.[11]

This program included research relevant to the present discussion such as particle adhesion, surface response to micrometeoroid impact, spacecraft and particle charging in interplanetary space, and the transport of released particles in the force fields present. Reference 11 comprises a summary of the early work. Of course, the complete analysis could never be tested because no flight measurements were possible. However, the combination of the early particle adhesion model and surface response model did compare well with limited ground simulation data.[12] Later adhesion work, with dust supplementing glass beads in a centrifuge, brought particle adhesion to its present status.[13] This author estimated the release of particles due to impact using the improved particle adhesion data.[14] A larger source due to spallation from the Orbiter thermal protection tiles was also modeled at that time.

Turning to the transport of particles into the field-of-view of an instrument on an Earth satellite, one must augment the forces with atmospheric drag and, in the satellite-centered coordinate system, centrifugal and Coriolis forces. The electrostatics is also different in Earth orbit. This type of analysis was performed for IRAS with the use of these various models.[8] The predictions of a particle in the field-of-view about 1% of the time led to a design change in the electronics to discriminate against objects only viewed for a short time. This strategy worked well for IRAS, but prevented any confirmation of the prediction. Simpson and Witteborn[15,16] considered the effect of particles on an infrared telescope as a part of their excellent assessment of the Shuttle environment. They concluded that sightings of particles would not seriously affect a Shuttle Infrared Telescope Facility (SIRTF) provided that vents and engine firings were properly controlled.[15] In their later publication[16], they considered the spallation estimates of Reference 14. This predicted environment would affect instruments with broad bandwidths (10 μm) at or below 10μm wavelength or with long integration times.

CONCLUSIONS

In general, most particle predictions are incompletely tested. For example, this author's prediction of the particles spalled from Shuttle tiles[14] is unproven. (See, for example Simpson and Witteborn.[16]) Simulations are difficult. Flight data are very uncontrolled (in the experimental sense). Scientific instruments are understandably designed, as possible, to be blind to these particles (e.g., IRAS and the Space Lab Infrared Telescope). Fast particles, such as those produced by spallation, may not be directly observed but add to the background brightness. The various postulated particle production mechanisms must be understood, and their relative contributions estimated. Nevertheless, there is a large body of data which must be more thoroughly analyzed, but only a few experimental models exist. Better scientific instruments, typically more sensitive to particles, are expected in the future. The analysis of an instrument's response to the particle environment, such as was carried out by Simpson and Witteborn[16] for SIRTF, requires better input. Contamination estimation procedures must be ready.

ACKNOWLEDGEMENT

The research described in this paper was carried out by the Jet Propulsion Laboratory, California Institute of Technology, under a contract with the National Aeronautics and Space Administration.

REFERENCES

1. D.F. Spencer, "A Consistent Mechanism for the Canopus Sensor Operation Observed on the Mariner C," internal document, Jet Propulsion Laboratory, 1964.
2. D.W. Schuerman, D.E. Beeson, and F. Giovane, Coronagraphic technique to infer the nature of the Skylab particulate environment, Appl. Optics, 16, 1591 (1977).
3. D.W. Schuerman and J.L. Weinberg, "Preliminary Study of Contaminant Particulates Around Skylab," Marshall Space Flight Center, NASA CR-2759, 1976.
4. D.W. Schuerman, "Skylab Contamination Results: S052 Particle Analysis and T025 Particle Analysis," Contractor Report, Marshall Space Flight Center, NAS8-32485, 1977.
5. K.S. Clifton and J.K. Owens, Optical contamination measurements on early Shuttle missions, Appl. Optics, 27, 603 (1988).
6. B.D. Green, G.K. Yates, M. Ahmadjian, and H. Miranda, The particulate environment around the Shuttle as determined by the PACS experiment, in "Optical System Contamination: Effects, Measurement, Control," P. Glassford, editor, Proc. SPIE, 777 (1987).
7. D.W. Schuerman and F. Giovane, Coronagraph: The ultimate device to monitor orbiting particulates in the Space Transport System (STS) environment, in "Spacecraft Contamination Environment," Proc. SPIE 338 (1983).
8. L.C. Andreozzi, W.R. Irace, and C.R. Maag, Contamination control of the infrared astronomical satellite, in "Optics in Adverse Environments II," Proc. SPIE 216 (1980).
9. A.R. Hoffman and R.C. Koukol, Particulate contamination control for the Viking and Voyager unmanned planetary spacecraft, in

"Proceedings of the USAF/NASA International Spacecraft Contamination Conference," J.M. Jemiola, editor, AFML-TR-78-190 or Marshall Space Flight Center, NASA-CP-2039, 1978.

10. J.K. Owens and K.S. Clifton, Camera/Photometer, in "STS-2 Induced Environment Contamination Monitor (IECM) - Quick-Look Report," E.R. Miller, editor, Marshall Space Flight Center, NASA TM-82457, 1982.
11. J. Barengoltz and D. Edgars, "The Relocation of Particulate Contamination During Spaceflight," Jet Propulsion Laboratory, NASA TM 33-737, 1975.
12. J.H. Goad, Jr., J.D. Battista, D.M. Robinson, and W.P. Chu, "Removal of Spacecraft-Surface Particulate Contaminants by Simulated Micrometeoroid Impacts," Langley Research Center, NASA TN D-7494, 1974.
13. J.B. Barengoltz, Particle adhesion to surfaces under vacuum, J. Spacecraft and Rockets, 26, No. 2, 103 (1989).
14. J. Barengoltz, Particulate release rates from Shuttle Orbiter surfaces due to meteoroid impact, J. Spacecraft and Rockets, 17, No. 1, 58 (1980).
15. J.P. Simpson and F.C. Witteborn, Effect of the Shuttle contaminant environment on a sensitive infrared telescope, Appl. Optics, 16, No. 8, 2051 (1977).
16. J.P. Simpson and F.C. Witteborn, Infrared radiation from the Space Shuttle contaminant environment, in "Proceedings of the USAF/NASA International Spacecraft Contamination Conference," J.M. Jemiola, editor, AFML-TR-78-190 or Marshall Space Flight Center, NASA-CP-2039, 1978.

PART II. PARTICLE DETECTION, IDENTIFICATION, ANALYSIS AND CHARACTERIZATION

STATISTICAL ASPECTS OF SURFACE PARTICLE COUNTING

B. Leslie, J. Pecen, K. Gross and A. Neukermans

Tencor Instruments
2400 Charleston Rd
Mountain View, CA. 94043

A method is described which characterizes the statistical variations in detecting and sizing surface defects, such as particles on monitor wafers, using monodisperse spheres. This variability is related to the systematic and random errors involved in the measurement process. In addition, these parameters determine the statistics with which defects are counted and a general technique is described to predict the mean and variance in counting any defect population.

INTRODUCTION

The detection and accurate measurement of light scattering defects on a surface such as an unpatterned silicon wafer is a process which has several sources of statistical variability. As such, it is not possible to obtain an exact count or sizing of these defects on a given sample. A user must be aware of these limitations and understand the statistical nature of the detection method before he can interpret the measurements with any degree of confidence. Failure to do this may lead to costly mistakes in the maintenance and control of process tools that this detection method is intended to monitor.

An example of this problem occurs when a user attempts to count the number of defects added to a monitor wafer when it is passed through a semiconductor process tool. For a typical application, where the number of defects is small, say less than 20, the number of reported defects can vary by as much as 15% for repeated scans of the substrate on the same instrument. This level of variability, though a natural consequence of the statistical nature of the measurement process, can be quite disconcerting to the user.

Another manifestation of this problem is in comparing 'Defect Maps' for repeated scans of a substrate by the same instrument. Although the total number of measured defects is approximately the same from scan to scan, it is commonly observed that as many as 30% of the defects may be reported on

Particles on Surfaces 3, Edited by K.L. Mittal
Plenum Press, New York, 1991

some maps but not others and their reported size may also vary from scan to scan. Even though the measuring instrument is operating properly, this unpredictable behavior can give serious concerns about its quality and performance.

The explanation for these phenomena is to be found in terms of the statistical variability in the signal detected from the light scattering defects. There are three main sources of variability: i) the imperfections in the detection sensitivity of the instrument itself, ii) the errors introduced by the spatial sampling of the surface under test, and iii) Poisson noise in the detected signals from the defects as well as the background scatter from the substrate under test. This paper will evaluate the influences from these sources of variability and their corresponding effect on the final count and size accuracy of the measured defects.

DEFECT DETECTION METHOD

The method by which light scattering defects are detected involves scanning the surface under test with a focussed laser spot. The spot is moved sequentially with respect to the substrate so that the entire surface is sampled by the beam. In a typical implementation as shown schematically in Figure 1, the laser spot is scanned in a line focus at high speed as the substrate is moved through this scanning beam at relatively slower velocity in a direction perpendicular to the beam motion. When the beam illuminates a defect on the surface, light is scattered away from the reflected beam so that it is intercepted by the collection optics. The scattered light is collected and directed onto a low noise photodetector. This signal is amplified and digitized at regular intervals by a high speed analog-to-digital converter so that the scattering amplitude is measured at a uniform array of points across the surface of the substrate under test.

SIGNAL BROADENING EFFECTS

The detected scattered light is a function of both the properties of the scattering source as well as of the instrument parameters. Variability in either of these components will result in a distribution of signal amplitudes. The instrument dependent variability can be assessed by measuring the response to a population of monodisperse polystyrene latex

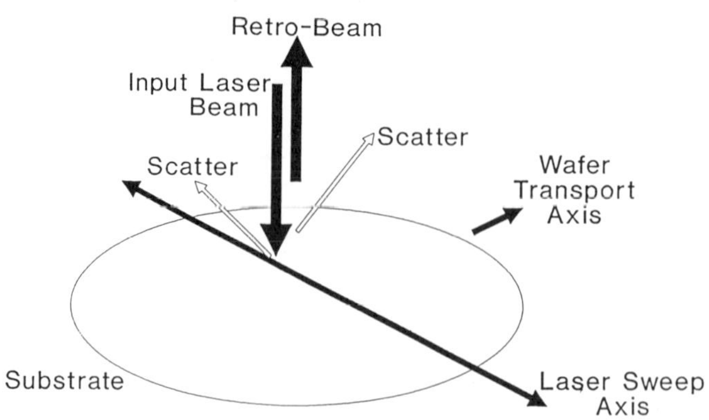

Figure 1. Scanning Technique to Detect Surface Defects on a Substrate Under Test.

(PSL) spheres on a clean substrate, such as a silicon wafer. These reference particles are routinely used to calibrate the detection sensitivity of this type of instrument[1]. The size of these particles is precisely controlled to better than 2% at the one sigma level[2,3] so that the scattering efficiency of each particle is essentially the same. Furthermore, since the size distribution of these particles is known, one can precisely predict the expected variation in the amount of light they scatter.

Figure 2 shows a typical distribution of the measured defect sizes from the calibration wafer described above. The width of the distribution is characterized by measuring its coefficient of variation (CV), which is defined as the ratio of the standard deviation of the distribution to its mean[4]. The measured CV is a combination of the instrument dependent broadening effect which is characterized by a CV, "V_i", and the sample dependent broadening which is characterized by a CV, "V_s". Since these are independent effects, the variance of the combined effect is the sum of the variances from the components[4]. The CV of the instrument dependent signal broadening can, therefore, be calculated from:

$$V_i = V^2 - V_s^2 \qquad (1)$$

where V is the measured CV for the reference particles.

Figure 3 shows the measured instrument dependent CV, as a function of the diameter of the PSL spheres on bare silicon using a Surfscan 5500. As reported by Neukermans et al.[5], this behavior can be separated into two limiting cases described by curves (A) and (B). Curve (A) is the limiting case for particles with diameters greater than about 0.3μm. It is typically less than 6% in magnitude and is independent of the size of the reference particles. There are two causes for this type of signal broadening: i) the variability in the detection sensitivity of the instrument, and ii) spatial sampling errors.

The first source of signal broadening is due to instrumental errors from spatial or temporal variations in sensitivity across the substrate. This can be caused by variations in the laser power, the shape and size of the laser beam on the surface, the collection efficiency of the detection

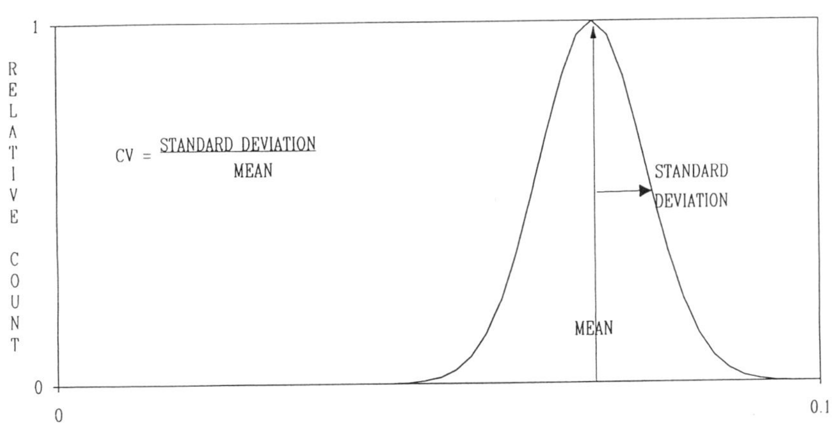

Figure 2. Signal Distribution for a Calibration Wafer with Monodisperse Spheres on a Silicon Substrate.

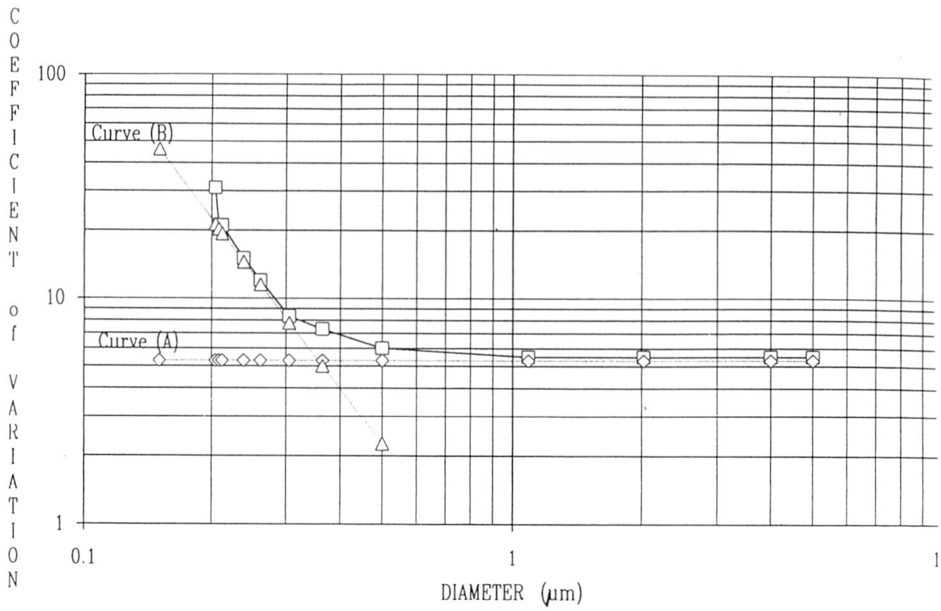

Figure 3. Coefficient of Variation Versus Particle Diameter(μm) for PSL Spheres on Bare Silicon, Curves (A) and (B) Indicate the Limiting Cases for Large and Small Defects, Respectively.

optics or electronic gain of the instrument as well as spatial or temporal variations in the sample pitches "dx" and "dy". For a Surfscan 5500, these effects can result in a broadening of the signal with a CV that ranges from 3% to 4%.

The second source of signal broadening is spatial sampling errors due to the finite distance between sampling points on the substrate. As the laser beam is swept over a small particle on the wafer surface, it will produce a scattered signal with a Gaussian profile which corresponds to the intensity distribution of the laser beam. This signal is sampled at regular intervals, as described above, and the peak amplitude is selected to represent the scattering efficiency for that defect. The peak signal occurs when the center of the beam is closest to the defect as shown in Figure 4. Since the particles are randomly positioned with respect to the sample points, the measured signal will vary according to the Gaussian distribution $\exp(-2\ (R/ro)^2)$ where ro is the radius of the laser spot and R is the distance of a given particle to the nearest sample point.

Monte Carlo modelling techniques predict an intrinsic broadening of the signal distribution as shown in Figure 5. For the Surfscan 5500, where the beam radius is 65μm and the sample pitches, in the X and Y directions, are 25μm and 33μm respectively, spatial sampling is expected to produce a 4.5% broadening of the signal distribution.

Curve (B) in Figure 3 increases in inverse proportion to the particle diameter raised to the power of 2.5 and dominates for diameters less than 0.3μm. This distribution arises due to Poisson noise in the scattered signal which occurs due to the quantized nature of the detected signal[6]. If the number of quanta in the signal for a given sample is "N", then repeated measurements of the same signal will show statistical variations due to the finite number of quanta comprising the signal. The standard deviation of this distribution is equal to the square root of the number of quanta. Since the CV is defined as the ratio of the standard deviation

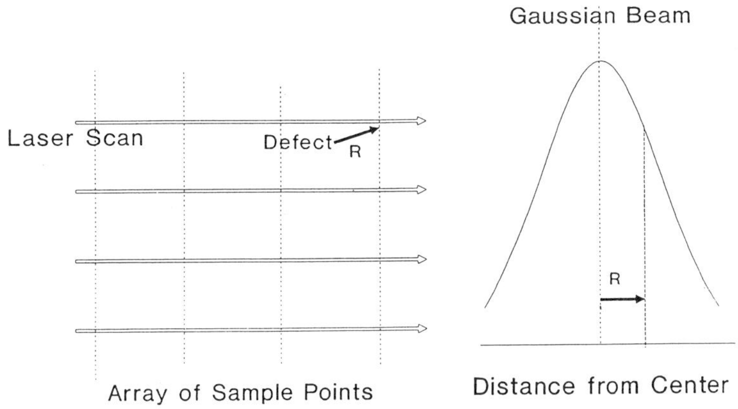

Figure 4. Array of Sample Points on the Substrate Under Test and the Intensity Distribution of the Laser Beam versus Position.

to the mean, the CV for a Poisson distribution is equal to $N^{-0.5}$. This effect will be present in all signals, but the relative noise becomes significant only when the number of quanta detected in a given measurement period is small.

In this case, the dominant source of Poisson noise is at the input of the photodetector where the number of quanta in the signal is the smallest. For a given particle of scattering cross-section,"s", the number of photoelectrons in the peak sample is given by,

$$N(s) = \frac{2 * s * P * G1 * G2 * \tau}{\pi * r0^2 * h\omega} \qquad (2)$$

where "P" is the power in the laser spot, "r0", the radius of the laser spot, "hω" is the energy per photon, "G1" is the collection efficiency of

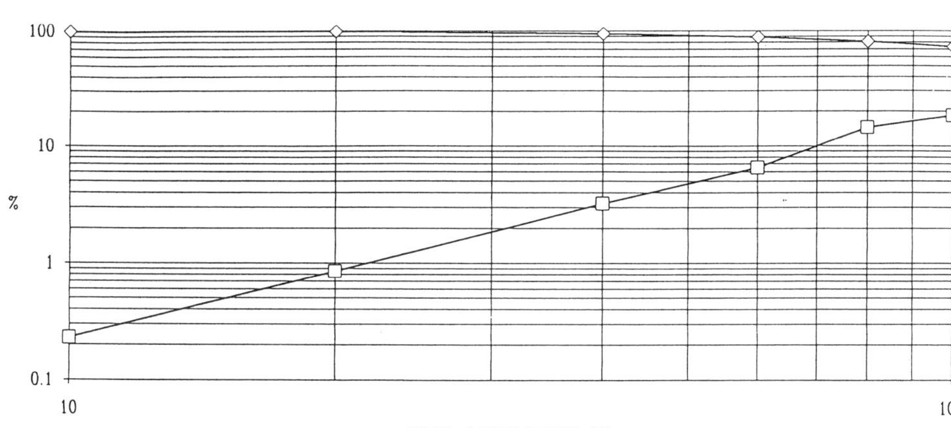

Figure 5. Signal Broadening,(CV), due to Spatial Sampling versus the Ratio of the Sample Pitch to the Beam Radius (%).

the optics,"G2" is the conversion efficiency of the photodetector and "τ" is the analog integration time per sample. For a 0.2μm PSL sphere, using 5mW of laser power in a 65μm spot at a wavelength of 633nm, about 16 photoelectrons are collected in a 25ns sample integration period. This primary signal will exhibit statistical variations with a CV of 25%. If the Poisson noise in the additional amplification stages is taken into account[7], the overall Poisson noise is increased by about 22% resulting in an overall CV of 30%. Furthermore, since the signal increases in proportion to the fifth power of the particle diameter, i.e. d^5, in this region, the Poisson noise contribution of the signal will follow a $d^{-2.5}$ trend for small particle sizes, which agrees with the observed slope of curve (B) in Figure 3.

Since each of these three effects described by curves (A) and (B) in Figure 3 is independent, their combined influence on the signal distribution is estimated by adding their individual variances[4]. The coefficient of variation of the measurement is, therefore, estimated as the square root of the sum of the squares of the individual CVs.

BACKGROUND NOISE

Another source of variability which can affect accurate particle counting at diameters less than about 0.3μm is the "haze" noise due to scattering from the surface of the substrate. Haze is a low level of background light which is diffusely scattered from imperfections at the wafer surface or in its sub-surface[8]. The average level of this scatter is relatively constant across the wafer, and is separated from particle data by filtering techniques. The instantaneous haze signal, however, has fluctuations due to Poisson noise in the same manner as previously described for the scatter signal. This noise has high frequency components in the same range as the particle data and will appear in the particle channel as a background count which can dominate at low signal levels.

In order to reject these spurious noise counts, one must employ a threshold which is set at a level, "T", sufficiently far above the noise to minimize the false count rate. Any signal which fails to exceed this level is rejected. The false count rate is estimated as the product of the total number of sample points taken on the wafer, "Ns", times the probability that a haze noise sample will exceed the threshold, "P(T)",

$$Nf = Ns * P(T) \tag{3}$$

In order to predict the false count rate and determine where the thresholds should be placed, one must estimate the amplitude of the haze noise in the particle channel. The average number of photoelectrons in the Haze signal, "N(H)", is derived in an analogous manner to the particle signal described by equation (2),

$$N(H) = \frac{P * H * R * G1 * G2 * \tau}{h\omega} \tag{4}$$

where "H" is the fraction of the reflected beam that is scattered by the surface haze and "R" is the reflectivity of the surface under test. Poisson noise in this signal results in a distribution of amplitudes with

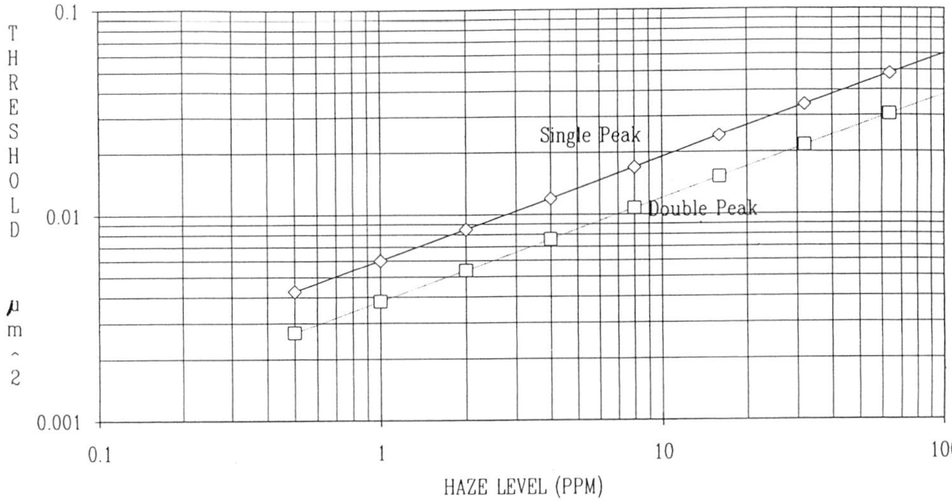

Figure 6. Minimum Detection Threshold (μm^2) versus Haze Level (PPM) for 'Single' and 'Double' Peak Noise Rejection Schemes.

a standard deviation equal to $N(H)^{0.5}$. If the Poisson noise in the additional gain stages is taken into account[6], the noise level is increased further by 22%. In order to compare this noise to an equivalent signal level in the particle channel, equation (4) is scaled by the ratio "s/N(s)" from equation (2), to give the standard deviation of the background noise level in the particle channel, "B(H)",

$$B(H) = 1.92 * r^2 \sqrt{\frac{H * R * h\omega}{P * G1 * G2 * \tau}} \qquad (5)$$

If we assume there are about 10^7 sample points per wafer, then the detection threshold must be placed about 5.2 times B(H) for a "Single Peak" thresholding technique to limit the number of false counts below one per wafer scan. Alternatively, a "Dual Peak" correlation technique[9] can be used to further improve the performance. In this scheme, the signal from the defect must exceed detection threshold in at least two neighboring sweeps to be counted as a valid defect. This allows the detection threshold to be reduced to only 3.3 times B(H) to obtain the same false count rate as before.

The minimum threshold limits for a false count rate of 1 in 10^7 are plotted as a function of thr haze level for both the 'Single' peak and 'Double' peak schemes in Figure 6. Both curves increase in proportion to the square root of the haze and represent the minimum detectable defect size for a surface with a particular haze level. The 'Double' peak technque provides a 40% improvement in the minimum detection limit over the corresponding 'Single' peak sensitivity.

COUNT ACCURACY

Having described the statistical variability in the measured scatter signal and the background noise as a function of the size of the defects, we will now consider the influence of these effects on the counting accuracy of the instrument.

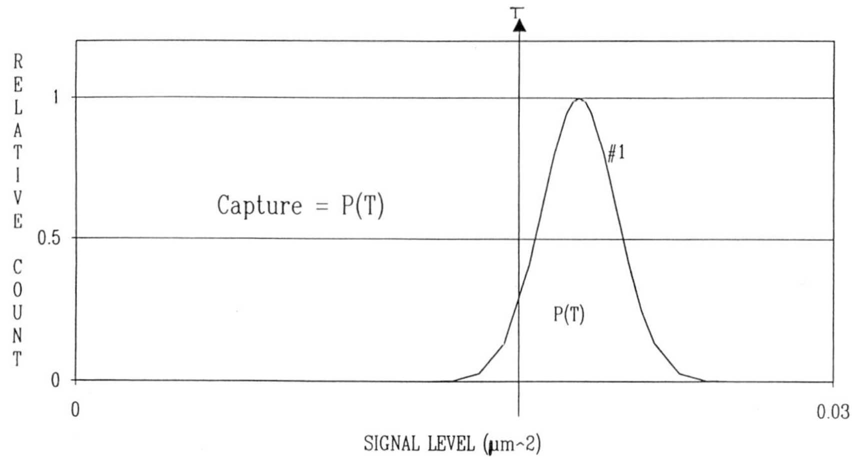

Figure 7. Position of Detection Threshold with Respect to Signal Distribution Determines capture Rate for a Population of Monodisperse Spheres.

As described above, a threshold must be used to reject false positive counts arising from the background noise in the haze from the substrate. However, this can result in the undesirable effect of also rejecting real defect counts. As we have seen previously, a defect of a given size will produce a range of scatter signals as described in Figure 3 for PSL spheres. If the detection threshold is placed within two or three standard deviations of the mean of the signal level for a given defect size, the signal from some of the defects will exceed the threshold and be counted while others will not and these will be rejected.

This effect is described by the term, "Capture Rate" for the given defect size. This is defined as the ratio of the number of defects detected by the instrument to the number of defects in the sample population of interest. For a population of monodisperse PSL spheres, the capture rate,

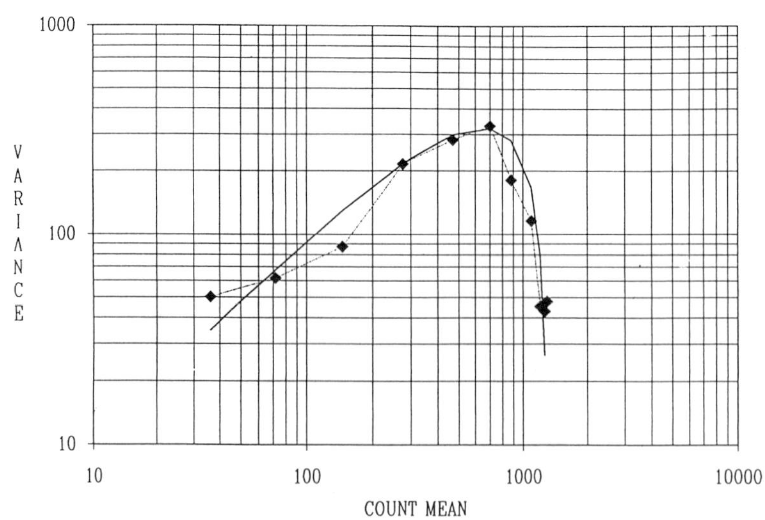

Figure 8. Variance versus Mean in Counting a population of 0.208µm PSL Spheres on Bare Silicon.

"C", is simply the area of the signal histogram that is above the threshold as depicted in Figure 7 where,

$$C = P(T) \tag{6}$$

The capture rate is a function of the size of the defect and is determined by the relative position of the collection threshold to the mean signal level and the width or CV of the signal distribution for that defect. Since an individual particle is either captured or not, the statistics in counting an assembly of monodisperse spheres will follow a binomial probability distribution as described by Cooper and Neukermans[10]. If the capture rate for an individual particle is "C" as defined above, then repeated measurements on a population of "N" spheres will give a distribution in the number counted. The mean and variance of this distribution is predicted from binomial statistics as,

$$\text{Mean} = N * C \tag{7a}$$

and

$$\text{Variance} = N * C * (1 - C) \tag{7b}$$

For example, for a population of 100 monodisperse spheres with the threshold set so that the capture rate is 90%, the number of particles detected will vary with a distribution characterized by a mean of 90 and a standard deviation of 3. This means that the total number of counts will vary in range of 84 to 96 for a confidence level of 95%.

This analysis was tested by scanning a population of approximately 1200 PSL spheres, 0.208μm in diameter on a bare silicon wafer using a Surfscan 5500. By moving the display threshold with respect to the signal histogram, the capture rate was varied in the range of 5% to 95%. The mean and variance of the particle count were calculated for 20 repeated measurements at each threshold and these results are displayed in Figure 8. The data show close agreement with the behavior predicted by Binomial statistics as described in equation (7). In particular, the

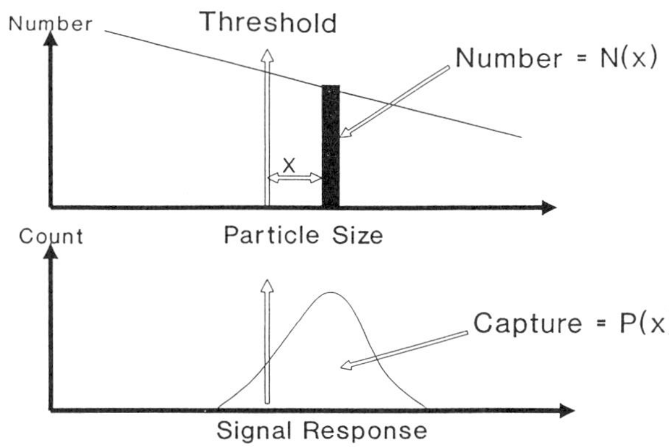

Figure 9. Statistical Analysis of a General Population of Defects in Terms of Sub-Populations of Monodisperse Particles.

variance is a parabolic function of capture which reaches a maximum for a capture of 50% and falls off to zero for a capture of either 0% or 100%.

This technique can be extended to counting the number of defects above a given size for a general population of particles as might be collected on a monitor wafer. The approach is outlined in Figure 9 where the defect population is divided into a series of narrow strips each of which can be treated as a sub-population of monodisperse scatterers. For example, for the sub-population which is at a distance "X" above the threshold, there are "N(X)" particles and their scattering response is characterized by a histogram as shown in Figure 9. The capture rate, "P(X)", is estimated as described previously for monodisperse spheres and the mean and variance in counting this sub-population are calculated from equation (7). The individual means are summed over the entire range of particles, to estimate the average number of defects counted. Similarly, since the sub-populations are independent, the individual variances in the sub-population count are summed over the entire range to estimate the variance in the total count. This gives the general formulae for the mean count and variance as,

$$\text{Mean Count} = \sum N(X) * P(x) \tag{8a}$$

and

$$\text{Variance} = \sum N(x) * P(X) * (1 - P(X)) \tag{8b}$$

For a natural population of particles with an inverse square distribution of size, equations (8a) and (8b) can be simplified to the form,

$$\text{Mean Count} = N0 * (1 + SDEV/(2*d)) \tag{9a}$$

and

$$\text{Variance} = 1.23 * N0 * (SDEV/d) \tag{9b}$$

where "N0" is the number of particles greater than size "d" and "SDEV" is the standard deviation of the signal histogram in units of an equivalent defect diameter, for particles of size "d". For example, consider a population with an inverse square distribution in size where there are 20 defects greater than 0.20μm in diameter. The standard deviation of the

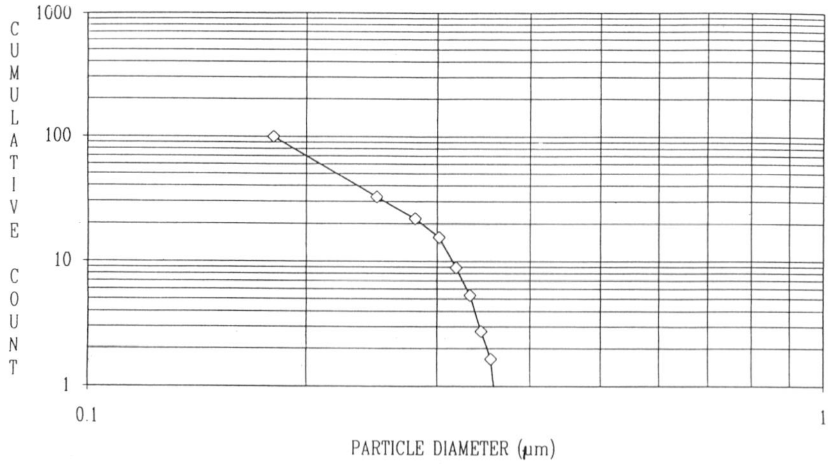

Figure 10. Cumulative Histogram of a Population of Particles on a 'Prime' Wafer versus Defect Size(μm).

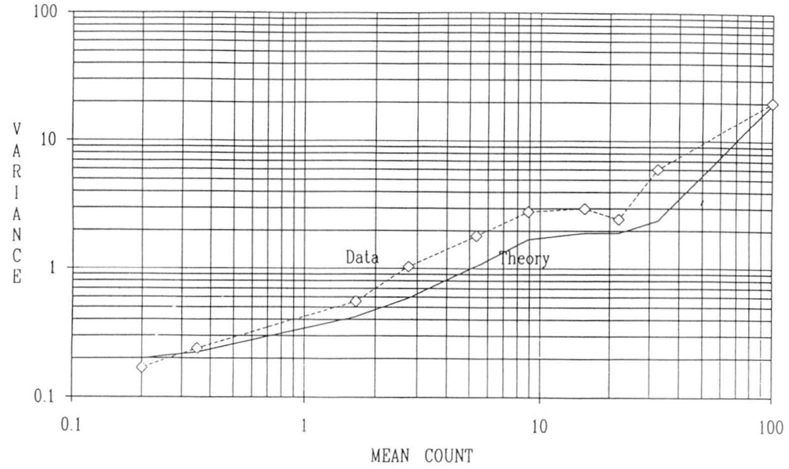

Figure 11. Variance versus Mean for Counting a Population of 'Native' defects on a 'Prime' Silicon Wafer.

signal histogram is 0.012µm in this region. Thus, repeated scans of this population will vary with an average count of 20.6 particles and a standard deviation of 1.1 particles. This implies that the number counted will be in the range of 18.4 to 22.8 particles at a 95% confidence level.

In addition, one can use this model to understand the large variability in the number of particles that "appear" and "disappear" in map-to-map comparisons of the same wafer, as discussed in the Introduction. For particles which are more than three standard deviations below the detection threshold, the capture rate is essentially zero so that they will never be detected by the instrument. For particles which are more than three standard deviations above the threshold, the capture rate is essentially unity so that they will always be counted by the instrument.

The particles that lie within these two bounds have a finite probability of producing signals that are either above or below the threshold with a probability that depends on the distance of the true particle size from the threshold. As such these particles will sometimes be seen on some scans but not on others. The number of particles expected to behave in this manner is determined by calculating the number of particles in the variable region which is plus or minus 3*SDEV around the threshold. For the example cited above with a natural population of 20 particles greater than 0.2µm, the number of particles may "appear" and "disappear" from map to map is 14 out of an average count of 21.

In order to verify this analysis, a series of measurements were made on a natural population of defects on a 'Prime' wafer using the Surfscan 5500. A cumulative histogram of the particle distribution is shown in Figure 10 and indicates a complex shape with at least two regions with slopes of 2.5 and 12. The display threshold was varied in the range of $0.006\mu m^2$ to $0.23\mu m^2$, and the number of defects greater than the threshold was counted for 20 repeated scans of the same wafer. The mean and variance of the distribution of counts are plotted in Figure 11 for each threshold. The results indicate a close agreement with the theoretical behavior which was calculated from equations (8a) and (8b).

CONCLUSIONS

The counting accuracy of a monitor surfscan tool can be characterized by measuring the coefficient of variation of the signal response for a series of monodisperse spheres.At high signal levels, the coefficient of variation in the signal response is small and is dominated by the combined effects of spatial sampling errors and variations in the instrument sensitivity. At low signal levels, the coefficient of variation increases sharply with smaller defect sizes due to Poisson noise in the signal and background light levels.

These parameters determine the capture rate for any population of monodisperse spheres and any given detection threshold. The counting statistics for monodisperse spheres follows a binomial distribution where the mean and variance in the total count are a simple function of the capture rate and number of defects. This analysis can be extended to evaluate the statistical accuracy in measuring a general population of defects. General formulae are derived which describe the mean and variance in counting any known distribution of defects at a given threshold.

REFERENCES

1. "Absolute Contamination Standard Application Note" VLSI Standards Inc. Mountain View, CA. (1990).
2. "Certified Particle Size Standards", Bulletin 92D, Duke Scientific, (1990).
3. "Standard Latex Particles", Japan Synthetic Rubber Co. Ltd, September (1989).
4. R. R. Sokal and F. J. Rohl, "Biometry", pp 58-60, Freeman,(1969).
5. A. Neukermans, S. Saadat, J. Pecen and G. Kren, Proc. S.P.I.E. (1987).
6. M. L. Boas, "Mathematical Methods in the Physical Sciences", pp 712- 718, Wiley, (1966).
7. R. W. Engstrom, "Statistical Theory of Noise in Photomultiplier Tubes" RCA Photomultiplier Handbook, (1980).
8. P.O. Hahn and M. Kerstan, Proc S.P.I.E. pp 172-181, (1988).
9. J. Pecen, A. Neukermans, G. Kren and L. Galbraith. Solid State Technology, pp 149-154, (1987).
10. D. W. Cooper and A. Neukermans, (to be published) (1991).

LIGHT SCATTERING BY SUBMICRON SPHERICAL PARTICLES

ON SEMICONDUCTOR SURFACES

E.J. Bawolek[a] and E.D. Hirleman[b]

[a]Department of Electrical and Computer Engineering
[b]Department of Mechanical and Aerospace Engineering
Arizona State University
Tempe, Arizona 85287-6106

In this work we report on the angle resolved light scattering characteristics of individual polystyrene spheres on silicon surfaces. A He-Ne laser (632.8 nm) focused to a 15 μm $1/e^2$ diameter was employed to illuminate 4.10 and 0.804 μm diameter spheres on optically smooth ($\sigma \ll \lambda$) silicon. Scattering was measured as a function of incident beam polarization for incident angles of 30, 45, and 75.3 degrees (Brewster's Angle). A ring/wedge photodetector array centered on the specular beam was employed for the scattering measurements. The detector intercepted light scattered up to 45 degrees from the specular beam. The results obtained are qualitatively explained with reference to the silicon surface reflectance which varies significantly as beam incident angle and polarization are changed. Conditions associated with high surface reflectance result in larger values for sphere scattering cross section. We propose that the scattering measurements can be modeled in terms of a first order theoretical model developed previously by one of the authors (Hirleman). Initial results show that the theory is useful for understanding trends and predicts the correct order of magnitude for the scattering cross section. Possible improvements to the model are discussed.

INTRODUCTION

Optical methods are widely employed for surface inspection and contamination detection in the integrated circuit and digital storage media industries. Commercial instruments are available which can scan a smooth (or patterned) surface with a laser and count individual scattering centers of submicron size by detecting (integrating) the light scattered by the feature. Although these instruments can exhibit high sensitivity to particulates, the calibration (i.e. the size vs. scatter relationship) must be determined experimentally for each specific substrate/thin film/contaminant combination. Prediction of the instrument response from first principles is a desirable goal. Calculation of the scattering behavior of a particle on a surface is possible by solving Maxwell's equations numerically[1].

Particles on Surfaces 3, Edited by K.L. Mittal
Plenum Press, New York, 1991

Angle resolved measurements of light scattered by particles on surfaces are becoming a topic of increasing importance because a need exists to thoroughly test and verify scattering prediction algorithms. The ideal algorithm would predict a particle's scattering response (intensity, polarization, and ellipticity vs. direction) as a function of its size, morphology, and refractive index. The substrate's optical properties and roughness would also be included. The total integrated scatter (TIS) obtained by commercial instruments is not sufficiently detailed for the purpose of testing predictions, although an accurate simulation would necessarily be capable of computing TIS. For example, Lee et. al.[2] have had good success in correlating scattering computations with measurements of TIS made on commercial instruments. However, because their experimental data contain no angular information, it is difficult to assess the precise degree to which their models fit the data. By performing and reporting angle resolved scattering measurements on sphere/surface combinations of interest to the semiconductor and storage media industries, a database necessary for the development of accurate simulation algorithms can be established. We anticipate that advances in computing power will make numerical simulation practical in the foreseeable future, producing two significant benefits: First, they will make possible self-calibrating contamination detectors which can accurately size particles based on fundamental physical principles. Second, they will improve the design of future contamination detectors by making it possible to predict the instrument performance before constructing a prototype.

In this paper we report on the angle resolved light scattering characteristics of individual polystyrene spheres on silicon surfaces. The results are obtained as a function of incident beam angle and polarization. By measuring individual spheres it is possible to separate the effect of the local surface scatter from that of the sphere. The data are important not only from the standpoint of supporting development of numerical models, but also for improved understanding of the performance of present-day contamination detectors.

THEORY

Light Scattering by Spherical Particles

The scattering of light by a sphere in free space can be calculated exactly by solving Maxwell's equations. The original solutions to this problem were independently derived by Lorenz and Mie. A contemporary interpretation of their formulations appears in van de Hulst[3]. Our coordinate notation follows van de Hulst's treatment and that of Weber and Hirleman[4]. A plane wave incident on a sphere in a uniform medium scatters energy in all directions according to the relation:

$$I(\theta_s, \phi_s) = I_0 F(\theta_s, \phi_s, \alpha)/k^2 r^2 \tag{1}$$

where I is the irradiance of the radiation scattered in the directions specified by θ_s and ϕ_s, I_0 is the irradiance incident on the particle, k is the wavenumber defined by $k = 2\pi/\lambda$ (λ = wavelength of illumination), r is the distance from the origin to the measurement point (assumed to be large enough to avoid near-field effects), and F is a scattering function given by:

$$F(\theta_s, \phi_s, \alpha) = i_1(\alpha, \theta_s)\sin^2\phi_s + i_2(\alpha, \theta_s)\cos^2\phi_s \tag{2}$$

where α is the dimensionless size parameter $\pi d/\lambda$, d is the sphere diameter, and i_1 and i_2 are the Lorenz-Mie scattering functions for perpendicular and parallel scattering polarizations. The differential scattering cross section $dC_{sc}(\alpha)$ for scattering by a sphere into a unit solid angle is obtained when both sides of Equation (1) are multiplied by r^2 and divided by I_0:

$$dC_{sc}(\alpha)/d\Omega = F(\theta_s,\phi_s,\alpha)/k^2 \tag{3}$$

where $dC_{sc}(\alpha)$ has dimensions of area and $d\Omega$ represents a differential solid angle (steradian).

Scattering from Spheres on Surfaces

The problem of predicting light scatter by a sphere in contact with a surface is far more difficult to solve than the idealized case of an isolated sphere in a uniform medium. Analytic solutions for the most general case have not (to our knowledge) been published in the literature. Bobbert and Vlieger[5] used a matrix formalism to combine Lorenz-Mie scattering theory with Debye potentials giving a mathematical description of how a plane surface modifies a spherical scattered wave. The results are analytic only in the case of a perfectly conducting plane. Videen[6] has derived the Mueller scattering matrices for a sphere on or near a plane surface for the case where the surface is a perfect conductor. His solution assumes that the scattered fields from the sphere are incident on the surface at near-normal incidence. Solution by numerical methods appears possible[1], but at this time still taxes the capabilities of available computers. A compromise approach to modeling the sphere-surface system is to assume that the observed scatter consists of rays which are directly scattered from the sphere into the detector and by rays that are reflected off the surface either before or after interacting with the sphere. This approach was followed by Nahm and Wolfe[7] as proposed originally by Young[8]. Weber and Hirleman[4] extended this approach with good results by explicitly including a term which accounts for attenuation by surface reflection. Lee et al.[2] provide a useful discussion of several of these approaches. Their study compares predictions of two modifications to Mie theory and numerical simulations. Knollenberg[9] takes a fundamentally different approach by computing the relative field intensities of standing waves produced when the incident beam undergoes reflection at the surface. The position of the particle within the resulting fringe pattern determines the extent to which energy can be coupled into the particle and subsequently scattered. The surface reflectivity influences this process by affecting the contrast of the standing wave pattern. Our approach follows that of Weber and Hirleman[4]. The method shares some similarity with Lee's[2] Model 1, but is simplified in that it does not include all of the scattering components of the Lee model. We approximate the particle on a surface response in terms of a two-component model in which:

$$dC_{sc} = dC_{sc,b} + R\,dC_{sc,f} \tag{4}$$

where R represents the specular reflectance of the surface, and the subscripts f and b on dC_{sc} represent forward and backscatter contributions as defined in Figure 1. Because Weber and Hirleman[4] studied the behavior of a particle on a nickel substrate, they were able to assume a constant value for R, independent of incident beam angle and polarization. The differential scattering cross sections reported in that study were integrated over the range of angles subtended by the ring elements of their photodetector. In this study, silicon substrates are employed. Unlike a metallic substrate, the surface reflectance of silicon varies strongly as the beam incident angle and polarization are varied. Figure 2 shows this behavior, calculated from electromagnetic theory[10], assuming a silicon refractive index of n=3.9, k=0.23. The values were selected for consistency with measurements we made of the

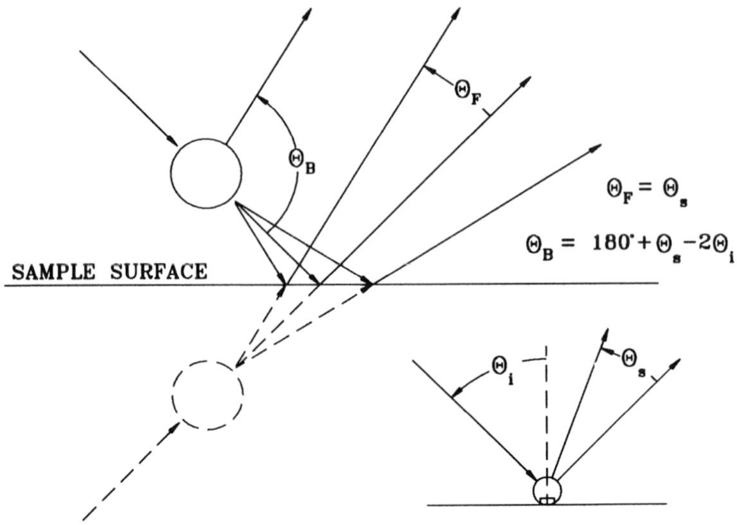

Fig. 1. Idealized schematic illustrating the angle definitions and scattering components assumed by the modified Lorenz-Mie model of Equation (4). The sphere is shown separated from the surface to facilitate explanation. Here θ_i is the angle the incident beam makes with the surface normal. The ray of interest makes an angle θ_s relative to the specular beam. The scatter into angle θ_s is assumed to be composed of two rays scattered at angles θ_b and θ_f as shown. The ray scattered at θ_f is modified by the surface reflectance, *R*.

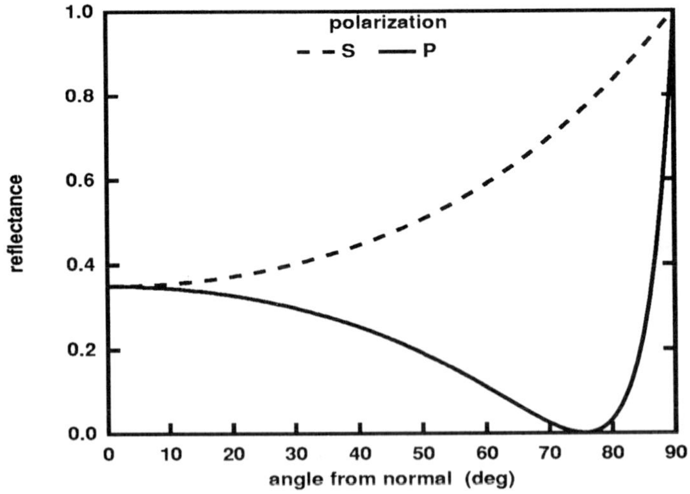

Fig. 2. Reflectance of silicon as a function of polarization and incident angle. The silicon refractive index assumed was n=3.9, k=0.23. S polarization corresponds to the electric field vector perpendicular to the plane defined by the incident and reflected beams; P polarization corresponds to the electric field vector parallel to the plane.

Brewster angle and the extinction coefficient on one of the silicon substrates. The exact value of the refractive index and extinction coefficient will vary slightly as the silicon carrier type and conductivity change, but our values are reasonable when compared with results of other researchers, for example Russo[11]. Because of the strong variation of silicon surface reflectance with polarization and angle, it is necessary to compute the value of *R* individually for each polarization and value of θ_i. Our computations have to date been confined to the plane defined by the incident and specular beams. We have not yet attempted to integrate our results over a finite detector area such as a ring element. Thus, we have left Equation (4) in differential form. It should be noted that in this form the model resembles a computation of a bidirectional reflectance distribution (BRDF) function, but should be extendable to ring detector geometries.

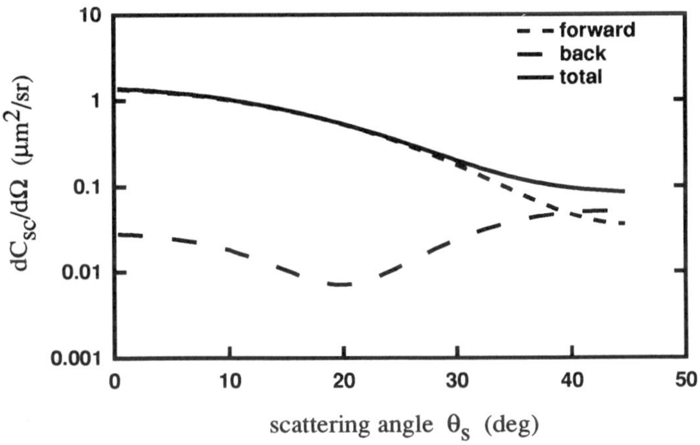

Fig. 3. Plot of differential scattering cross section $dC_{sc}/d\Omega$ as a function of scattering angle θ_s using the modified Lorenz-Mie theory of Equation (4) for scattering by a 0.804 μm polystyrene sphere on a silicon surface. The calculation assumes an incident beam angle of 45 degrees, S polarization and λ = 632.8 nm. The refractive index of the sphere is taken as n=1.59; k=0.0. The three curves represent the three terms of Equation (4).

Figure 3 shows an example calculation for a polystyrene sphere on silicon substrate. Examination of this figure shows that the forward scattering component dominates for scattering angles up to 38 degrees. The model therefore predicts that variations in the value of *R* can have a significant effect on the scattering behavior of a particle. For example, near the Brewster angle, the rays scattered at small angles would be highly attenuated, reducing the scattering cross section. At large scattering angles the backscatter contribution dominates the behavior. Thus, its contribution to the scattering behavior cannot be ignored.

EXPERIMENTAL

Apparatus

The equipment used in this research is shown diagrammatically in Fig. 4. This apparatus is an enhancement of a system developed by Weber and Hirleman[4]. The apparatus consists of a light source, polarization rotator, focusing optics, photodiode array, positioning stages, stepper motor controller, and data acquisition microcomputer. To prevent particle contamination of samples, the optical components and sample stage are located in a 100% vertical laminar flow clean enclosure continuously purged with HEPA filtered air. The environment inside the enclosure was measured according to Federal Standard 209D using a Climet model CI-8060 particle counter and found to be better than Class 10 under static conditions.

The optical system utilizes a linearly polarized He-Ne laser operating at 632.8 nm with nominal power output of 5 mW and $1/e^2$ diameter of 0.64 mm. A polarizing beamsplitter is employed to provide adjustable

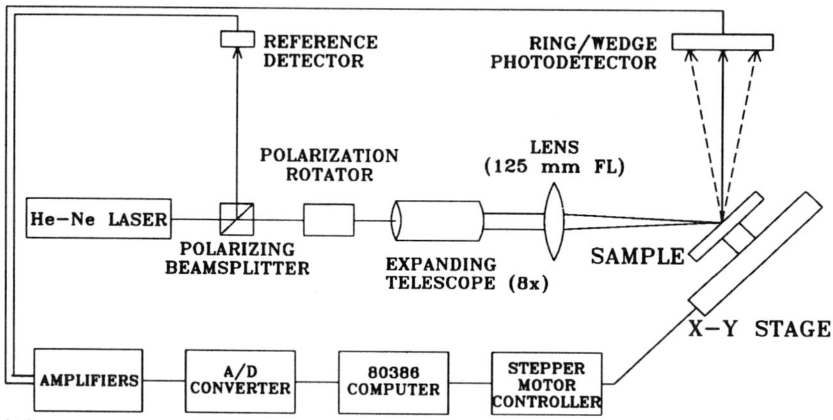

(a)

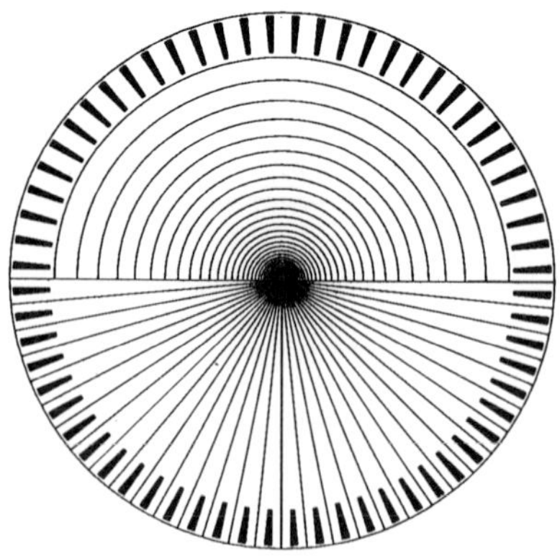

(b)

Fig. 4. Schematic of the experimental apparatus used in these studies.
(a) Equipment configuration.
(b) Schematic detail of ring/wedge photodetector.

attenuation of the laser beam and to divert a small portion of the beam energy to a reference detector. All scattering data are normalized to the reference detector output. The main beam exiting the beamsplitter passes to the polarization rotator, which permits continuous variation of the polarization between S (electric field vector perpendicular to plane of incidence) and P (electric field vector parallel to plane of incidence). Extinction measurements were employed to establish the initial settings of the polarization rotator, and to establish proper setting of the apparatus at Brewster's angle. The expanding telescope produces a Gaussian output beam with a $1/e^2$ diameter varying between approximately 1 and 7 mm. A 125 mm focal length lens is used to focus the beam onto the sample surface, producing a spot with a minimum diameter of 15 µm. By changing the configuration of the components, it is possible to produce various angles of incidence on the sample. For these studies, the test surface was oriented such that the surface normal was either 30, 45, or 75.3 degrees relative to the incident beam.

A photodiode array consisting of 32 rings and 32 wedges was employed to collect the scattering data. The center of the array is a circular element with a diameter of 178 µm. The remaining rings are half-circles whose areas increase in proportion to the radial distance from the wafer center. The other half of the detector array is composed of 32 identical wedge-shaped detectors equally spaced around the semicircle. The photodiode array was aligned such that the specular beam was incident on the center ring detector with the detector surface perpendicular to the specular beam in all instances. The detector was calibrated by means of a white light source with variable intensity, cross referenced to a Newport Model 835 Optical Power Meter. Individual calibration curves were developed for each ring and wedge relating the detector output to absolute illumination intensity. Although data were collected from the wedges in all of our studies, in this paper we report results based on analysis of the ring outputs only.

The positioning stages of Fig. 4 moved the test surface in a Cartesian coordinate system. Software was developed and implemented on an 80386 based microcomputer to control stage motion and simultaneous data acquisition. Samples were incrementally scanned in X and Y directions to interrogate a selected surface region. Reduced area scans with step resolutions as small as 1 µm were employed to measure individual spheres. Mechanical hysteresis was minimized by beginning scans from a "home" position (i.e. top to bottom). The data acquisition system consisted of parallel analog preamplifiers multiplexed into a 12-bit analog to digital converter (ADC) board inside the microcomputer. The digitized detector signals and reference detector voltage were recorded in conjunction with stage location to provide the data necessary for determining the scattering cross sections.

Sample Description and Preparation

The samples used in this study were 0.804 and 4.10 µm diameter polystyrene spheres marketed by Duke Scientific Corp. These spheres are intended for use as size calibration standards. The reported standard deviations for the sphere sizes are 0.6% and 4.9%, respectively. The substrates were monocrystalline silicon (bare silicon) cleaved from standard (100) orientation p-type silicon test wafers. These wafers have a smooth, virtually featureless surface (Fig. 5). The cleaving process was employed to produce a sample size which was suitable for our apparatus. The experimental data were taken from spheres deposited on the polished wafer surface (as opposed to deposited on the cleaved surface). Sample substrates were cleaved prior to sphere deposition.

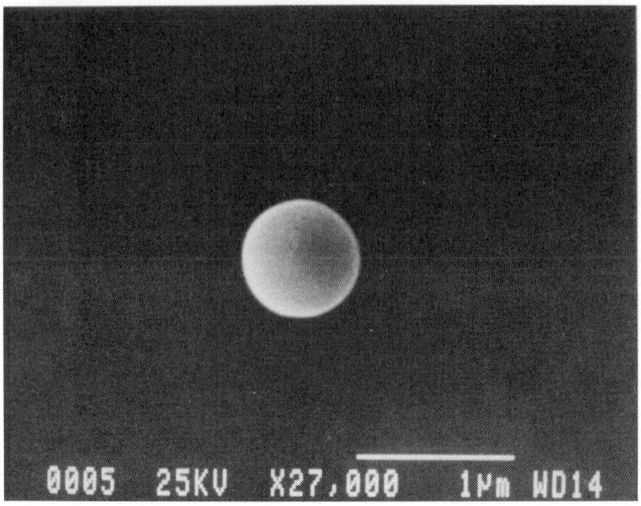

Fig. 5. Scanning electron micrograph of a typical 0.804 µm sphere deposited on bare silicon substrate employed in this study. The surface roughness of the silicon (σ) is much smaller than the illuminating wavelength (λ).

Spheres were deposited onto the substrate surfaces by a spin coating technique. The spheres, supplied by Duke Scientific, are in aqueous suspension with high concentrations (1×10^8 and 7.5×10^{10} per ml for the 4.10 and 0.804 µm spheres, respectively). These solutions were diluted 1:100 and 1:500 respectively (by volume) by adding 0.2 µm filtered reverse osmosis deionized (RODI) water. One drop of methanol ($\approx$ 0.06 ml) was added to approximately 30 ml of solution to aid dispersion. These solutions were dynamically dispensed from an eyedropper while the substrate was spun at 3000 RPM. Five to seven drops were typically dispensed in a line progressing from the outside of the sample toward the center. The sample was dried by continuing the spin process for 30 seconds. The areal density of the deposited spheres was estimated by counting individual spheres within several microscope fields. The measured density was 0.2 per mm^2 for the 4.10 µm spheres and 23 per mm^2 for the 0.804 µm spheres. The fields examined were virtually free from spurious contaminants, despite the fact that the process of cleaving could potentially produce particulates. Part of the reason for this is that the spin process washes away larger sized particles. Another factor is the high areal density of the deposited spheres: Some unwanted contaminants undoubtedly exist on our samples, but they are much lower in number than the target spheres. In the limiting case, our experiments require only the small region within the illuminated spot to be contaminant-free to yield useful data. The microscope fields we examined easily met this criterion.

Procedure

Experiments were typically run in two parts: First, a "global" scan was performed in which an area of approximately 1 to 6 mm^2 was examined in steps of 10 to 15 µm. An imaging algorithm was applied to the acquired data which helped us to identify regions in which one or more of the ring/wedge detectors showed response above the general background level. Second, features were selected for "detail" scans in which a region of 100 μm x 100 μm (sometimes smaller) was examined in steps of 1 μm. These detail scans were centered on the spheres and subsequently analyzed to produce the scattering cross section data reported in the next section. The detail scans were analyzed in two parts to separate the sphere scattering signal from that of the surrounding surface. In performing the data analysis, it was necessary to discard data from the innermost ring elements which were saturated by the specular beam. Depending on polarization, sample to detector distance, and incident angle, six to twelve of the innermost rings were blinded. This placed a limit on the minimum angle θ_s for which scattering data could be obtained. The maximum θ_s was a function of the detector to sample distance. In our work this distance was 15.0 mm for incident angles of 45 and 75.3 degrees. The maximum θ_s captured with this configuration was 45 degrees. With an incident angle of 30 degrees the instrument geometry limited the minimum sample to detector distance to 20.0 mm. The maximum θ_s in this case was 37 degrees.

Scattering data are reported as differential cross sections $dC_{sc}/d\Omega$, obtained by subtracting the background surface scatter from the peak scattering signal. The coordinate system necessitated by our detector does not allow our data to conveniently mesh into the bidirectional reflectance distribution function (BRDF) employed by some researchers to report light scattering by particle contaminated surfaces [7,8]. The use of the differential cross section is more closely aligned to comparison with measurements provided by contamination analyzers marketed to the semiconductor industry [2,12].

Although the detector was calibrated to absolute intensity, computation of the scattering cross section requires that the incident beam power and diameter be accurately known. For these reasons the total beam power incident to the wafer surface was measured prior to each series of experiments using the Newport power meter. The beam diameter was obtained by fitting a Gaussian function to the response data recorded from one or more detector elements as the stage position was varied. For large particles, the response is a convolution of the scattering cross section and the beam intensity functions. However, small particles can be regarded as an impulse function, and the response curve is a direct measure of the Gaussian beam diameter. Since the incident beam was at an angle to the sample surface the response function is a two-dimensional surface with elliptical contours. Figure 6 shows a typical response function.

The instrument sample stages had good positional repeatability. This made it possible to return to a specific location or feature and perform replicate measurements of its scattering behavior. It was also possible to switch polarization and remeasure the scattering of individual spheres. However, since our samples had no grids or reference markings, it was not possible to return to a specific location after the sample had been removed from the stage. Thus, we could not correlate the measured scattering signatures with optical images of specific spheres. In order to minimize the possibility of incorrectly including a sphere doublet or dirt particle in our data sets we typically measured five to ten samples at each set of experimental conditions and rejected statistical outliers from the data.

RESULTS AND DISCUSSION

Figure 7 shows the results of the scattering cross section measurements made on seven different 4.10 μm spheres deposited on bare

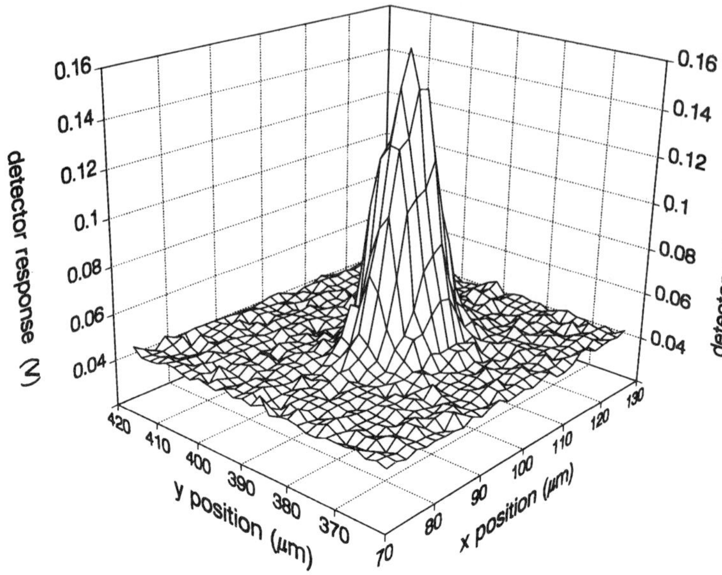

Fig. 6. Response surface plot for a single ring element (Ring 24, 17 degrees) as the laser beam at 30 degrees incidence was scanned over a 0.804 μm sphere on bare silicon substrate. The beam was measured to be 15 μm by fitting a Gaussian curve to a cross section of the data taken in the y direction through the widest portion of the curve. The elliptical shape of the contour is due to the non-normal incidence of the beam.

silicon substrate. The positions of the scattering minima and the cross section magnitudes are in general agreement with previous measurements made by Weber and Hirleman[4] on similarly sized spheres on nickel substrate. An exact correlation is not observed due to differences in experimental conditions, specifically beam polarization and substrate reflectivity. The data also show the variability between scattering signatures of individual spheres. The increased variability in the 36 to 40 degree region is not completely understood, but may be attributable to sensitivity of the results to the presence of a sharp minimum in the scattering function, which would shift with varying sphere size. Repeat measurements of a single sample showed less variability than the sample to sample measurements.

Figure 8 shows the differential scattering cross sections for 0.804 μm spheres on bare silicon substrates as a function of beam incident angle and polarization. These plots are averages of several samples. Examination of the figure reveals several key features: In the case of S polarization the scattering cross section is greatest for illumination at the Brewster angle, and decreases as the incident angle decreases. The situation is reversed for P polarization. In this case the scattering cross section is lowest for Brewster angle incidence. These results can be understood in terms of the reflectance behavior of silicon: For S polarization, the surface reflectance is 77% at the Brewster angle, diminishing to 40% at 30 degree incidence. The forward scattered light near the specular beam is reflected more efficiently near the Brewster angle, resulting in a larger scattering cross section. With increasing angular deviation from the specular beam the reflectivity

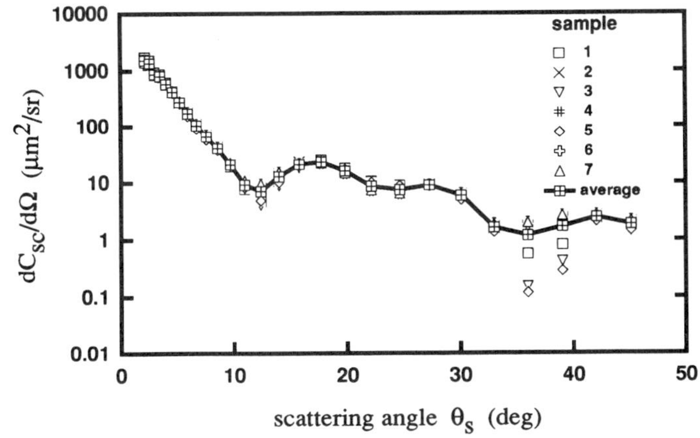

Fig. 7. Plot of differential scattering cross section $dC_{sc}/d\Omega$ for 4.10 μm polystyrene spheres on bare silicon substrate as a function of mean scattering angle θ_s for several samples. These measurements were made with a beam incidence of 45 degrees and P polarization.

diminishes and the curves tend to approach each other asymptotically. The situation is inverted for P polarization. Here, the surface reflectance is nearly zero at the Brewster angle. Thus, the specular beam and scattered beams near the specular are significantly attenuated. This can be seen in Fig. 8b where the cross sections reported for $\theta_s \le 20°$ and Brewster's angle incidence are nearly one order of magnitude smaller than those associated with other incident angles. At larger values of θ_s the associated silicon surface reflectance increases and the effect on the scattering cross section is less. The influence of substrate reflectance on scattering cross section has been observed previously by Galbraith and Neukermans[12], although their work was confined to measurements of total integrated scatter and did not include effects of incident beam angle and polarization.

The experimental results obtained on 4.10 μm spheres on bare silicon with Brewster angle incident illumination are compared with theory in Figure 9. The results agree in a qualitative sense, i.e. they exhibit similar trends and are of the same order of magnitude. Comparison between the S and P polarization cases shows the model correctly predicts a damping of the oscillations in the sphere scatter for $\theta_s \le 20°$. However, the detailed predictions do not match the experimental results. We postulate that several factors may be involved. In examining the figures it is obvious that the predicted minima are sharper than those actually observed. This is due in part to the fact that our detector necessarily integrates over a finite range of θ_s. The experimental minima may be further smeared by the fact that the measured spheres cannot be perfectly uniform, but instead exhibit a range of sizes. The discrepancy between theory and experiment at large angles is magnified because our current model does not integrate around the cone of scattered radiation intercepted by the ring detectors. We suspect that additional scattering components outside the plane of reflection are contributing a smoothing effect to the experimental data.

Comparisons between theory and experiment for 0.804 μm spheres appear in Figure 10. The lack of detailed agreement between theory and experiment suggests that the model deficiencies are more significant for

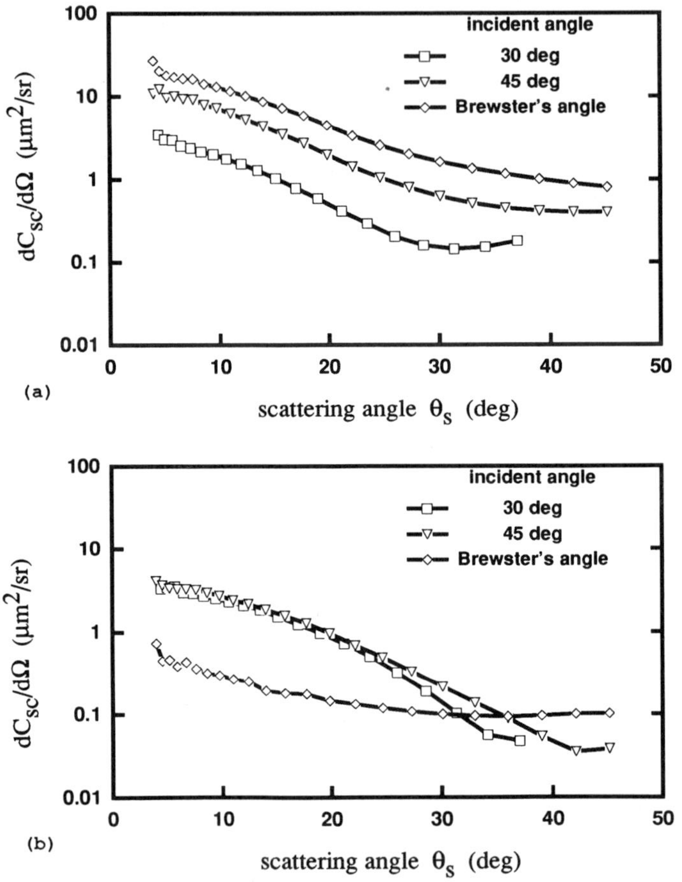

Fig. 8. Plot of the differential scattering cross section $dC_{sc}/d\Omega$ vs. mean scattering angle θ_s for 0.804 μm spheres on bare silicon substrate.

(a) S polarization

(b) P polarization

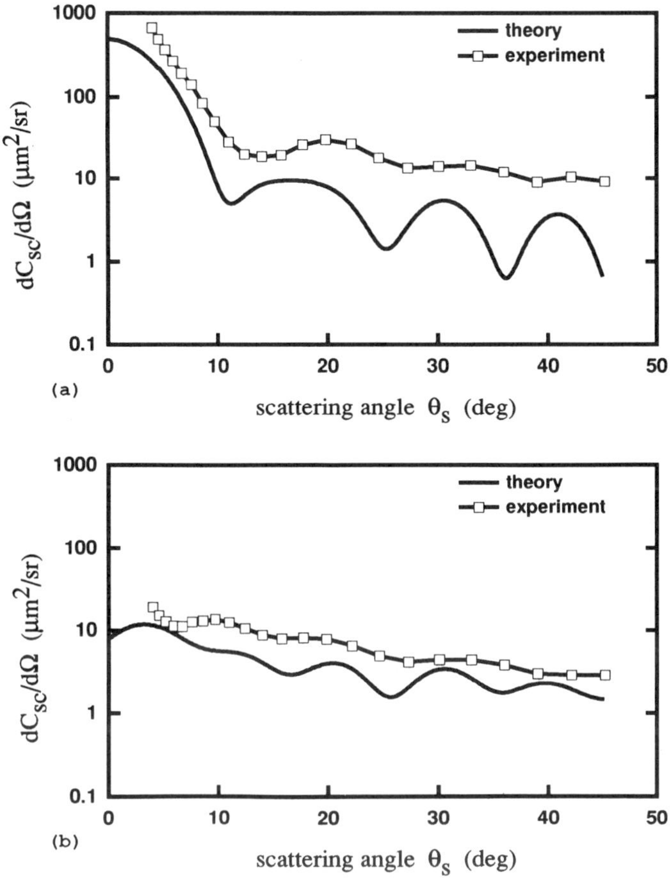

Fig. 9. Comparison between experiment and the modified Lorenz-Mie theory of Equation (4) for 4.10 µm spheres on bare silicon at 75.3 degree incidence.
(a) S polarization.
(b) P polarization.

smaller spheres. Another possibility is that standing wave effects as described by Knollenberg[9] are significant for this sphere size, and are influencing the scattering behavior. Our data are not extensive enough totest this hypothesis. A series of measurements over a range closely spaced incident beam angles could potentially resolve this question, but is beyond the capability of our current apparatus.

One trend in the data is evident in Fig. 10: The minimum in the experimental result occurs at a smaller θ_s than predicted by theory. This difference implies that the apparent size of the particle is larger than its true size. This was also observed by Weber and Hirleman[4], who speculated that the image of the sphere reflected by the sample surface was contributing to the scattering, giving a result in which the apparent size is increased. This effect would also explain in part why the predicted scattering cross sections are slightly lower than those actually observed. Because the sphere is imaged on a surface which attenuates reflected beams, the effective scattering by the sphere/image

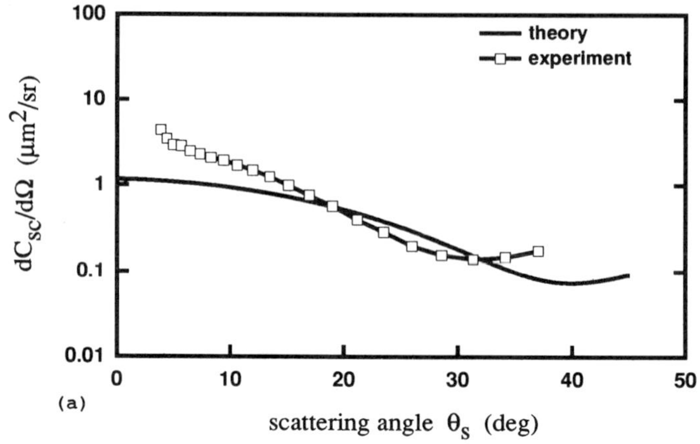

(a)

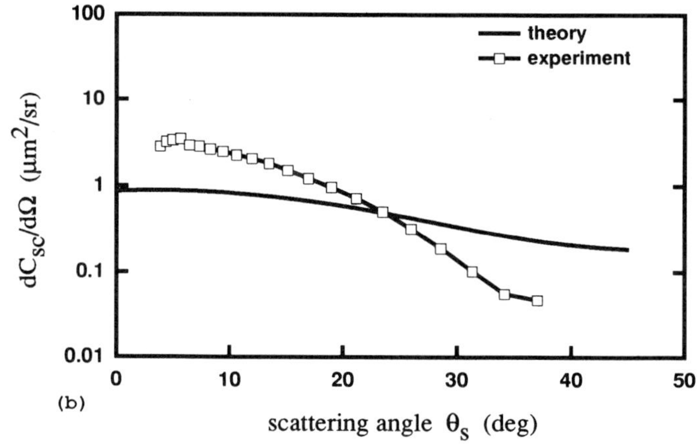

(b)

Fig. 10. Comparison between experiment and the modified Lorenz-Mie theory of Equation (4) for 0.804 µm spheres on bare silicon at 30 degree incidence.
(a) S polarization.
(b) P polarization.

combination will be more complicated than would result from simply doubling the sphere volume. Thus, corrections to the model to incorporate this effect will require detailed computations.

CONCLUSIONS

We have experimentally measured angle-resolved light scatter by submicron polystyrene spheres on silicon surfaces. The differential scattering cross sections of the spheres were shown to vary with the beam incident angle and polarization. Variations in scattering cross section were qualitatively consistent with variations in surface reflectance, with high surface reflectance resulting in larger apparent sphere scatter. A modification of Lorenz-Mie scattering theory was introduced to model the scattering behavior of the spheres on the silicon surface. The agreement between theory and experiment was good for the 4.10 µm diameter sphere, but less accurate for the 0.804 µm spheres. These studies imply that our model will be improved if the axial angular dependence of the surface reflectance outside the BRDF plane and additional sphere image effects are taken into consideration.

ACKNOWLEDGEMENTS

This work was supported by the Semiconductor Research Corporation contract 89-MJ-153 and by IBM Corporation. We thank Mr. C. W. Teutsch of Intel Corporation for supplying the substrates and Mr. James B. Mohr of Motorola for assistance in preparing samples for SEM study. We also want to acknowledge helpful discussions with our SRC Mentors and industrial contacts: Dr. Samuel Batchelder (IBM Corporation), Mr. Christopher Teutsch (Intel Corporation), Dr. Edward Hall (Motorola) and Mr. Brad W. Scheer (VLSI Standards).

REFERENCES

1. G. L. Wojcik, D. K. Vaughan, and L. K. Galbraith, SPIE Proc., 774, 21 (1987).
2. H. S. Lee, S. Chae, Y. Ye, D. Pui, and G. L. Wojcik, Aerosol Science and Technology 14, 177 (1991).
3. H. C. Van de Hulst, "Light Scattering by Small Particles," pp. 114-130, Dover, New York, 1981.
4. D. C. Weber and E. D. Hirleman, Applied Optics 27, 4019 (1988).
5. P. A. Bobbert and J. Vlieger, Physica 137A, 209 (1986).
6. G. Videen, J. Opt. Soc. Am. A 8, 483 (1991).
7. K. B. Nahm and W. L. Wolfe, SPIE Proc. 675, 295 (1986).
8. R. P. Young, Opt. Eng. 15, 516 (1976).
9. R. G. Knollenberg, Proc. Institute of Environmental Sciences, p. 501 (1986).
10. David K. Cheng, "Field and Wave Electromagnetics," pp. 352-361, Addison-Wesley, Reading, Massachusetts, 1983.
11. O. L. Russo, J. Electrochem. Soc. 127, 953 (1980).
12. L. Galbraith and A. Neukermans, SPIE Proc. 774, 13 (1987).

BRDF MEASUREMENTS AND MIE SCATTERING ANALYSIS OF SPHERICAL PARTICLES ON OPTICAL SURFACES

Robert J. Champetier

Eos Technologies Inc.
606 Wilshire Blvd., Suite 700
Santa Monica, CA 90401

Michael Sobelman and Julio R. Blanco

Department of Physics and Astronomy
California State University Northridge
Northridge, CA 91330

We have used Mie scattering intensity predictions to compute the bidirectional reflectance distribution function (BRDF) of arbitrary dust size distributions on optical surfaces. This modeling approach is supported by experimental data on spherical particles with known size and optical properties, deposited on glass and mirrors and whose BRDF was measured with polarizers and analyzers. Our conclusions confirm previous published results on modeling the effects of particulate distributions on optical scatter.

INTRODUCTION

Our interest lies in measuring and analyzing the changes in surface scattering brought about by the deposition of dust on optical surfaces. This issue affects a broad range of instruments, notably optical telescopes and other imaging sensors, and indeed has been the subject of entire technical sessions.[1] During the last two decades, the desire to improve the stray light performance of refractive, and especially reflective, optical sensors has spurred interest in smoother mirrors, scatter measurement techniques, control of particulates and other contamination, determination of surface dust counts and the resulting effects. In this paper, we present measurements of scatter caused by latex spheres deposited on glass and mirror surfaces, and show that useful analyses can be made using computations based on Mie theory,[2] but that care is necessary when applying such calculations to surface Bidirectional Reflectance Distribution Function (BRDF).

BACKGROUND

The BRDF of a surface describes the concentration of reflectance per

Particles on Surfaces 3, Edited by K.L. Mittal
Plenum Press, New York, 1991

unit solid angle, under specified restrictions such as uniformity, isotropy, and infinitesimal measurement parameters.[3] In spite of these limitations, and perhaps due to a lack of acceptance for appropriate nomenclature, the name BRDF persists in general practice and consequently is used here as well. The definition given below (Equation 1), from ref. 3, describes the BRDF, f_r, in terms of the incident irradiance, E_i, and the reflected radiance, L_r, in the usual coordinate system, as depicted in Figure 1. The units are 1/steradian.

$$f_r(\theta_i,\phi_i;\theta_r,\phi_r) \equiv \frac{dL_r(\theta_i,\phi_i;\theta_r,\phi_r;E_i)}{dE_i(\theta_i,\phi_i)} \equiv \frac{dL_r(\theta_i,\phi_i;\theta_r,\phi_r;E_i)}{L_i(\theta_i,\phi_i)\cdot\cos\theta_i\cdot d\omega_i}. \tag{1}$$

Most measurements use laser sources which illuminate a spot on the sample smaller than both the sample and the area viewed by the detector. For flat samples presenting a geometrically smooth surface, we may then approximate the true BRDF using measurements of power, Φ (in watts), in Equation (2),

$$BRDF=\frac{\Phi_r}{\Phi_i\cdot\omega_r\cdot\cos\theta_r}, \tag{2}$$

where again ω_r is the solid angle subtended by the detector.

This is the formula which is in general use.[4] Note that for the case of a perfectly isotropic and geometrically flat diffuser, the measured signal should vary as the cosine of the angle of reflection, θ_r, since the projected area of the illuminated spot diminishes with the view factor $\cos\theta_r$, and vanishes at grazing observation. As a result, the BRDF remains constant for such a surface. We shall see later that certain problems

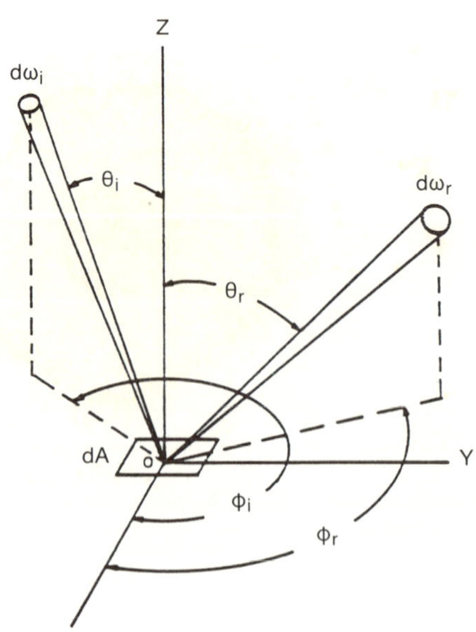

Figure 1. Geometry of incident and reflected beams after ref. 3.

arise when applying this approach to discrete scatterers of finite size such as latex spheres or dust.

The characterization of surfaces in terms of BRDF saw its start in the late 1960's, both with respect to definitions,[3,5] and measurements (for instance, see Arnold and Beard).[6] Coarse estimates for dust scatter were done[7] by assuming isotropic scatterers with a projected area equivalent to that of the dust. The BRDF is simply that of a perfect diffuser scaled down to the appropriate covering fraction of the dust. The covering fraction, CF, also called the obscuration ratio, is defined as the fractional area covered by particulates, usually the cumulative projected area of the dust. It is thus equal to the probability that a random point on the surface be shaded by a particle of dust. Using the diffuse approximation, and considering an example of a cleanliness level 500 dust, we get, BRDF of perfect diffuser = $1/\pi$, BRDF for dust with CF = 0.003 is $BRDF_{dust} = 0.003/\pi = 0.001$.

The cleanliness level is a MIL-STD-1246B designation[8] in common use to describe the size distribution and count of dust on surfaces. For example, a CF of 0.003 corresponds nominally to the level 500. Since there are no appropriate reported data to compare with theory, only a crude, order of magnitude agreement can be noted with the model predictions of Young[9] for scatter angles 20 to 40°, or the measurements reported in the literature from 10 to 30°.[10] No valid comparison is feasible with Carosso[11] where glass substrates were used instead of mirrors, or with Chen and Hedgeland,[12] who used black glass. Close agreement results from the model of Ma, Fong and Lee[13] in the range 10 to 50°, which qualitatively agrees with the data of Williams and Lockie.[14] The diffuse model is expected to fail for angles less than about 10°, as no allowance is made for the known strong forward scatter of the large particles making up the dust.

Some simple experiments revealed the presence of the resonance lobes for the case of latex spheres on mirrors, and this pointed to a possible use of Mie's Theory. Young[9] pursued this approach experimentally and developed a simple model by adding two contributions: the power directly scattered by each particle toward the detector, and that forward scattered and reflected by the mirror or other substrate. The scatter contributions of particles of various sizes and counts were added so as to reflect actual dust distributions of interest. This approach, or a variation of it, forms the basis for other models currently in use.

The data presented here were collected for the purpose of establishing a basis for modeling dust scatter using Mie computations.[7] The choice of dielectric spheres with a narrow size distribution was made so as to test the resemblance of computed scatter distributions for isolated spheres, with BRDF measurements of spheres deposited on anti-reflection (AR) coated glass, and on a metallized mirror. These results were not reported earlier, as there appeared then to be limited interest in such studies. The current calculations were carried out with a more recent code described by Bohren and Huffman,[15] which is used in our own model, now in the process of being fully tested. This code was checked against that of Dave[16] as far as the Mie intensities are concerned, with very good agreement.

EXPERIMENTAL

BRDF Scatterometer

This study compares theory with data, and required some care in sample preparation and measurement technique. The BRDF apparatus was in routine use and its results had been compared to those of other laboratories to assure consistency. For this study, a helium neon laser operating at 632.8 nm was used as a source of visible light. The laser was some 10 m away from the sample, and lenses with focal lengths of about 6 m were used to place the beam waist, about 3 mm in diameter, at the sample position, and along a horizontal path. Figure 2 is a sketch of the optics. The power reflected by the chopper went to a reference detector.

The samples were typically 5 cm in diameter, had sharp edges, and were held from the back surface by a mechanical stage. The detector was fastened to a boom on a turntable, and could be scanned about the vertical axis of the turntable. The entire sample zone was within a class 100 per Fed-Std-209B clean air bench. Drift in the entire source, detector and signal readout with phase lock amplifiers, was minimized and checked routinely. The detector consisted of an aperture stop subtending a maximum 1.62° at the sample, an imaging lens, a field stop limiting the viewed circular area to a diameter of 26 mm at the 675 mm distance, and an expansion space ahead of the photomultiplier tube so that the cathode received a defocused beam. The detector assembly was mounted on a swivel holder so as to achieve proper aim and centering on the samples. This photometric system exhibited a flat response across the entire field of illumination in the sample plane. The linearity of the tube and amplifier was verified over some 6 orders of magnitude using combinations of calibrated neutral density filters.

Sample Preparation

The samples were of two types: glass, antireflection coated on both sides, and first surface aluminized mirrors. The BTDF (where the T stands for transmittance) of the glass at normal incidence is shown in Figure 3. In this figure, the beam is incident horizontally, and the polarizer and analyzer are set horizontal, H-H, and vertical, V-V. In order to prepare surfaces with latex sphere populations as free of clumping as possible, the spheres, which were 1.1 μm in diameter, were first ultrasonically shaken while in a liquid suspension. Then they were passed on to a nebulizer of a

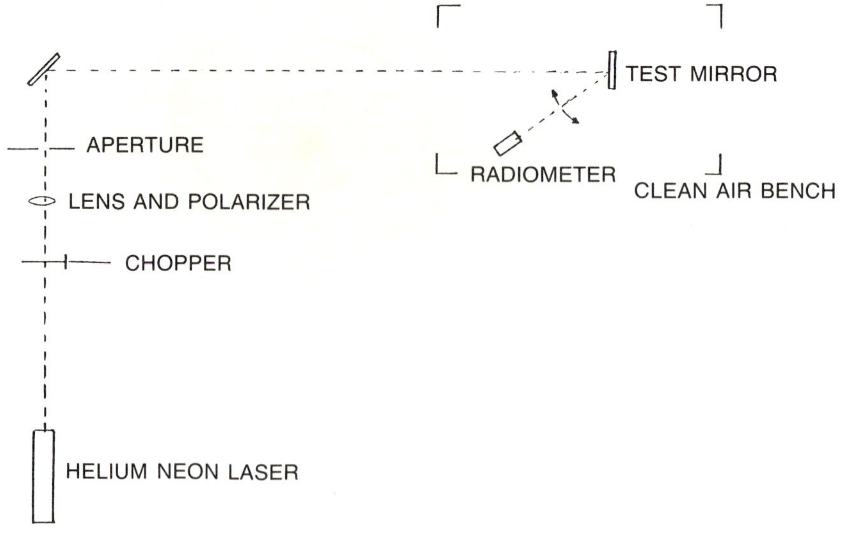

Figure 2. Apparatus for BRDF measurements.

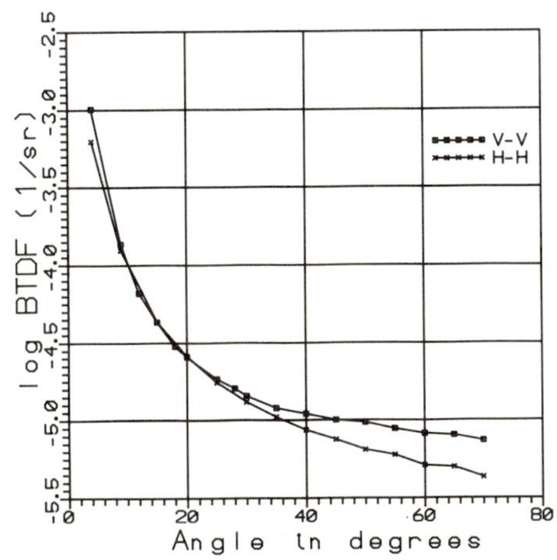

Figure 3. BTDF of clean AR coated glass. Horizontally incident beam, horizontal polarizer and analyzer (H-H) and vertical polarizer and analyzer (V-V), normal incidence, 632.8 nm. Angle is measured from the transmitted specular beam.

Babington type schematically shown in Figure 4, which was used to generate a fine mist. The nebulizer was set so that the probability of two latex spheres per droplet was low. It was placed well above the samples so that the liquid would evaporate on the way down. This device resulted in very uniform surface densities with no more than the statistically expected doublets or triplets of the latex. Particle counts were first made using witness mirrors, but later done with the actual samples using a microscope.

BRDF Calibrations

The reference samples for calibrating the BRDF measurements were made from Eastman's White Standard, an ultra-pure form of barium sulfate, $BaSO_4$,

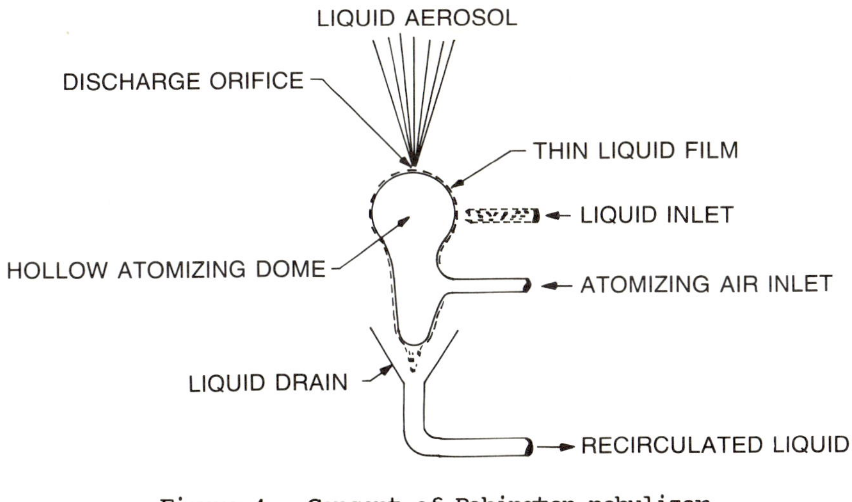

Figure 4. Concept of Babington nebulizer.

with well known reflectance properties. The powder was placed in a cavity several millimeters deep and pressed flat by hand using a glass plate.

Two methods were used to calibrate the BRDF measurements. The first was to illuminate (632.8 nm) the barium sulfate at normal incidence (along the horizontal beam path) and perform polar scans about the vertical axis of the detector stage. Figure 5a shows such scans for the incident beam polarized horizontally and the analyzer in front of the detector set horizontally, H-H, and vertically, H-V. Figure 5b shows a similar comparison for the case in which the incident beam is vertically polarized, V-V and V-H. Note the higher signals for matching polarizer orientations H-H and V-V than for the crossed cases. This indicates that our pressed barium sulfate was not a perfect depolarizer and, therefore, not a perfect diffuser. This point is also evident from the decrease in BRDF toward 90°.

The absolute values of the BRDF's were obtained using an instrumental constant, which converts the signal in volts to BRDF in sr^{-1}. This constant is determined by integrating the raw signals over the hemisphere above the sample and setting the sum for both analyzer polarizations equal to the known reflectance of barium sulfate. The figures show the data after the instrumental constant has been factored in. It is worthwhile to note that the average of the H-V combination from 6 to 30° is 0.1518, while for the V-H pair, it is 0.1511, or a difference of 0.5%. The slight difference between the two other scans, (H-H) and (V-V), may therefore be real rather than apparatus error. These latter two scans differ in the geometry of the scan plane in relation to the plane of incident polarization. The resulting values for normal incidence and an 8° observation angle are:

V-V	0.185	H-H	0.196
V-H	0.151	H-V	0.152
sum	0.336	sum	0.348

overall average : 0.342, difference V vs. H : 3.4%.

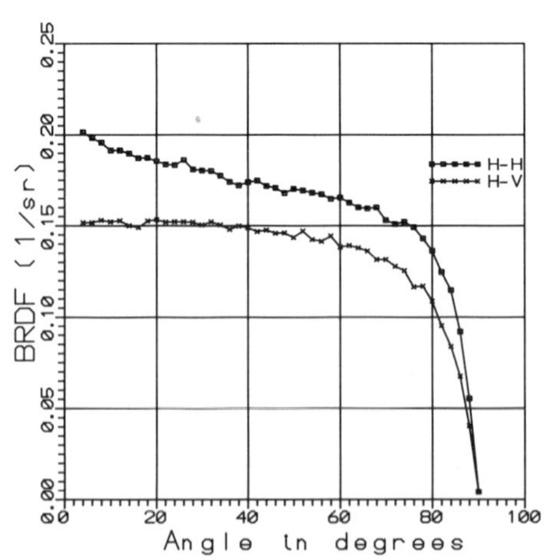

Figure 5a. BRDF of $BaSO_4$. Horizontally incident beam, horizontal polarizer and analyzer (H-H), horizontal polarizer and vertical analyzer (H-V), normal incidence, 632.8 nm. Angle is measured from the surface normal.

The second calibration is obtained by measuring the ratio of the scattered beam signal to the specular signal. This ratio requires attenuation with neutral density filters for the 100% beam, and these filters were checked with the same equipment. This calibration was carried out for one polarization combination only, for normal incidence and an 8° detector angle. The agreement with the integration approach above was within 3.3%. The integration method is thought to be more reliable in this case.

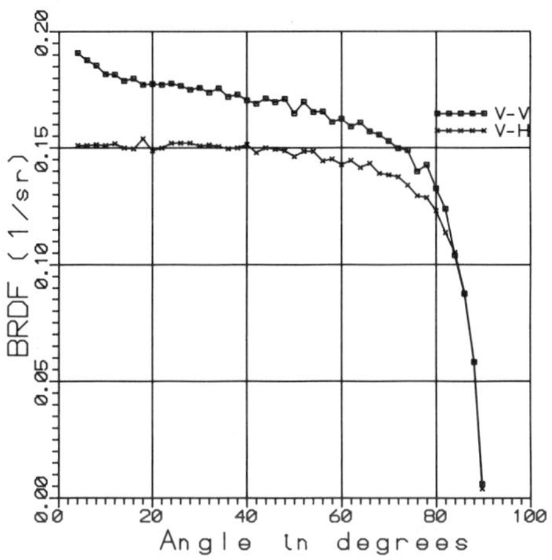

Figure 5b. BRDF of $BaSO_4$. Horizontally incident beam, vertical polarizer and horizontal analyzer (V-H) and vertical polarizer and analyzer (V-V), normal incidence, 632.8 nm. Angle is measured from surface normal.

Latex Spheres on Glass

The spheres had a diameter of 1.099 μm and an index of refraction of 1.5905 at the 632.8 nm wavelength, as reported by the supplier. Initial attempts to find good agreement between measured and calculated scattering patterns and absolute levels were marginally successful. Several experiments were then run so as to rule out systematic errors, including the positioning of the spheres on the incident side of the glass as well as the side where the beam exits. Concerns over humidity effects were tested by warming the samples to 43°C during measurements. No systematic trends were seen and all the data were in good agreement.

Since significant discrepancies in the diameter of the latex spheres were measured with an electron microscope, the possibility of our spheres departing from the claimed values was considered. Experimental diameters and indices were then determined by optimizing the fit of the data measured on glass, to Mie calculations, using three criteria. These criteria were applied simultaneously and were: (1) the position of the first minimum in the pattern, (2) the position of the second minimum for the V-V case of vertical polarizer and analyzer, and (3) the ratio of the two scattered polarized intensities as a function of angle. The optimum fit was found for a diameter of 1.163 μm and an index of 1.6050, which leads to a size parameter $2\pi a/\lambda$ of 5.775. The BRDF (actually BTDF) measurements are shown in Figures 6a and 6b, along with the computations. Note the quantitative

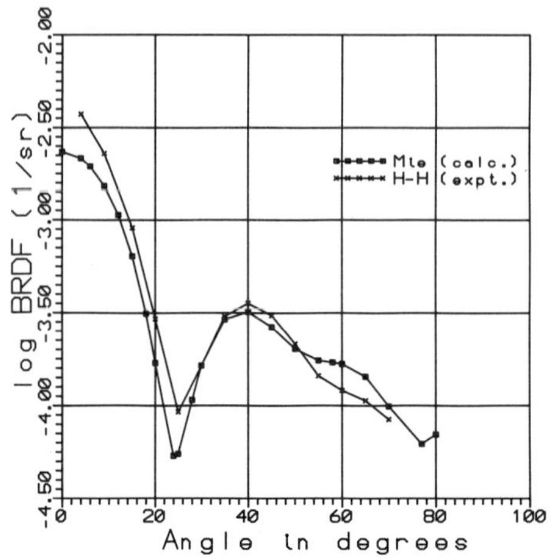

Figure 6a. BRDF (actually BTDF) of Latex on Anti-Reflection Coated Glass. Horizontally incident beam, horizontal polarizer and analyzer (H-H), normal incidence, 632.8 nm. Angle is measured from transmitted specular beam.

agreement up to large scatter angles. The slight disagreement at angles less than 7° may be due to the contribution from the clean glass alone, whose BTDF is comparable to or higher than the BRDF of the spheres alone.

This part of the experiment showed that quantitative agreement was achieved between theory and measurements, and therefore provided a further check on the validity of our BRDF measurements. It should be pointed out that the expression used for Figures 6a and 6b differs from those in general use by not including a $1/\cos\theta_r$ coefficient for the measured and

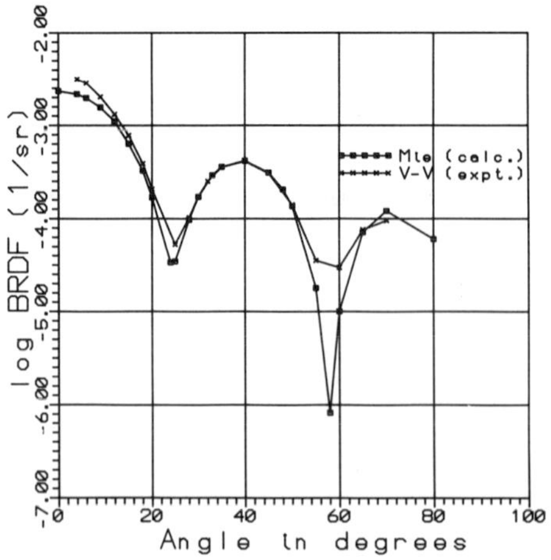

Figure 6b. BRDF (actually BTDF) of Latex on Anti-Reflection Coated Glass. Horizontally incident beam, vertical polarizer and analyzer (V-V), normal incidence, 632.8 nm. Angle measured from transmitted specular beam.

predicted BRDF values. Mie solutions, in Bohren and Huffman,[15] are given in terms of the parameter S_{11}, a dimensionless coefficient which relates the incident intensity, I_i, (called irradiance E in radiometric terminology) to the scattered intensity, I_s, by the expression

$$I_s = \frac{S_{11} I_i}{k^2 r^2} \tag{3}$$

for the unpolarized case, where k is $2\pi/\lambda$ and r is the distance to the detector. The subscript s, not r, is normally used for opaque surfaces.

We may, if we wish, create an expression for BRDF based on this irradiance and treat the spherical scatterers on the glass surface as equivalent elements of area dA with no thickness. For the case of a finite beam incident on a surface at angle θ_i, with beam cross-sectional area A_0, illuminating an area A on the surface equal to $A_0/\cos\theta_i$, and a surface density of spheres D, the number N of spheres illuminated is AD, or $A_0D/\cos\theta_i$. The irradiance on each true sphere is not altered as the beam incidence is changed, unlike a true flat surface, where the irradiance decreases as $\cos\theta_i$. The expression becomes, after some manipulation,

$$f_r = \frac{DAS_{11}}{k^2 \cos\theta_s}, \tag{4}$$

where θ_s is the scattering angle.

However, the possibility of a discontinuity arises at $\theta_s = 90°$ where discrete particles are used rather than a surface. At a grazing observation angle, the illuminated area of a flat surface vanishes as $\cos\theta_s$, causing any measured signal to also vanish. For a perfectly isotropic diffuser, the BRDF remains constant with θ_r. The measured signal for such a sample is proportional to BRDF$\cdot\cos\theta_r$, and therefore decreases with the $\cos\theta_r$ factor, and vanishes at θ_r=90°.

For particles, however, no such disappearance occurs and the usual BRDF expression, Equation (2), leads to a discontinuity at 90°. Spherical particles, whether distributed in a plane or a volume, each intercepts a fixed fraction of the illuminating beam, regardless of the angle of incidence (constant irradiance). Similarly, the detector "sees" a fixed cross-sectional area for these spheres, independent of its angle to the plane. We assume a low particle surface density, so that the particles do not shadow one another.

There are two cases of interest where the incidence angle to the plane is varied. The first assumes a finite beam incident on a large plane with particles. The number of illuminated particles increases as $1/\cos\theta_i$, just as the beam footprint does. The second case postulates a large beam and a small sample, where the number of lit particles remains fixed and each particle receives a constant irradiance.

These observations serve to point out that BRDF terminology is suitable for surfaces but not, in the strict sense, to particulates. We may for convenience employ such expressions, either with or without cosine terms, but care is required in using the results correctly. For cases where large angles are considered, the standard Mie scattering expressions are quite reliable, as they do not include terms relating to the projected area of the source elements.

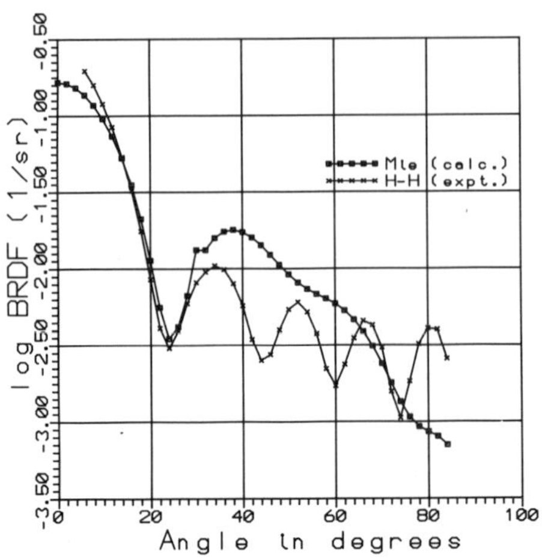

Figure 7a. BRDF of Latex on Aluminum. Horizontally incident beam, horizontal polarizer and analyzer (H-H), normal incidence, 632.8 nm. Angle is measured from the reflected specular beam.

Latex Spheres on Aluminized Mirrors

The same spheres and deposition apparatus were used to prepare samples on first surface aluminized mirrors, in order to check the extent to which Mie computations might be applicable. The data are given in Figures 7a and 7b using the same methods and data computations as for the glass substrates. The angle of incidence was 3°. The BRDF of the clean substrate in this case was considerably lower, by typically two orders of magnitude, than that of the spheres. The surface density averaged 0.001178 spheres per μm^2 of mirror surface.

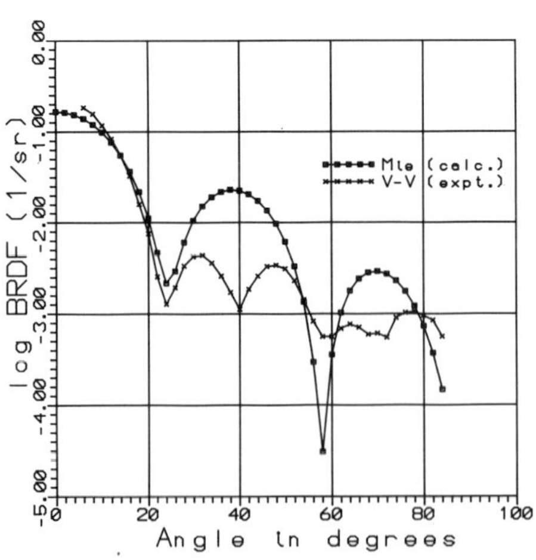

Figure 7b. BRDF of Latex on Aluminum. Horizontally incident beam, vertical polarizer and analyzer (V-V), normal incidence, 632.8 nm. Angle is measured from the reflected specular beam.

The fit to the theory appears to be best in the angular range 10 to 30°. At larger angles, we observe a general order of magnitude trend of agreement and markedly different resonance features. This difference in qualitative features is not surprising if one considers that the assumptions required for a valid application of the Mie analysis are violated when the particle is brought in contact with a conducting surface. It is tempting to state that agreement at large angles between measurements and a Mie model may be largely fortuitous. It is significant, however, that most applications of such analyses may stress only angles less than 20 degrees where the agreement is better. Clearly, more investigations would be required for the 0 to 10° range.

MODEL DEVELOPMENT

The model we are developing is based on a simple summation of Mie scattering patterns for an arbitrarily specified population of spheres representing actual dust particles. Refinements have not yet been implemented regarding optimum optical constants and dissection of large, irregular and typically fibrous dust and lint particles into a suitable equivalent set of smaller spheres. These improvements must await the availability of precise BRDF measurements on well characterized dust populations.

It is expected, however, that the shape differences between actual dust and ideal spheres may not have a major deleterious effect on the usefulness of a model for engineering approximations. Young[9] has reported significant success in matching his model to data. In checking our calculations, we have computed the BRDF to match Young's test distributions which are described in his original report, and we have obtained good agreement. We have also exercised variations in the sampling of particle sizes, so as to avoid bias due solely to the use of insufficient computations.

The effect of particle shape has been investigated by numerous authors over the last 30 years.[17-33] The similarity with sphere scattering appears to be adequate to support approximation schemes of interest here. To illustrate the use of such a model we have computed the BRDF for hypothetical sphere distributions. Figures 8a and 8b show the BRDF for a distribution with the slope of -0.926 and level 1000 in the usual MIL-STD-1246B format, which implies a maximum particle size of 1000 μm. For strictly spherical particles, the CF is 12.28%. The contributions from the spheres alone are shown, spanning the range 0 to 180°, so as to display both forward scattered (0 to 90°), and back scattered (90 to 180°) levels. Contributions from both may be expected to add to the BRDF of an actual mirror, but the precise weighting of these will require further experiments. Several authors[34-37] have analytically investigated the scattering by a sphere in proximity to a surface, and application to dust scatter modeling may soon be verifiable.

The BRDF levels corresponding to arbitrarily defined size increments have also been computed and are shown in Figures 8a and 8b as well as the whole distribution. Of these subsets, the dominant one is due to the 1 to 30 μm range, and the larger sizes are ranked in decreasing order in these calculations. In order to gauge the importance of the various particulate sizes in causing scatter, but on a size distribution more representative of actually observed dust populations on exposed optics, the same calculations

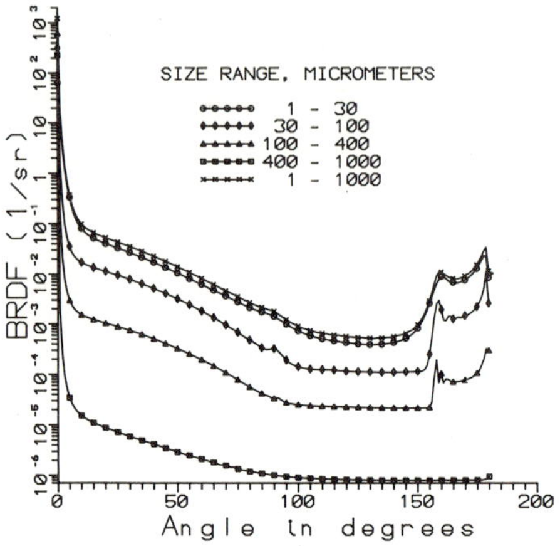

Figure 8a. Predicted BRDF for dust alone, MIL-STD-1246B with a slope of -0.926, cleanliness level of 1000, particle sizes 1-1000 μm, wavelength = 514.5 nm, index of refraction n = 1.5 - i0.0004, covering fraction = 12.28%, calculated from 0 to 180° at 1° intervals.

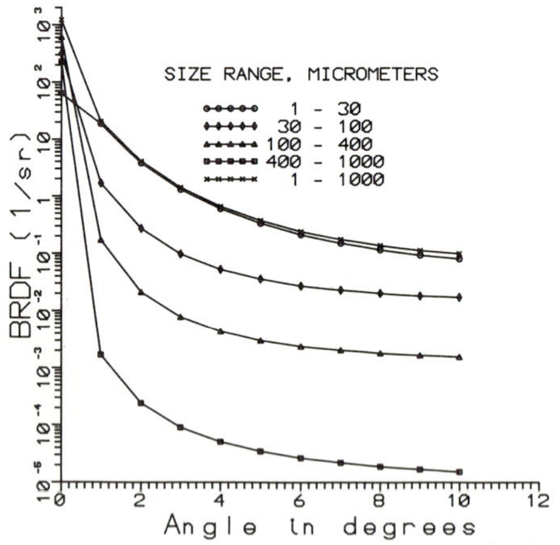

Figure 8b. Predicted BRDF for dust alone, MIL-STD-1246B with a slope of -0.926, cleanliness level of 1000, particle sizes 1-1000 μm, wavelength = 514.5 nm, index of refraction n = 1.5 - i0.0004, covering fraction = 12.28%, calculated from 0 to 10° at 1° intervals.

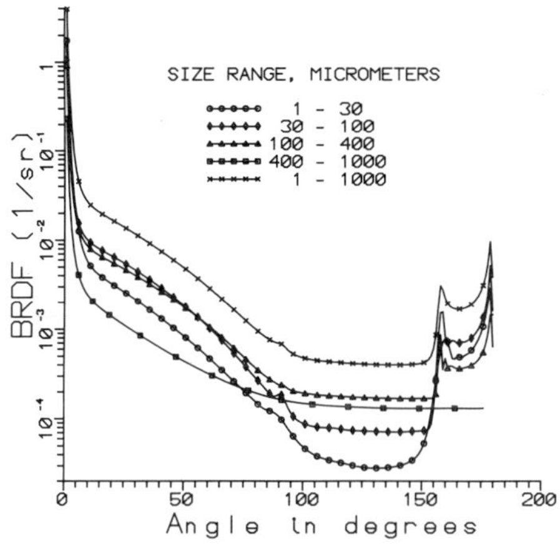

Figure 9a. Predicted BRDF for dust alone, MIL-STD-1246B with a slope of -0.4 with size distribution truncated past 1000 μm diameter, wavelength = 514.5 nm, index of refraction n = 1.5 - i0.0004, covering fraction = 12.28%, calculated from 0 to 180° at 1° intervals.

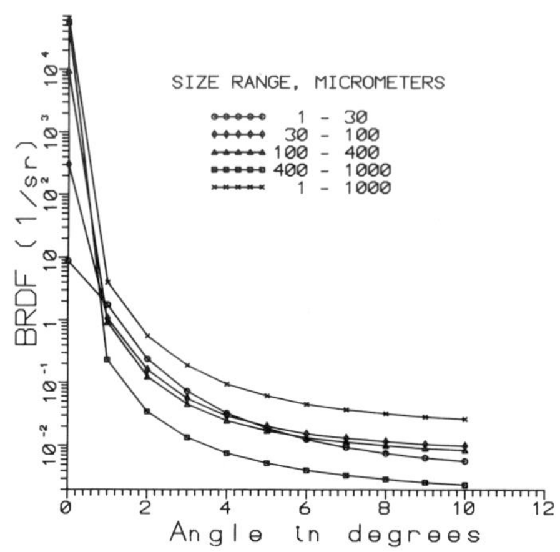

Figure 9b. Predicted BRDF for dust alone, MIL-STD-1246B with a slope of -0.4 with size distribution truncated past 1000 μm diameter, wavelength = 514.5 nm, index of refraction n = 1.5 - i0.0004, covering fraction = 12.28%, calculated from 0 to 10° at 1° intervals.

were performed for a slope of -0.4, but otherwise in the same MIL-STD-1246B format. The absolute counts were adjusted so as to result in precisely the same CF as in the first case with a slope of -0.926. The latter case is shown in Figures 9a and 9b. Note the change in ranking, with the 1 to 30 μm sizes no longer dominating the BRDF levels.

Clearly, the inference that the largest particles are spheres is unrealistic and leads to exaggerated forward scatter at the small angles, particularly less than 2°. With that caveat spelled out, the trend still appears to be that the larger particles cause most of the scatter near 0°. Experimental evidence supports the stated trends, when single large pieces of lint were observed to completely dominate mirror scatter.

It should be mentioned that the computations shown in Figures 8a,b and 9a,b were performed by an ELXSI System 6400 (SR13.0) computer. The program language used was FORTRAN and the mathematical subroutines were selected from the IMSL Math/Library.[38] The BRDF calculations were lengthy (depending on the particle size parameter) though the FORTRAN code developed was not optimized for efficiency. Each BRDF curve shown in these figures takes several days of CPU time when calculated at intervals of 1°, as these were.

CONCLUSIONS

Our experimental work indicates that there is ample justification for basing models of dust on Mie computational approximations. More work is unquestionably needed to refine these models, particularly for the large particles in typical distributions, which dominate the scatter near 0° and are the least well approximated by spherical analogs. Caution has been pointed out in the formulation of BRDF expressions starting with Mie results, particularly for cases of non-normal incidence and large angle viewing. The radiometric nomenclature so useful for surfaces breaks down when particulates on surfaces are considered.

ACKNOWLEDGEMENTS

The authors wish to thank Norman Stahlberg and Robert Giguere for their most able assistance in the early phases of this work. JRB acknowledges partial support by the California State University at Northridge.

REFERENCES

1. J.C. Stover, Editor, "Scatter from Optical Components," Proc. SPIE 1165, 1-459 (1989).
2. G. Mie, Ann. Physik 25, 377 (1908).
3. F.E. Nicodemus, J.C. Richmond, and J.J. Hsia, NBS Monograph 160, (1977); F.E. Nicodemus, Appl.Opt. 9, 1474 (1970); Appl. Opt. 5, 715 (1966); Appl. Opt. 4, 767 (1965).
4. J.C. Stover, paper No. 1165-01 in ref. 1.
5. D.B. Judd, J. Opt. Soc. Am. 57, 445 (1967).
6. C.B. Arnold and J.L. Beard, paper 1165-09 in ref. 1.
7. R.J. Champetier, unpublished data (mid 1970's).
8. MIL-STD-1246B, "Product Cleanliness Levels and Contamination Control Program", Sept. 1989.
9. R.P. Young, Opt. Eng. 15, 516 (1976); American Institute of Aeronautics and Astronautics, paper No. 75-667 presented at the AIAA 10th Thermophysics Conference at Denver, Colorado in 1975.
10. "Final Report - Investigation of Contamination on Optical

Surveillance Systems," RADC-TR-87-230, Contract Number F30602-85-C-0279, 39-45 (1987); report can be requested from RADC (OCSE), Griffiss AFB, NY 13441-5700.
11. P.A. Carosso and N.J.P. Carosso, Appl. Opt. 25, 1230 (1986).
12. P.T. Chen and R.J. Hedgeland, paper 1165-26 in ref. 1.
13. P.T. Ma, M.C. Fong and A.L. Lee, paper 1165-33 in ref.1.
14. V.L. Williams and R.T. Lockie, Opt. Eng., 18, 152 (1978).
15. C.F. Bohren and D.R. Huffman, "Absorption and Scattering of Light by Small Particles," Wiley, New York, 1983.
16. J.V. Dave, IBM Report No. 320-3237 (1968); a copy of this report may be obtained from IBM Palo Alto Scientific Center, 1530 Page Mill Road, Palo Alto, CA 94304.
17. J.M. Greenberg, N.E. Pedersen and J.C. Pedersen, J. Appl. Phys., 32, 233 (1961).
18. B. Donn and R.S. Powell, in "Proceedings of the Interdisciplinary Conference" held at Clarkson College of Technology in 1962, M. Kerker, Editor, p. 151, MacMillan, New York, 1963.
19. J.R. Hodkinson, in "Proceedings of the Interdisciplinary Conference" held at Clarkson College of Technology in 1962, M. Kerker, Editor, p. 87, MacMillan, New York, 1963.
20. A. Mugnai and W.J. Wiscombe, Appl. Opt., 25, 1235 (1986).
21. A. Mugnai and W.J. Wiscombe, Appl. Opt. 28, 3061 (1989).
22. R.J. Perry, A.J. Hunt and D.R. Huffman, Appl. Opt. 17, 2700 (1978).
23. P. Latimer, A. Brunsting, B.E. Pyle and Carey Moore, Appl. Opt. 17, 3152 (1978).
24. R.M. Welsh and S.K. Cox, Appl. Opt. 17, 3159 (1978).
25. R.G. Pinnick, D.E. Carroll and D.J. Hofmann, Appl. Opt. 15, 384 (1976).
26. N.K. Mehta, A. Grimison and A. McB. Block, Appl. Opt. 9, 2031 (1971).
27. A.C. Holland and G. Gagne, Appl. Opt. 9, 1113 (1970).
28. D.H. Napper and R.H. Ottewill, in "Proceedings of the Interdisciplinary Conference" held at Clarkson College of Technology in 1962, M. Kerker, Editor, p. 377, MacMillan, New York, 1963.
29. I. Kirmaci and G. Ward, Appl. Opt. 18, 3328 (1979).
30. D.W. Schuerman, Editor, "Light Scattering by Irregularly Shaped Particles," Plenum Press, New York, (1980).
31. W.Wiscombe and A. Mugnai, in "Light Scattering by Irregularly Shaped Particles," D.W. Schuerman, Editor, p. 141, Plenum Press, New York, (1980).
32. R.H. Zerull, R.H. Giese, S. Schwill and K. Weiss in "Light Scattering by Irregularly Shaped Particles," D.W. Schuerman, Editor, p. 273, Plenum Press, New York, (1980).
33. J.B. Pollack and J.N. Cuzzi in "Light Scattering by Irregularly Shaped Particles," D.W. Schuerman, Editor, p. 113, Plenum Press, New York, (1980).
34. P.A. Bobbert and J. Vlieger, Physica, 137A, 209 (1986).
35. P.A. Bobbert, J. Vlieger and R. Greef, Physica 137A, 243 (1986).
36. D.C. Weber and E. D. Hirleman, Appl. Opt. 27, 4019 (1988).
37. G.L. Wojick, D.K. Vaughan and L.K. Galbraith, in "Lasers in Microlithography," D.J. Ehrlich, J.Y. Tsao and J.S. Batchelder, Editors, Proc. SPIE 774, 21 (1987).
38. FORTRAN subroutines for mathematical applications, version 1.1 (1989); for more information contact IMSL, Customer Relations, 2500 ParkWest Tower One, 2500 CityWest Boulevard, Houston, TX 77042-3020, telephone (713) 782-6060.

IDENTIFICATION OF SURFACE CONTAMINANTS USING INFRARED MICRO-PROFILING*

D. S. Blair and K. J. Ward

Process Optimization and Control Division
Sandia National Laboratories
Albuquerque, New Mexico, 87185

Infrared micro-profiling is the combination of infrared microspectroscopy with precise microscope stage movements. It can provide molecular and spatial information for a variety of samples as small as 10 µm in diameter. To illustrate the technique, different contaminant materials, including a cellulose acetate fiber, oils deposited from a fingerprint, and a thin film of solder flux residue, were infrared micro-profiled. An integrated absorbance data reduction technique commonly used in gas chromatography/FT-IR applications was applied to the micro-profiling data. This technique organizes the vast amount of data generated, enabling the user to plot the results in 3-dimensional projections, allowing extraction of relevant spatial information. A method of coadding spectra from different pixel elements is presented, providing higher quality spectra without increasing data acquisition time. This procedure improves spectral signal-to-noise which aids in the identification of unknown contaminants.

INTRODUCTION

Infrared spectroscopy (IR) is a vibrational spectroscopic technique used for the nondestructive identification of molecular species. It provides information about the molecular structure of a compound by determining the frequency and quantity of light it absorbs in the infrared region. The resultant IR absorption spectrum is characteristic of the compound. It is one of the compound's physical properties, like its boiling point or density. In fact, with the exception of optical isomers, no two compounds have identical infrared absorption spectra. The molecular information that IR spectroscopy provides distinguishes it from other analytical techniques that yield information about elemental composition such as x-ray fluorescence, atomic absorption, and emission spectroscopy. In general, due to quantum selection rules and vibrational absorption frequencies, IR spectroscopy is used for the identification of

*This work was performed at Sandia National Laboratories supported by the U.S. Department of Energy under Contract Number DE-AC04-76-DP00789.

Particles on Surfaces 3, Edited by K.L. Mittal
Plenum Press, New York, 1991

organic materials. However, many inorganics also exhibit IR detectable molecular vibrations and therefore can be identified by this technique[1].

IR spectroscopy can be performed on molecular species in any physical state. Therefore, gases, liquids, and solids can all be easily sampled. However, IR spectroscopy may not be as useful for the identification of dilute solutions of less than millimolar concentrations due to low absorbance intensity. Complex mixtures can be difficult to identify due to the large number of absorption bands present in the spectrum. Also, metals and metallic alloys cannot be identified using IR spectroscopy since these materials are not infrared active.

Dispersive spectrometers were among the first commercially available infrared spectrometers. These instruments compare a beam of infrared light transmitted through or reflected by a sample to a parallel beam not exposed to the sample. The infrared spectrum is obtained and recorded by scanning successive frequencies. However, with the ready availability of powerful, but affordable computers, many research and analytical laboratories now use Fourier-transform infrared spectrometers. These spectrometers examine the entire frequency range of infrared light simultaneously by interferometry. The pathlength of two beams of infrared light is varied with time during each scan. A time dependent pathlength difference between the two beams causes phase shifts of one beam relative to the other. When the two beams combine, a constructive and destructive interference pattern, known as an interferogram, is formed. A Fourier transform is used to mathematically change the time dependent interferogram into a frequency dependent spectrum. The result is a measure of the intensity reaching the detector as a function of frequency. This is equivalent to a single beam spectrum obtained with dispersive infrared spectroscopy. However, Fourier-transform infrared spectroscopy (FT-IR) is an improvement over dispersive infrared spectroscopy. It can increase spectral quality while reducing acquisition time by examining all frequencies simultaneously. For more information about FT-IR, the reader is referred to the comprehensive text on the subject by Griffiths and de Haseth[2].

Microscopy/FT-IR became a reality after the development of high sensitivity detectors, commercial availability of high-intensity infrared sources, and improvements were made in the throughput efficiency of infrared spectrometers. These instruments make it possible to examine samples as small as 10 μm in diameter. The size limitation is a physical constraint based on the diffraction limit of light in the infrared region at 1000 cm^{-1}. However, a contaminant of this size may be considered macroscopic in the microelectronics industry. Miniaturization of semiconductor devices has created a demand for the detection and identification of contaminants in the tenths of a micron range. In our laboratory, though, microscopy/FT-IR has proven itself extremely useful for detecting contaminants on microelectronic components that have originated from processing equipment, personnel, and storage facilities because many such contaminants are larger than the 10 μm limit.

Infrared micro-profiling further enhances microscopy/FT-IR. It combines infrared microspectroscopy with controlled microscope stage movements and can be used to both locate and help identify chemical species. Like traditional IR microspectroscopy, IR micro-profiling can be used to examine contaminants present as particles, droplets, smears, fibers, crystals, inclusions, bubbles, or additives[1]. Recently, it was used to map, chemically identify and spatially resolve, a fiber contaminant on a microelectronic device[3]. It has also been used to identify individual polymers present in a multilayer polymer structure[4], and a fluid inclusion[5]. As will be shown, the usefulness of IR micro-profiling includes the ability to detect contaminants not visible with standard optical microscopy. Furthermore, even though data collection time is increased, the "hands-on" time of the spectroscopist is greatly reduced.

During IR micro-profiling, a complete infrared spectrum is obtained at each stage position or pixel element. This creates a large number of spectra that must be manipulated so that relevant information can be extracted to produce a spatial image. One approach for examining the data is to create infrared maps that show the spectral regions of interest for each pixel organized on a grid scaled to the sampled area[5]. Another approach is to compress the raw spectral data to a single intensity value at each pixel and plot it as a function of position in an axonometric projection. This allows efficient examination of the data collected using a 3-dimensional representation of the image. The compression can be accomplished by measuring the intensity of a single absorption frequency, by integrating the total absorbance over a specified spectral range, or by a procedure known as Grâm-Schmidt reconstruction[3].

Three examples will be presented to illustrate the usefulness of IR micro-profiling for contaminant identification. The examples are a cellulose acetate fiber, a fingerprint, and a thin film of rosin-based solder flux. Three-dimensional axonometric projections of the images will be presented. These result from compression of the spectral data at each pixel to a single value by integrating the absorbance over a spectral range containing absorption frequencies characteristic of the contaminant. A brief discussion will also be presented on coaddition of spectra from different pixel elements that contain spectral information about the contaminant. This procedure improves spectral quality, increases the signal-to-noise ratio (S/N) and provides a good quality resultant spectrum for subsequent library searches that helps in the identification of unknown contaminants.

EXPERIMENTAL

Infrared spectra were obtained using a Spectra-Tech (Stamford, CT) IR-PLAN infrared microscope interfaced to a Nicolet 7199 FT-IR (Madison, WI). The IR-PLAN is an adaptation of an Olympus (Lake Success, NY) research grade optical microscope. It allows the operator to visually align and mask the sample area before IR spectroscopy is performed. The spectra reported here were all obtained with the microscope in reflectance mode. Infrared radiation was focused at the sample by a totally reflecting 15X Cassegrainian objective (numerical aperture, 0.58). Redundant Aperturing™ was used to collect spectra of high spatial purity and photometric accuracy[5]. The system is equipped with a narrow band mercury-cadmium-telluride (MCT) detector having a 0.25 x 0.25 mm element and a spectral range of 4000-800 cm^{-1}. The microscope was coupled with a Spectra-Tech motorized micropositioning stage that provided movement in two dimensions. X and Y stage step sizes were varied from 20 to 80 μm with a spatial precision of 1 μm. Software to drive the stage was written by the authors using Nicolet FORTRAN to comunicate to the stage via an RS-232 port of the Nicolet 1280. A square aperture with adjustable blades masked each pixel area consistent with the step sizes.

The collection of infrared data and movement of the motorized stage was fully automated. A Nicolet 1280 computer, with a Fast Fourier Transform (FFT) co-processor, was interfaced to a stage controller which moved the stage an assigned distance after each spectrum was collected. Coadditions of 64 scans/spectrum at 8 cm^{-1} resolution and Happ-Genzel apodization were collected at each pixel with an acquisition time of 20 seconds/spectrum. The total data acquisition time for collecting 400 spectra for each of the three examples to be presented was approximately 2.5 hours. For image reconstruction, each spectrum was reduced to a single intensity value by integrating the infrared absorbance in a specified spectral region and plotting it as a function of spatial location. The 3-dimensional axonometric projections were made by transferring the integrated Nicolet absorbance files to a DOS based PC. The data were loaded into a 3-dimensional array and plotted using Axum™, a technical graphics and data analysis software package commercially available from TriMetrix Inc. (Seattle, WA).

RESULTS AND DISCUSSION

To illustrate the capabilities of IR micro-profiling, a single cellulose acetate fiber on a gold coated glass slide was mounted on the microscope stage. A square 20 x 20 μm aperture masked each pixel area with X, Y stage step sizes of 20 μm. Four hundred pixels were collected. Data were collected, processed, and stored sequentially on the Nicolet computer disk. After examining individual spectra it was determined that the C-H stretching region, 2800 - 3000 cm^{-1}, was distinctive to the fiber and would be the spectral region used for integration. The integrated intensity values were plotted as a function of pixel position in an axonometric projection, Fig. 1. The pixels that exhibit infrared absorptions in the spectral region of interest can be seen to move diagonally along the projection from the lower left to the upper right corner. Figure 1 can be used to estimate the fiber's diameter, 40 μm, and length, 450 μm. However, visible inspection of the sample under 150X magnification indicated that the entire length of the fiber was not profiled. This is confirmed by the significant image intensity at the upper right edge of the figure.

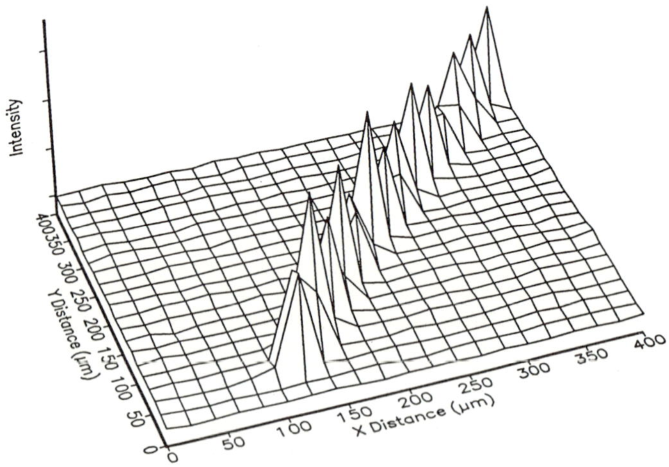

Figure 1. Axonometric projection of cellulose acetate fiber infrared micro-profiled using a 20 μm aperture, 20 μm step size.

Naturally occurring hand oils can easily be deposited on electronic components and their presence may adversely affect the component's performance. Therefore, a sample fingerprint on a gold coated glass slide was IR micro-profiled to further illustrate the usefulness of the technique. The print was visually inspected at 150X magnification. The oily regions of the print were approximately 240 μm wide and separated by approximatly 100 μm of clean, non-oil deposited areas. To adequately map the entire print with a 100 x 100 μm aperture and 100 μm step size would have resulted in the generation of over a million spectra and required over 8 months for data collection with our current system. Therefore, it was decided that only the upper left portion of the print would be imaged. A square 80 x 80 μm aperture was used to mask each pixel area concomitant with 80 μm stage movements. This resulted in the collection of 400 pixels. After the data were collected, processed, and stored, individual spectra were examined. Since all oils deposited were of interest, the entire C-H stretch region, 2800 - 3000 cm^{-1}, was again chosen as the spectral region to be integrated. Any absorbance intensity in this region was considered diagnostic for the presence of oils. The integrated intensity values for each pixel are plotted as a function of position in the axonometric projection, Fig. 2. Pixels corresponding to infrared spectra that contain absorptions in the C-H stretch range can be seen as

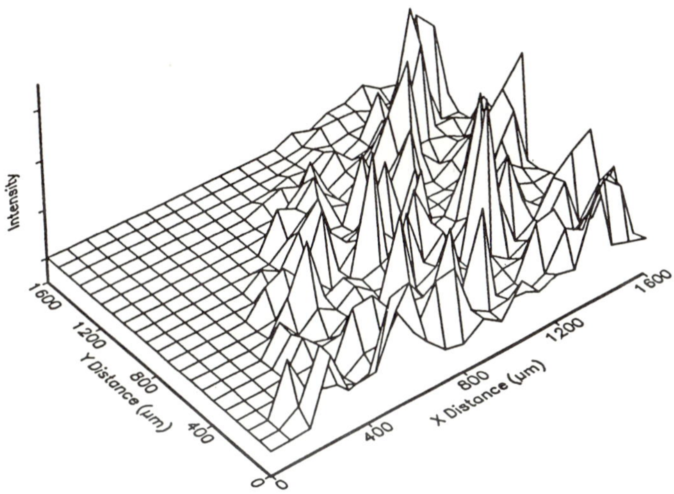

Figure 2. Axonometric projection of fingerprint on gold slide infrared micro-profiled using 80 μm aperture and 80 μm step size.

ridges in the projection that form four concentric curves. The valleys indicate regions of no IR absorption in the diagnostic range, i.e no oils deposited. These results illustrate that IR micro-profiling can be used to generate a visual image characteristic of a contaminant based on the chemistry of the contaminant. Furthermore, if sufficient computer disk space and data collection time had been allocated, the entire fingerprint could have been mapped.

Rosin-based solder flux is used in some electronics applications. However, the complete removal of flux after its use may be imperative to assure no subsequent reaction or degradation of printed circuit board or component materials. Therefore, to determine the effectiveness of IR micro-profiling in detecting the presence of solder flux, a thin film of Kester 197 Solder Flux (Litton/Kester Solder, Des Plaines, IL.) was applied to a copper coupon. It was dried and the coupon was then mounted on the microscope stage. Four hundred pixels were collected over a 400 x 400 μm area, mapped with 20 μm step sizes and 20 x 20 μm square aperture. The data were collected, processed, and stored. Examination of individual spectra revealed that absorbances characteristic of the compound were present in both the carbon-hydrogen, C-H, and carbonyl, C=0, stretch regions at 2800-3000 and 1700-1800 cm^{-1}, respectively. Therefore, either region could be used for integration. Data reduction and the generation of 3-dimensional projections of both regions yielded similar results. Figure 3 shows the axonometric projection of the integrated IR intensities as a function of pixel position for the C-H stretch region. Pixels with infrared absorption in this region are present in the upper left corner of the projection. The large intensity values on the edge suggest that the contamination continues past the area imaged. This was confirmed by a visual inspection of the sample under 150X magnification. Also, a spot that was not optically visible and measuring approximately 120 by 80 μm can be seen centered at (175, 300 μm). This confirms that IR micro-profiling is useful for the location and identification of small contaminants that are not visibly apparent.

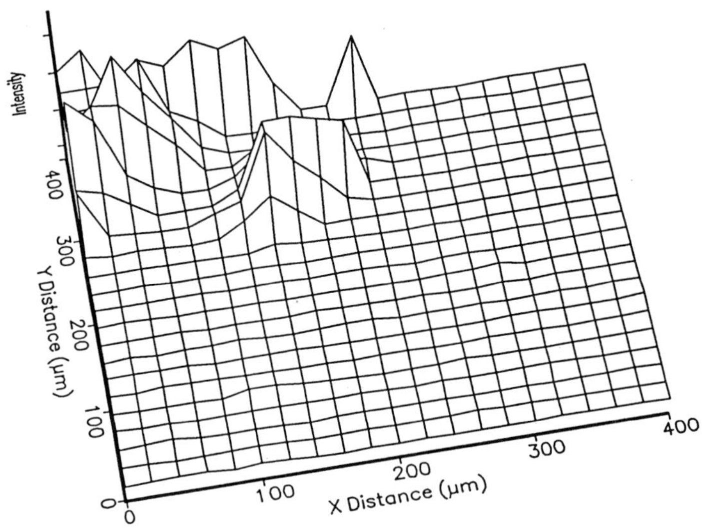

Figure 3. Axonometric projection of rosin-based solder flux on copper coupon infrared micro-profiled using a 20 μm aperture, 20 μm step size.

Identification of an unknown contaminant is often performed by matching the infrared spectrum of the unknown with that of known reference material. The ability to match spectra is a direct function of their quality. One measure of quality is the spectrum's S/N. This can be improved during data collection by either increasing the number of scans per spectrum at each pixel or by increasing pixel size. However, increasing the number of scans at each pixel could make the technique time prohibitive. Furthermore, increasing the pixel size would decrease spatial resolution. Therefore, we examined spectral averaging of data from multiple pixels as a post-data collection technique to obtain a higher quality resultant spectrum. This technique is commonly used in gas chromatography/FT-IR applications where a single compound elutes from a gas chromatographic (GC) column over a period of time. A number of infrared spectra may be collected in the time it takes a single compound to leave the column. This results in low absorption intensity spectra due to the relatively low sample concentration associated with each sp[Atrum. Spectra corresponding to a single compound eluting from the GC column are coadded to produce a single spectrum of higher S/N. It has been established that coadding spectra with integrated intensity values equal to or greater than 40% of the largest value results in an increased S/N[7].

This same approach was followed in coadding spectra from various pixel elements. Spectrum A in Fig. 4 is the spectrum of the highest intensity integrated absorbance pixel from the solder flux residue in Fig. 3. The major features of the spectrum indicate that the solder flux is a carboxylate salt with the absorption frequencies related to the O-H stretching vibration at 3380 cm^{-1}, C-H stretch at 2923 and 2860 cm^{-1}, and C=O stretch between 1715 and 1680 cm^{-1}. The sinusoidal type wave present between 2800 and 1850 cm^{-1} is characteristic of interference fringes that appear in spectra of thin samples with parallel faces. These fringes can be eliminated if thicker samples and/or samples with nonparallel sides are used. Spectrum B, Fig. 4, results from averaging spectra that correspond to pixels with integrated absorbance values greater than 40% of the maximum intensity after baseline correction. Examination of Spectrum B reveals that the fringing pattern is no longer apparent and the spectrum appears less noisy than Spectrum A. Reduction of the fringing pattern is due to variance in sample thickness across the mapped area; in some spectra no fringing pattern is present, in others the pattern has a different period. Therefore, averaging these spectra reduces overall fringe amplitude. Decrease in noise level is confirmed by a decrease of 64% in the root-mean-square (rms) noise between 2200 and 2000 cm^{-1}.

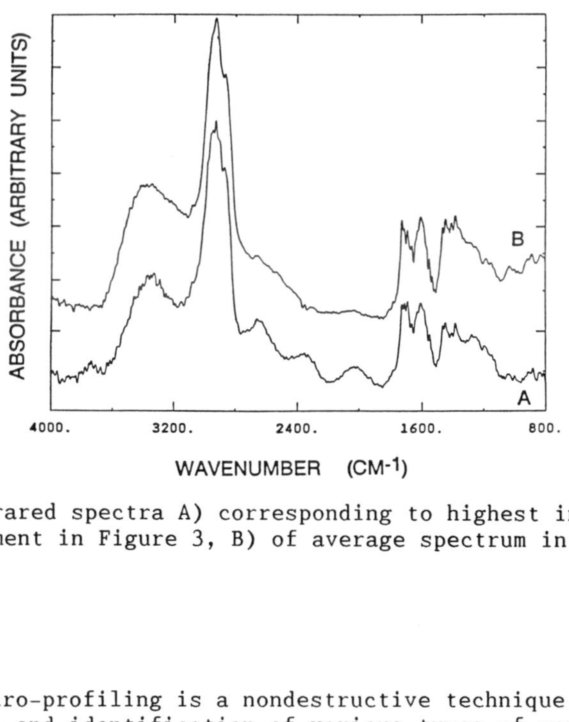

Figure 4. Infrared spectra A) corresponding to highest intensity pixel element in Figure 3, B) of average spectrum in Figure 3.

CONCLUSIONS

Infrared micro-profiling is a nondestructive technique that is useful for the location and identification of various types of contaminants common to the microelectronics industry. Furthermore, it can be used to detect contaminants that are not optically apparent. The large amount of spectral information that is generated can be successfully managed by compressing spectra for each pixel element to a single intensity value and plotting these values as a function of pixel position. Therefore, the availablity of versatile and flexible software facilitates the usefulness of micro-profiling. Also, spectral quality can be improved, reducing rms noise 64%, by coadding spectra from different pixel elements with integrated absorbance values exceeding 40% of the maximum intensity.

ACKNOWLEDGEMENTS

The authors would like to gratefully acknowledge the technical contributions of G. W. Koepp for assistance in software development and implementation.

REFERENCES

1. K.J. Ward and David R. Tallant, in "Particles on Surfaces 2: Detection, Adhesion and Removal," K.L. Mittal, editor, p.115, Plenum Press, New York, 1989.

2. P.R. Griffths and J.A. deHaseth, "Fourier Transform Infrared Spectrometry," Wiley & Sons, 1986.

3. K.J. Ward, SPIE Proc. 1145, 212 (1989).

4. M.A. Harthcock and S.C. Atkin, in "Infrared Microspectroscopy: Theory and Applications," R.G. Messerschmidt and M.A. Harthcock, Eds.,p.21, Marcel Dekker, New York, 1988.

5. H.J. Milledge and M.J. Mendelssohn, in " Analytical Applications of Spectroscopy," C.S. Creaser and A.M.C. Davies, editors, p.217, The Royal Society of Chemistry, Norwich, England, 1988.

6. R.G. Messerschmidt,in "Infrared Microspectroscopy: Theory and Applications," R.G. Messerschmidt and M.A. Harthcock, Eds., p.1, Marcel Dekker, New York, 1988.

7. C.C. Johnson and L.T. Taylor, Anal. Chem., 56, 2643 (1984).

ANALYSIS OF PARTICLES ON SURFACES BY TOTAL REFLECTION X-RAY FLUORESCENCE SPECTROMETRY

H. Schwenke, J. Knoth and U. Weisbrod

GKSS-Forschungszentrum
Institut für Physik
Postfach 1160, 2054 Geesthacht, Germany

The capability of Total Reflection X-ray Fluorescence spectrometry (TXRF) for the analysis of the elemental composition of contaminants in and on surfaces is described. Instrumental aspects are discussed in this context. Detection limits better than 10^{10} at/cm^2 or 0.2 pg on a surface area of 2 mm in diameter have been obtained. The unique sensitivity of TXRF allows the determination of metals in airborne particles at a concentration level as low as 10 ng/m^3. The question as to whether contaminants are located on, or embedded in, a surface has been answered by recording the fluorescence intensities as a function of the incident angle of the primary beam.

INTRODUCTION

The exploitation of the total reflection phenomenon in X-ray Fluorescence (XRF) spectrometry has proven to be beneficial for enhancing the detection performance of this traditional analytical technique[1]. Total Reflection X-ray Fluorescence (TXRF) spectrometry has considerably expanded the application range of XRF in trace element analysis down to ppb- or pg-levels[2,3]. The potential of TXRF concerning near surface layers in the nm range, i.e., the examination of trace element contaminants on wafer surfaces, has only been perceived recently[4,5]. Moreover, it was claimed that this technique provides a novel means for measuring concentration profiles along the normal to a flat interface[6]. In this paper we first outline the basic physical and instrumental characteristics of TXRF. Then we turn to the analysis of airborne particles, collected on quartz flats which were used as targets in a size separating impactor. Finally we demonstrate the capability of the TXRF method to differentiate between particulate and film type contaminants.

TOTAL REFLECTION X-RAY FLUORESCENCE SPECTROMETRY (TXRF)

TXRF makes use of total reflection of the primary X-ray beam at grazing incidence. The reflecting surfaces are used either as sample carriers (Fig. 1), or may themselves be the object of investigation. The fluorescence radiation is detected by an energy dispersive detector.

Particles on Surfaces 3, Edited by K.L. Mittal
Plenum Press, New York, 1991

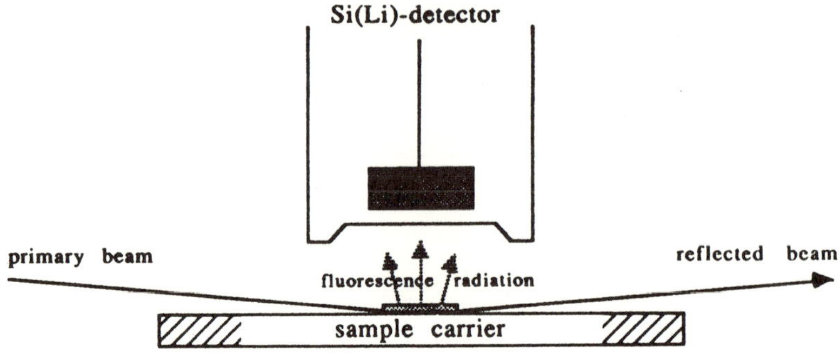

Fig. 1. Arrangement for Total Reflection X-ray Fluorescence Analysis.

Total reflection occurs at grazing angles below a critical value. Fig. 2 shows the reflectivity of X-rays as a function of the angle of incidence. At low angles the reflectivity approximates unity and it drops down beyond the critical angle. The high reflectivity in the total reflection mode results in an extremely low energy transfer (Fig. 3) from the incident beam into the irradiated substrate, because most of the energy is reflected and does not penetrate through the interface. Using an X-ray reflector as the sample support this low energy transfer leads to a large reduction in the background, normally contributed by the support, whereas the sample is exposed to the full, in fact even the doubled, intensity of the exciting beam.

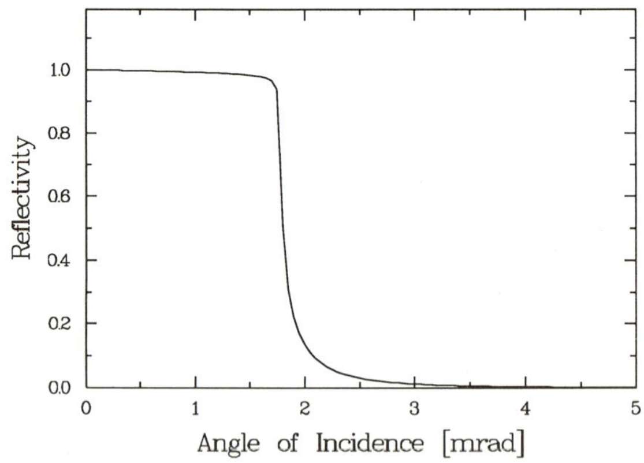

Fig. 2. Reflectivity for 17.5 keV X-rays on quartz.

Fig. 4 shows the design of a commercially available TXRF-instrument[7]. The beam of an X-ray tube is directed at grazing incidence over an X-ray mirror onto the polished surfaces of a sample carrier, e.g., quartz. The fluorescence radiation of the sample over the mass range typically from ng to µg is detected by a Si(Li)-detector. Fig. 5 shows a photograph of the instrument, which comprises two tubes and two sets of excitation optics which allow coverage of the entire periodic table, as well as a sample changer and the usual equipment for unattended operation. The detection limits are displayed in Fig. 6.

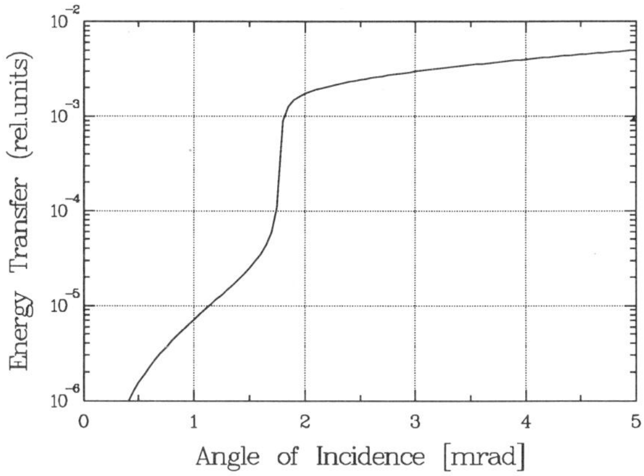

Fig. 3. Energy transfer of a 17.5 keV-beam to a quartz surface as a function of the incident angle.

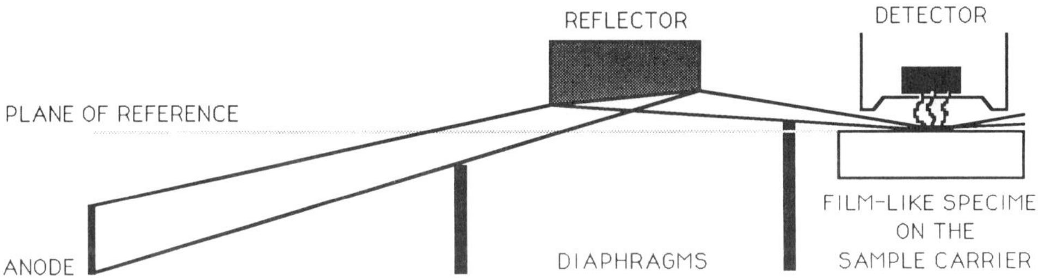

Fig. 4. Path of the X-rays in a commercially available TXRF-instrument.

Fig. 5. Photograph of a TXRF-instrument.

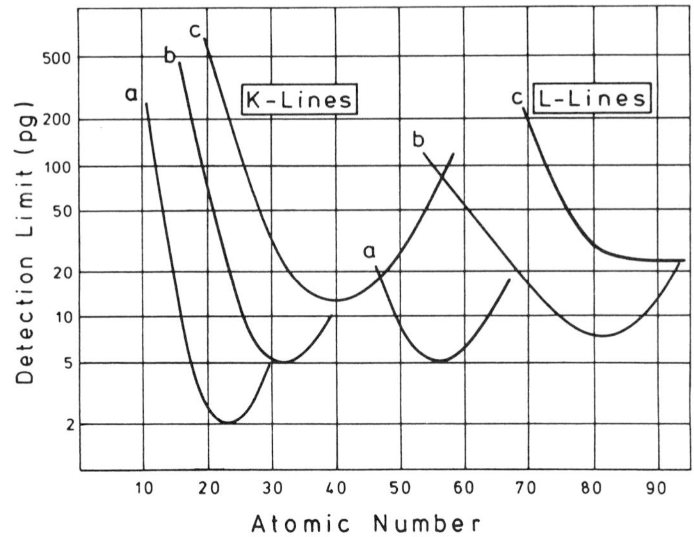

Fig. 6. Detection limits of a commercially available instrument using a Mo tube (curve b) or a W tube (curve a: W-L excitation, curve c: white spectrum excitation).

PARTICULATE TYPE CONTAMINANTS

From the broad range of application in trace element analysis e.g. in the environmental field[3] or in the purity control of reagents[8] we have selected the analysis of airborne particles as an example here. In this particular case polished quartz discs which are used as sample carriers for TXRF are also suited as targets in the various stages of a size separating impactor (Fig. 7). The particles, which are adherent to the surface, are directly ready for quantitative analysis.

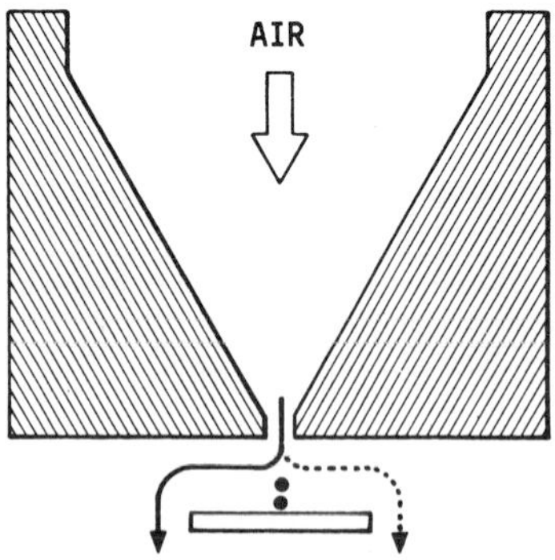

Fig. 7. Airborne particles impacting with TXRF-sample carriers.

Fig. 8 shows two different size separated fractions of particles collected on the surface of X-ray mirrors, which were used in a second step as sample carriers in TXRF. With a sampling time of 30 minutes in the impactor a detection limit e.g. for lead in the 1 - 2 µm-fraction of 1 ng/m^3 has been obtained[9,10].

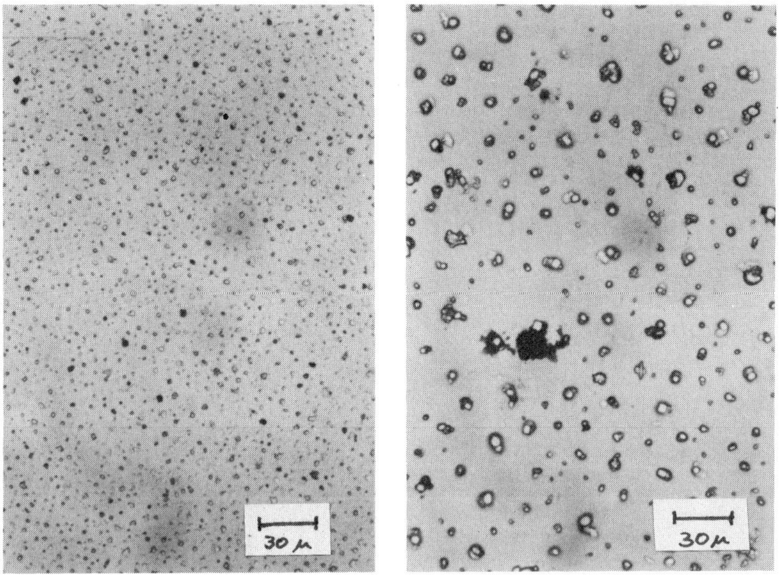

Fig. 8. Photographs of airborne particles collected in two sizes on X-ray mirrors used as targets in different stages of a size separating impactor (see Fig. 7).

FILM TYPE CONTAMINANTS

Up to this point we have only dealt with the particulate type contaminants on surfaces. In order to analyze the film type, and to differentiate one type of contamination from another, we need to take a closer look at the features of the total reflection technique. Total reflection is not only characterized by high reflectivity, but also by low penetration depth (Fig. 9) which makes TXRF well suited for the analysis of thin films in the nanometer-region.

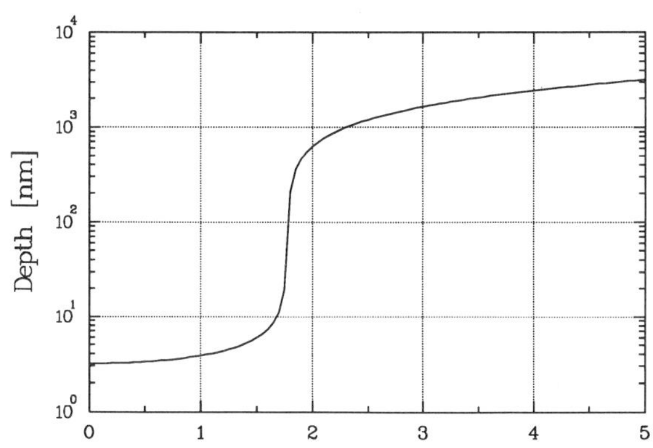

Fig. 9. Penetration depth of 17.5 keV X-rays in quartz as a function of the incident angle.

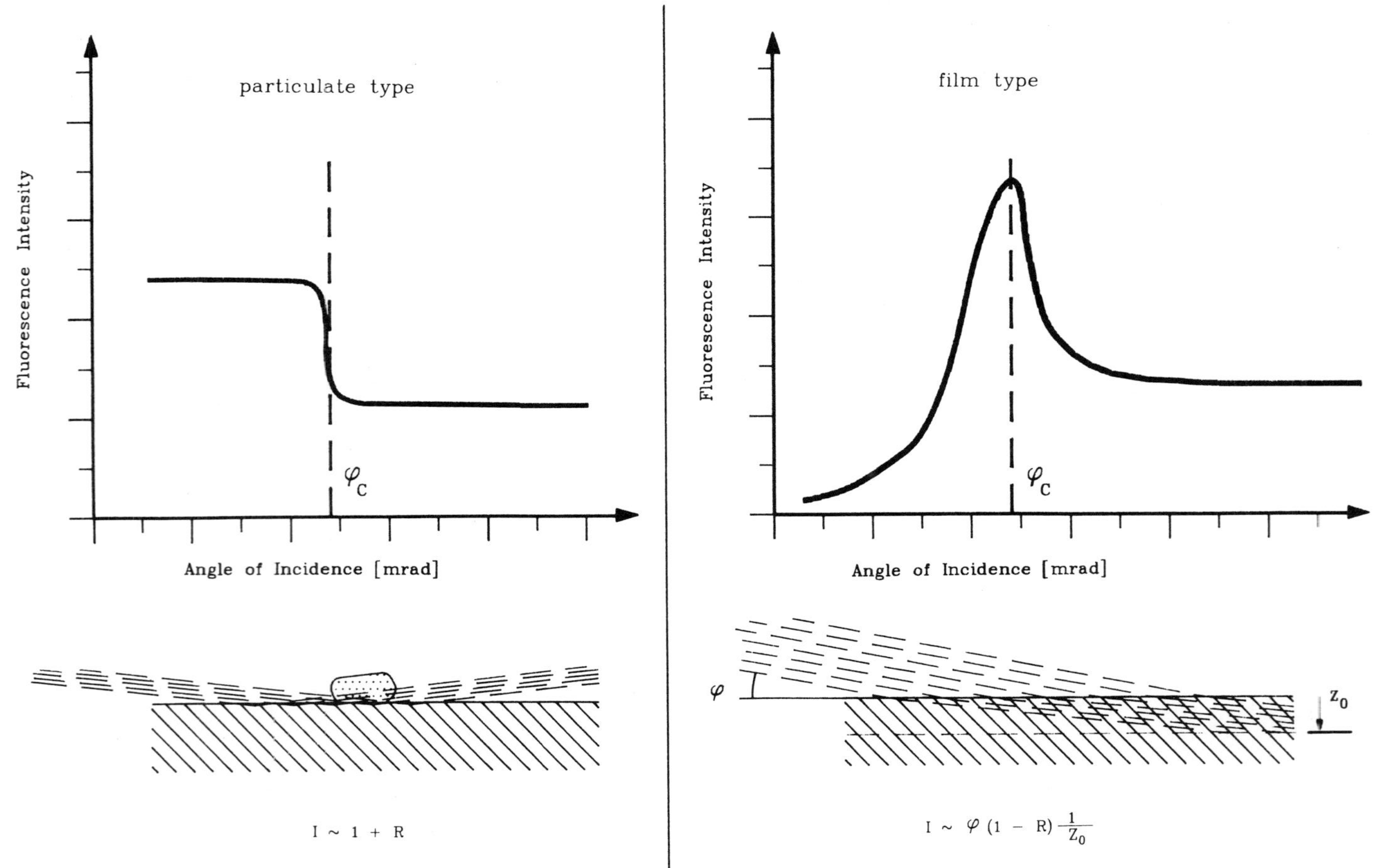

Fig. 10. The primary intensity for particulate compared to film type contamination as a function of the incident angle.

The induced fluorescence intensity of atoms which are constituent parts of a near surface layer is, however, completely different from the fluorescence radiation of atoms, which are located clearly above a surface (Fig. 10). Simple considerations may illustrate the difference. The intensity which excites the atoms located above an interface is given by $I = I_0 (1 + R)$, where I_0 denotes the primary intensity and R the reflectivity. These atoms are excited by the full primary beam as well as by the reflected beam. Surface atoms, on the other hand, are excited merely by the unreflected portion (1 - R) multiplied by two factors which take into account the field compression due to refraction and the angle dependent irradiance. In the case of film type contaminants, as opposed to particulates, Fig. 10 exhibits that below the critical angle there is a strong dependence of the fluorescence signal on the angle of incidence.

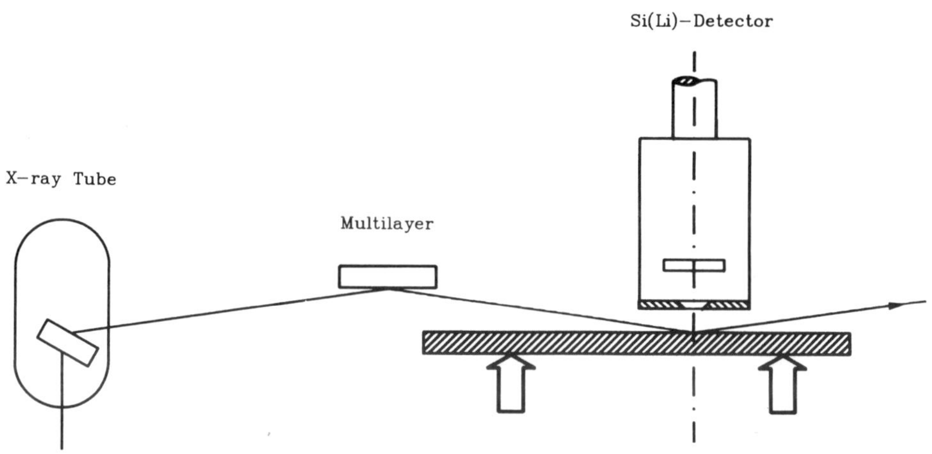

Fig. 11. Design of a TXRF-instrument for surface analysis. The examined surface is aligned by means of stepping motors which are controlled by the fluorescence intensity.

In order to cope with this complication which does not occur in the examination of the particulate type contaminant (right part of Fig. 10), we were compelled to extend the development of TXRF-instrumentation and quantitation software. On the instrumental side we improved, among other things, the angle setting to an absolute accuracy of 0.05 mrad by means of stepping motors which are controlled by a software package using the fluorescence signals from the Si(Li)-detector. In addition, we established spectral purity by a multilayer mirror in the path of the X-rays (Fig. 11). As a side effect we reduced in this way the detection limits further by an order of magnitude, i.e., from 3 pg (Fig. 6) to 0.2 pg for Nickel (Fig. 12).

Further considerations about the difference in fluorescence yields of particulate and film type contaminants lead to the perception of a transition zone between these types, which may not be covered by the simplified presentation in Fig. 10. Actually, for instance, it would appear impossible to decide whether crystals or other material agglomerations in nanometer dimensions belong to a surface or should these be treated as separately located on top of it.

Fortunately, however, the situation can be fully resolved by means of calculations by applying a common formalism to both cases on the basis of classical electrodynamics. The propagation of electromagnetic waves at an interface between two media characterized by different refractive indices is described by the Fresnel relations for transmission and reflection[11].

For grazing incidence, the transmission coefficient t at the interface of the adjacent media ν and $\nu+1$ is

$$t_{\nu,\nu+1} = \frac{2\,\Theta_\nu}{\Theta_\nu + \Theta_{\nu+1}}$$

and the reflection coefficient r

$$r_{\nu,\nu+1} = \frac{\Theta_\nu - \Theta_{\nu+1}}{\Theta_\nu + \Theta_{\nu+1}}$$

with the angles Θ_ν of the incident and $\Theta_{\nu+1}$ of the refracted beam.

The complex refractive index is given by

$$n = 1 - \delta + i\beta.$$

Using Snell's law the refracted angle Θ_ν is of the form

$$\Theta_\nu = \sqrt{\Theta^2_o - 2\delta_\nu + 2i\beta_\nu}$$

where Θ_o denotes the glancing incident angle.

Owing to their complex formulation the Fresnel relations provide, in addition, information about the phase-relations of the wavefronts which proceed from the respective interfaces. Recursive computing techniques for the resulting intensities are basically known from the publications of Parratt, and other authors since 1954[12,13]. They provide a complete description of the primary intensity close to any reflecting interface including effects of interferences. In this paper we do not intend to consider the details of these calculations, as this will be done elsewhere. We intend to discuss only two results.

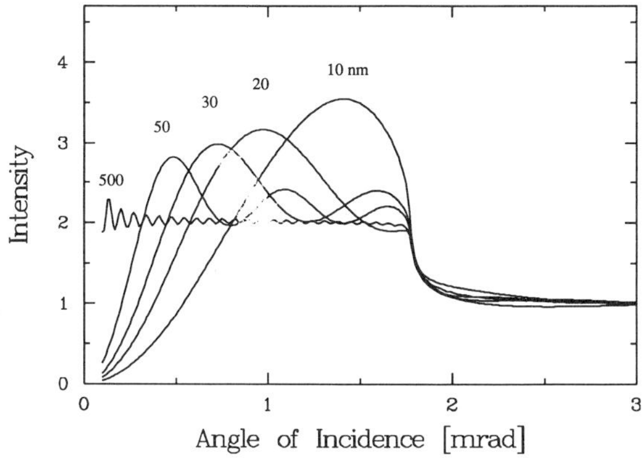

Fig. 12. Detection limit for nickel in 1000 s counting time obtained with a prototype instrument (according to Fig. 11) which is characterised by a multilayer monochromator and computer controlled angle setting.

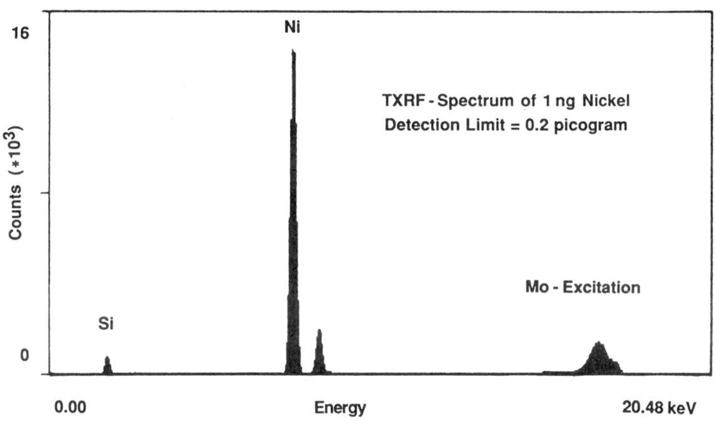

Fig. 13. The intensity of the primary field (17.5 keV) in air above a quartz surface as a function of the incident angle integrated over various distances (from 10 to 500 nm) perpendicular to the air/quartz interface.

Fig. 13 illustrates the field intensity integrated over various distances perpendicular to a quartz surface as a function of the incident angle of the primary beam for the system vacuum/quartz. It turns out that the previously shown fluorescence intensities (Fig. 10), which were derived from basic physical considerations, in fact represent limiting cases of the more general description on the basis of the Fresnel equations. The Fresnel formalism describes also standing wave phenomena, which are exemplified in Fig. 13 for the size dependent fluorescence of particles on a surface.

Comparing Fig. 10 with Fig. 13 it can be seen that TXRF definitely assigns particles bigger than about 500 nm to the particulate type contaminant; whereas conglomerates on a surface smaller than approximately 10 nm cannot be distinguished from the film type contamination using this technique.

Fig. 14 shows a TXRF-spectrum of iron contamination on a silicon wafer surface. The fluorescence signal of iron was measured at grazing incidence of the primary beam as a function of the incident angle. The experimental data in Fig. 15 are displayed as a sum of the two components which represent the different types of contamination. In this way, we were able not only to quantify the total amount of iron atoms per area, but also to differentiate between the two contamination types within the already mentioned limits.

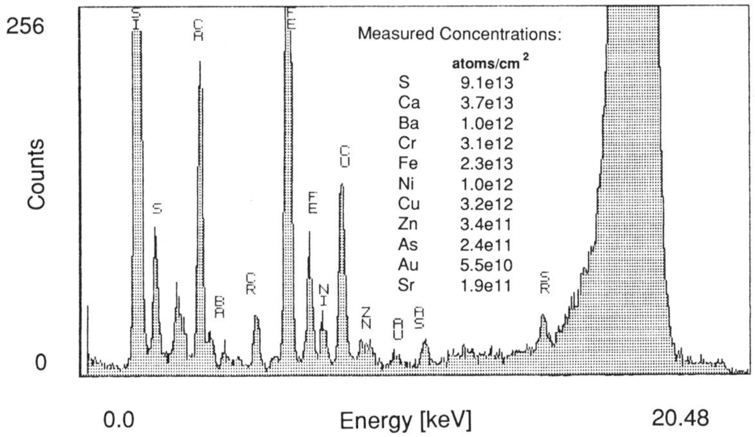

Fig. 14. Analysis results and X-ray spectrum of a contaminated silicon wafer surface (excitation energy 17.5 keV, counting time 1000 s).

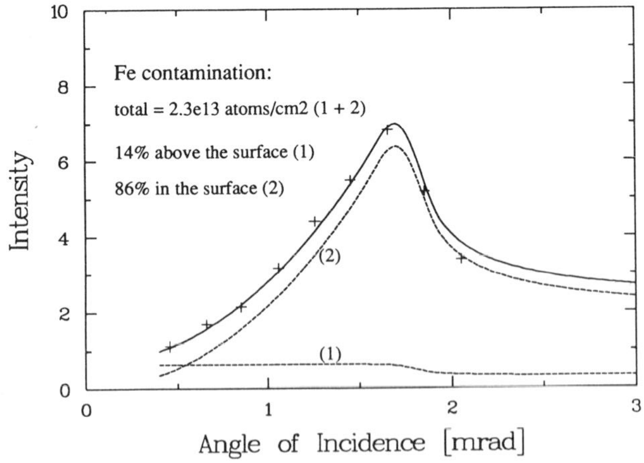

Fig. 15. Quantitative differentiation between film and particulate type iron contaminants on a silicon wafer. The measured values (+) are fitted by the solid line curve, which represents the sum of the calculated particulate component (dashed curve 1 : $0.3 \cdot 10^{13}$ atoms/cm^2) and the contribution of the film type contaminant (dashed curve 2 : $2.0 \cdot 10^{13}$ atoms/cm^2).

CONCLUSION

Totally reflecting sample carriers can be used in XRF in order to obtain detection limits in the pg- or ppb level. TXRF is also a surface sensitive method with a probing depth in the nanometer range. A special feature is the capability of finding out in a nondestructive manner whether contaminants on surfaces are of the particulate or film type. We have presented the technique primarily under the aspect: "Particles on Surfaces". However, it can be applied equally as well to the examination of silicon wafers or the analysis of near surface layers in alloys or of thin films on any substrate.

REFERENCES

1. Y. Yoneda and T. Horiuchi, Rev. Sci. Instrum., 42, 1069 (1971).
2. J. Knoth and H. Schwenke, Fresenius Z. Anal. Chem., 301, 7 (1980).
3. A. Prange, Spectrochimica Acta, 44B, No. 5, 437 (1989).
4. P. Eichinger, H.J. Rath and H. Schwenke, in "Semiconductor Fabrication: Technology and Meterology", ASTM STP 990, D.C. Gupta, Editor, American Society for Testing and Materials, 1988.
5. H. Schwenke, W. Berneike, J. Knoth and U. Weisbrod, Adv. X-ray Analysis, 32, 105 (1989).
6. W.B. Yun and J.M. Bloch, J. Appl. Phys. 68(4), 1421 (1990).
7. Technical Note: EXTRA II, Rich. Seifert & Co., Bogenstr. 41, 2070 Ahrensburg, Germany.
8. A. Prange, K. Kramer and U. Reus, Spectrochimica Acta, in press.
9. B. Schneider, Spectrochimica Acta, 44B No 5, 519 (1989).
10. A. Prange, Adv. X-ray Analysis, 32, 211 (1989).
11. M. Born and E. Wolf, "Principles of Optics", Pergamon Press, New York (1965).
12. L.G. Parratt, Phys. Rev., 95, 359 (1954).
13. B. Pardo, T. Megademini and J.M. André, Revue Phys. Appl., 23, 1579 (1988).

CHARACTERIZATION OF SURFACE METAL PARTICULATE CONTAMINATION USING THE POLYSILICON CHEMICAL VAPOR DEPOSITION PROCESS

William A. Syverson
IBM General Technology Division
Essex Junction, Vermont 05452

Paul J. Schubring
Department of Mechanical Engineering
Rensselear Polytechnic Institute
Troy, New York 12180

For many semiconductor wafer processes and, in particular, polysilicon chemical vapor deposition (CVD), metallic contamination is a known problem. Metallics can act as localized catalysts during the CVD process, which results in an accelerated, non-uniform polysilicon deposition that is "spike-like" in shape and appearance. These "spike" defects are known to cause interlevel electrical shorts in integrated circuits. With a 400-nm deposition, the spikes range from 0.25 to 2.0 μm in height and are often uniquely characteristic of the contamination from the metallic element tested.

This paper describes the experimental findings that characterize the formation of polysilicon spikes for many metallic elements. Polysilicon CVD, the analytical technique used in the study, is described, and contamination by 22 elements is characterized. The experimental findings are then used to compare some commonly used process chemicals such as sulfuric acid and some photoresist developers.

INTRODUCTION

At IBM's semiconductor facility near Burlington, Vermont, there have been many occasions when defects found in polysilicon chemical vapor deposition (CVD) films result in substantial device-yield degradation. One of the most unique and damaging defects are interlevel electrical shorts caused by locally enhanced polysilicon depositions, or "spikes." The spikes were found to extend well above the base polysilicon film and into the IC films above the base. Depending upon the spike's location within a device, a spike as small as 0.25 μm in height can cause an electrical failure. Figures 1 through 4 are dark-field optical and scanning electron micrographs (SEM) of a family of contaminants found in the sulfuric acid used to clean wafers. Figure 5 is an SEM of a residual polysilicon spike that caused an electrical failure of the device.

Particles on Surfaces 3, Edited by K.L. Mittal
Plenum Press, New York, 1991

Figure 1. Dark-field optical micrograph of a circular defect, X1000.

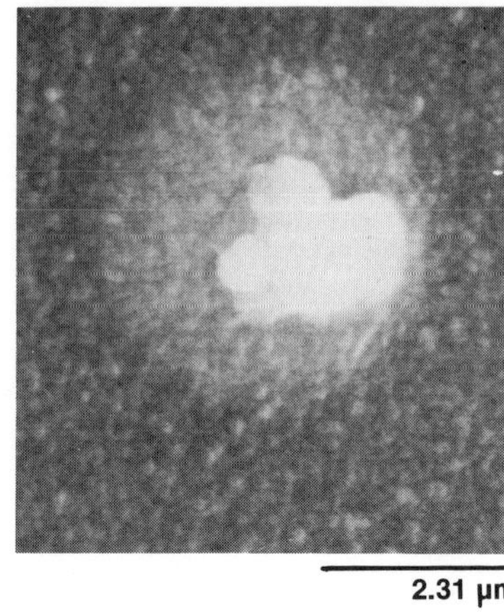

Figure 2. SEM micrograph of a bulk sulfural circular defect.

Figure 3. Dark-field optical micrograph of a polysilicon spike, X1000.

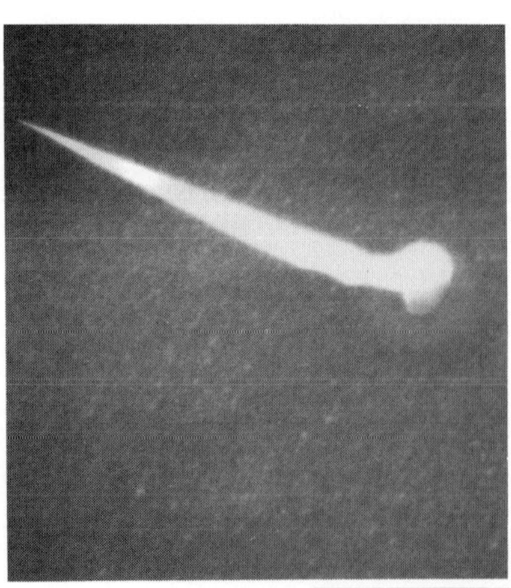

Figure 4. SEM micrograph of a polysilicon spike.

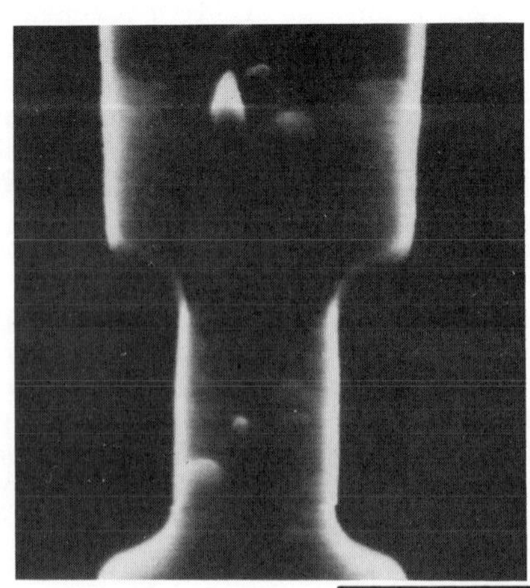

Figure 5. SEM micrograph of a polysilicon spike inside a device.

During several investigations as to the cause of similar polysilicon spikes, several distinct spike formations were observed. Therefore, an effort was undertaken to determine which metallic contaminants would result in related polysilicon defects. This report summarizes the findings to date.

EXPERIMENTAL TECHNIQUE

The method used to characterize individual metals is summarized below. Since a silicon dioxide surface is common to many semiconductor wafer processes, oxide wafers were selected as the experimental vehicle.

SAMPLING TECHNIQUE

Using a 45-nm silicon dioxide covered wafer:

1. Deposit a single laboratory purity grade metal particulate source on the wafer surface.

   Method of deposition:

   - Solids: scratch on the wafer surface.
   - Powders: sprinkle onto the wafer surface.
   - Hard solids: low-pressure evaporation.
   - Liquids: dip/air dry.

2. Deposit 400 nm of LPCVD polysilicon.

   - Temperature: 620 °C
   - Process Time: 110 minutes
   - Pressure: 0.25 torr
   - Gas: SiH_4

3. Inspect and photograph.

   - Dark-Field Microscope
   - Scanning Electron Microscope

To prevent contamination of subsequent product runs, all quartzware and process hardware were replaced upon completion of the deposition test runs.

RESULTS AND DISCUSSION

Nearly all metallic elements tested resulted in some type of polysilicon spike after deposition; many of the spikes also appear to be uniquely characteristic of the specific metal element being tested. One of the most significant findings is that each of the three commonly used alkali-metal photoresist developers resulted in a variety of shapes and sizes after polysilicon deposition. Of the samples tested, only lead, carbon, and the organic photoresist developer tetra methyl ammonium hydroxide (TMAH) did not produce polysilicon spikes. Table I summarizes the metallic elements evaluated and the contaminant sources (i.e., solid, low-pressure evaporation, etc.) and gives a brief description of some distinguishing spike characteristics that result from each element. SEM micrographs of the spikes are included in the appendices.

Table I. Description of polysilicon spikes resulting from contamination.

Contaminant Source	Method of Dep.	Height of Spikes (μm)	Appearance and Shape of Spikes
Aluminum	Evap	1.0	Conical in shape, convoluted.
Barium	Power	1.0	Conical in shape.
Calcium	Powder	0.5-1.0	Straight, cylindrical in shape.
Chromium	Evap	0.7-1.0	Conical and horn-like in shape.
Cobalt	Powder	0.2-0.5	Elliptical in shape.
Copper	Solid	10.0	Long, thin, highly refractive.
Gold	Solid	1.0-3.0	Cylindrical, bent, "halo" at base.
Iron	Solid	1.0	Conical in shape.
Magnesium	Solid	0.3-0.4	High density, twisted in shape.
Manganese	Powder	1.0	High density, conical in shape.
Molybdenum	Powder	2.0-5.0	Straight, cylindrical in shape.
Nickel	Powder	2.0-3.0	Conical, uniform in shape.
Platinum	Solid	0.5-1.0	Conical in shape, convoluted.
Silver	Solid	0.5	Conical, uniform in shape.
Tantalum	Powder	5.0	Cylindrical, rounded tips.
Tin	Solid	3.0	Thin, mono-crystalline, rounded ends.
Titanium	Solid	0.5-1.0	Conical in shape.
Tungsten	Powder	0.5-1.0	Conical in shape.
Zinc	Solid	0.5-1.0	Conical in shape.
KCl	Powder	1.0	Conical in shape.
KOH	Solid	0.3-0.5	A variety; conical in shape, convoluted.
NaOH	Liquid	1.0-3.0	A variety but cylindrical in shape.
$NaSiO_3$	Liquid	0.5	A variety but conical in shape.
NaCl	Powder	0.5-1.0	Conical in shape.
Carbon	Solid	---	No spikes.
Lead	Solid	---	No spikes.
TMAH	Liquid	---	No spikes.

LPCVD conditions: 620 °C, 0.25 torr

- For aluminum, the solid source did not yield any spikes, possibly due to the inherent presence of an aluminum-oxide film covering the source. Therefore, aluminum was sputtered onto the wafer surface.
- Two KOH samples were processed in the same run; both wafers contained spikes. The liquid source is the actual solution used in the manufacturing process.

SPECIFIC EXAMPLES

Because of their small size, many of the observed spikes are undetectable when using standard bright-field or dark-field optical microscopy; therefore, SEMs were required to fully characterize the wafer contamination. Unless a specifically sensitive process step, much like polysilicon deposition, follows the occurrence of the contamination, the resultant defects could easily remain undetected.

The SEM micrographs in the appendices show examples of the characteristic polysilicon spikes unique to each of the contaminants evaluated. The spikes are identifiable by length and configuration. Many of the elements are common impurities in the wet chemicals used for semiconductor processing and, as such, are controlled to low (10 ppb) metallic levels. However, titanium, zinc ,aluminum, iron and nickel are also common materials found throughout the environment and tooling of all semiconductor fabrication areas.

CONCLUSION

We have shown that nearly all metallic contamination species tested with the polysilicon CVD technique result in a unique and observable defect or "spikes" and that the shape and size of many of these defects can be characteristic of the specific contamination itself. Finally, we have demonstrated that many common photoresist developers may contain a host of unwanted metallic contaminants that deposit on wafer surfaces. It is our hope that this paper may be used as a quick reference guide for identifying potential elemental contaminants of various processes.

ACKNOWLEDGEMENTS

We thank M. Bouffard, W. King, E. Poplawski, I. Stone and many others at IBM Burlington for their assistance in this work.

1. Aluminum

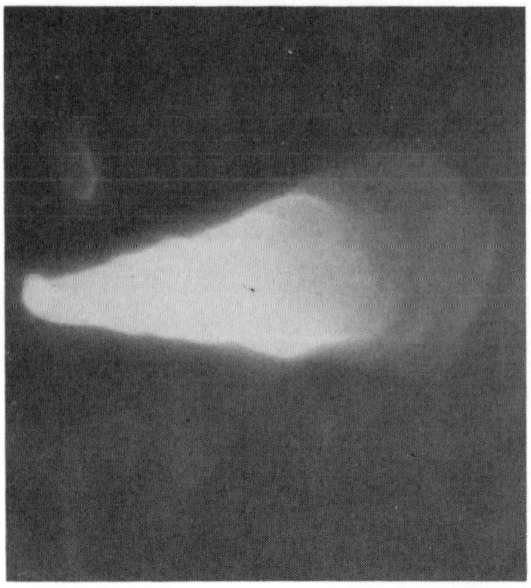

0.50 μm

2. Barium

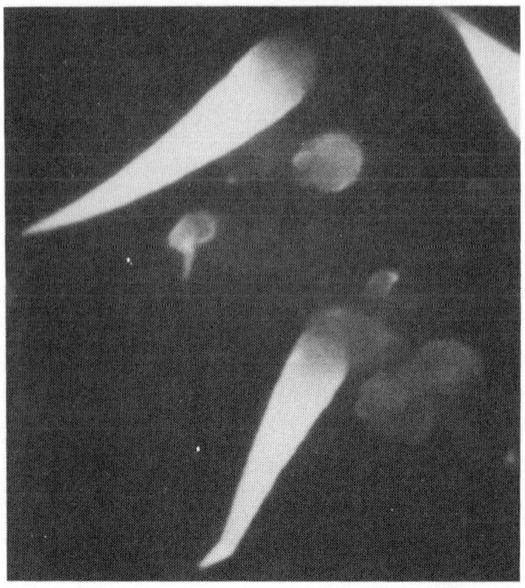

0.75 μm

3. Calcium

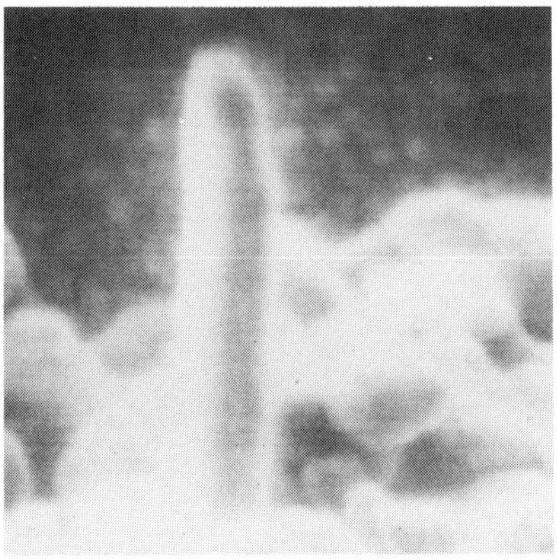

0.50 μm

4. Chromium

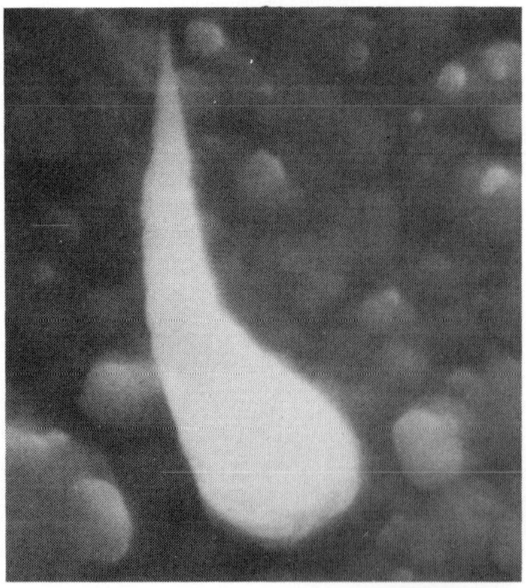

0.86 μm

5. Cobalt

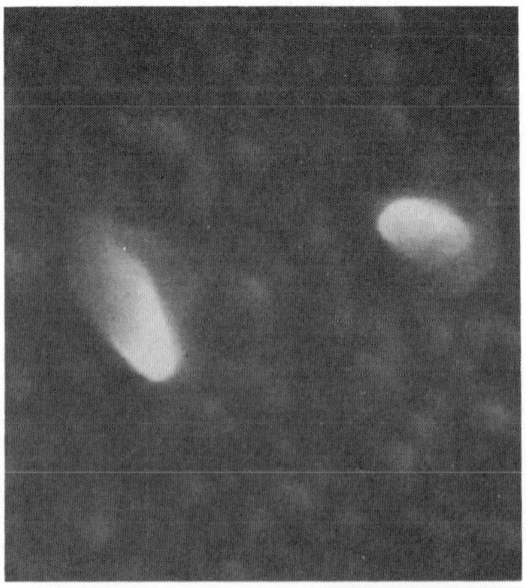

0.50 μm

Appendix A: SEM micrographs of polysilicon spikes formed by different metal elements deposited on a SiO_2 wafer surface.

6. Copper

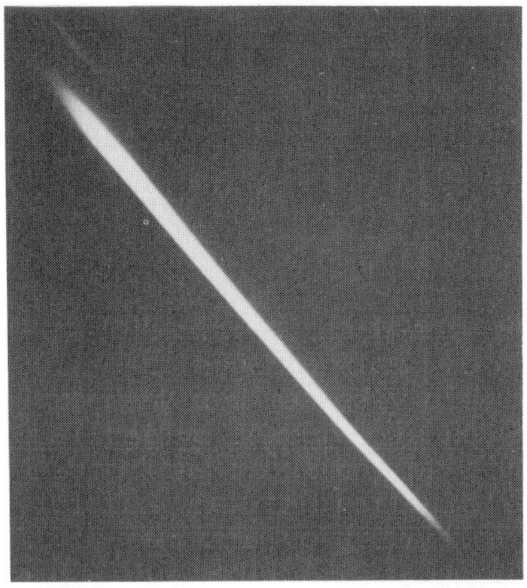

6.0 µm

7. Gold

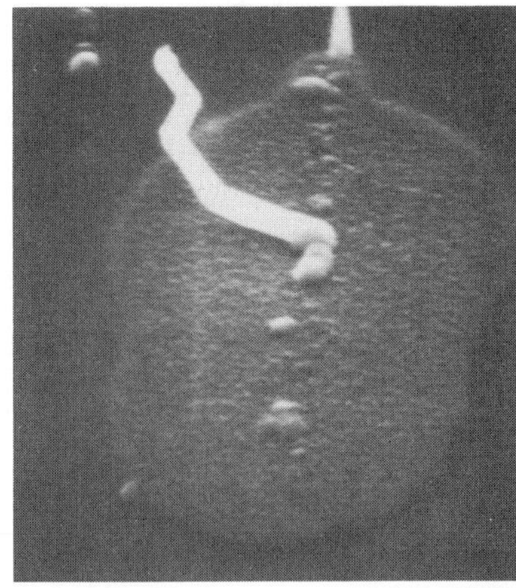

4.3 µm

8. Iron

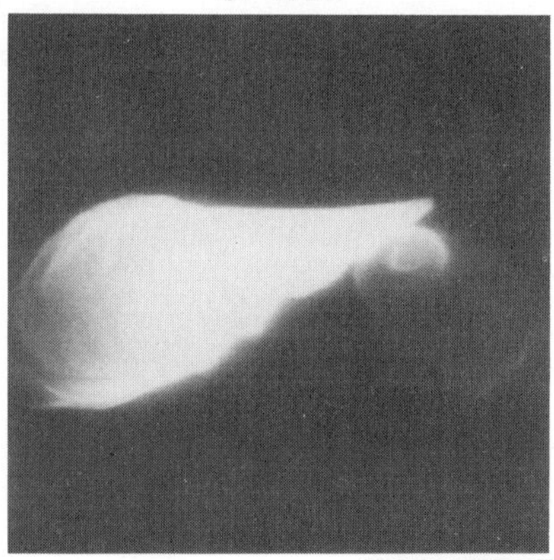

0.50 µm

9. Magnesium

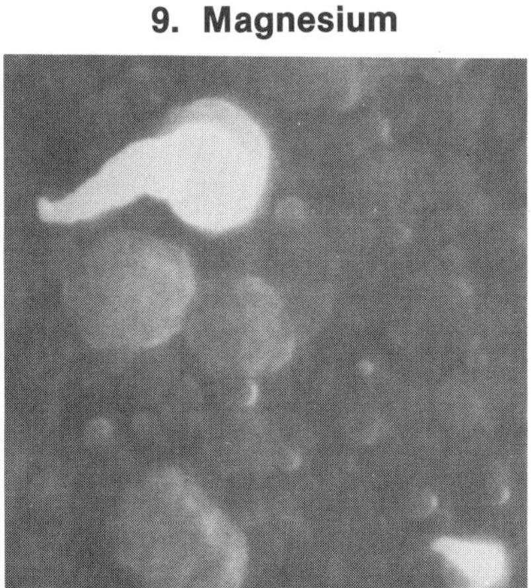

0.86 µm

10. Manganese

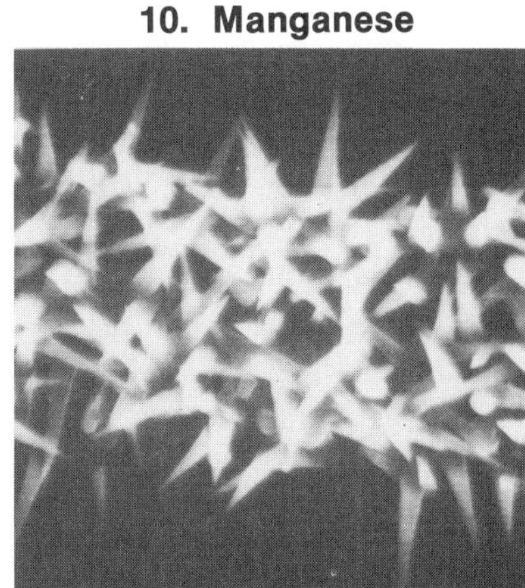

1.76 µm

Appendix A (cont.): SEM micrographs of polysilicon spikes formed by different metal elements deposited on a SiO_2 wafer surface.

11. Molybdenum

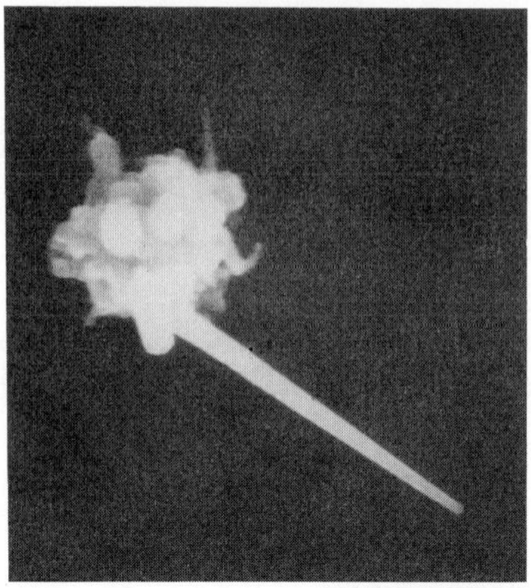

3.0 μm

12. Nickel

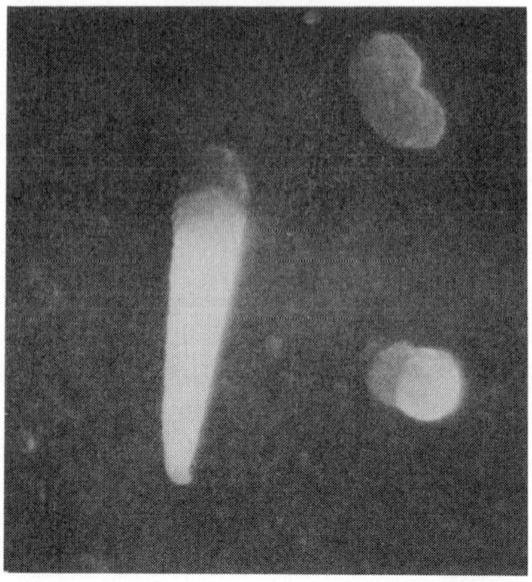

2.00 μm

13. Platinum

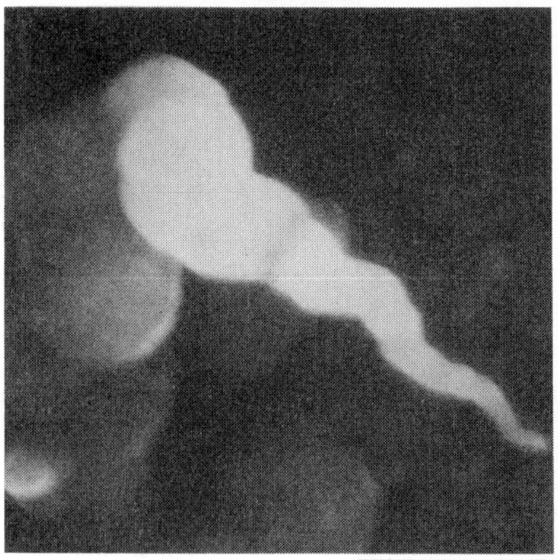

0.86 μm

14. Silver

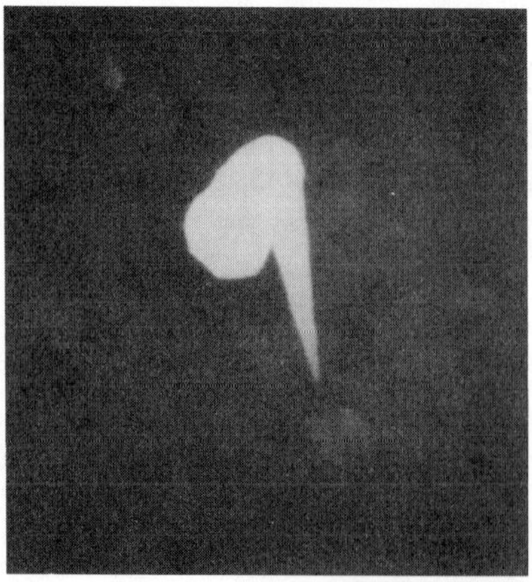

1.50 μm

15. Tantalum

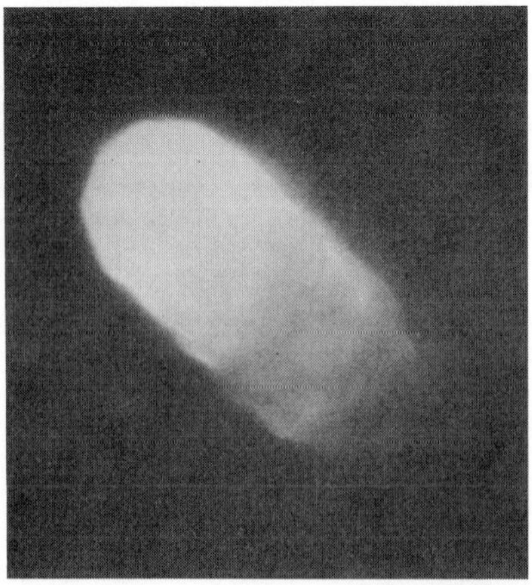

0.50 μm

Appendix A (cont.): SEM micrographs of polysilicon spikes formed by different metal elements deposited on a SiO_2 wafer surface.

16. Tin

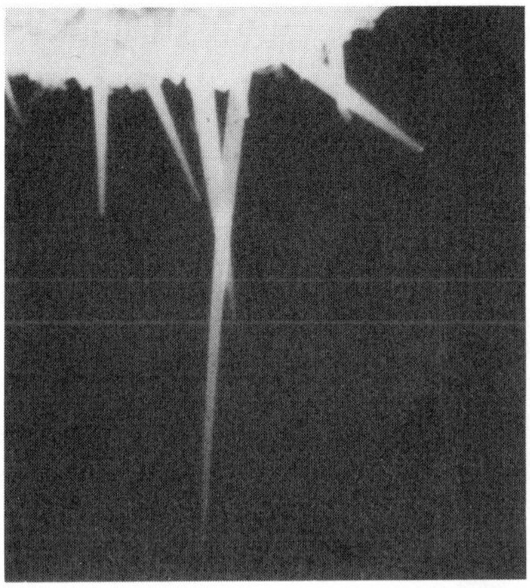

3.8 µm

17. Titanium

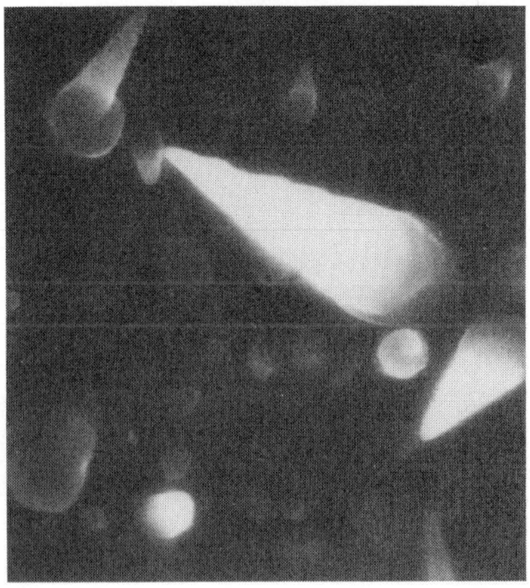

0.50 µm

18. Tungsten

0.86 µm

19. Zinc

1.00 µm

Appendix A (cont.): SEM micrographs of polysilicon spikes formed by different metal elements deposited on a SiO_2 wafer surface.

20. Sodium Hydroxide

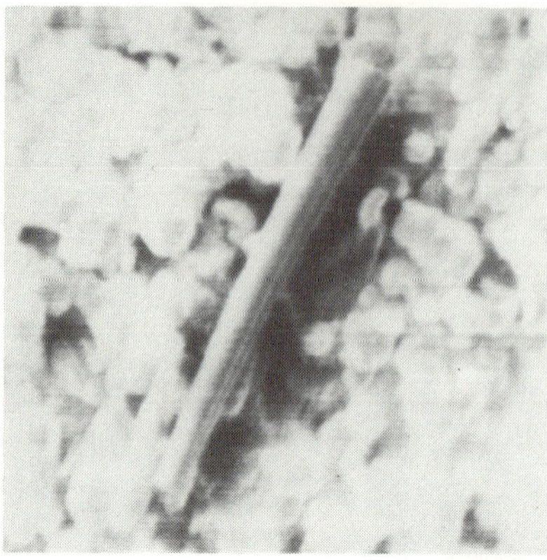

2.00 μm

21. Sodium Metasilicate

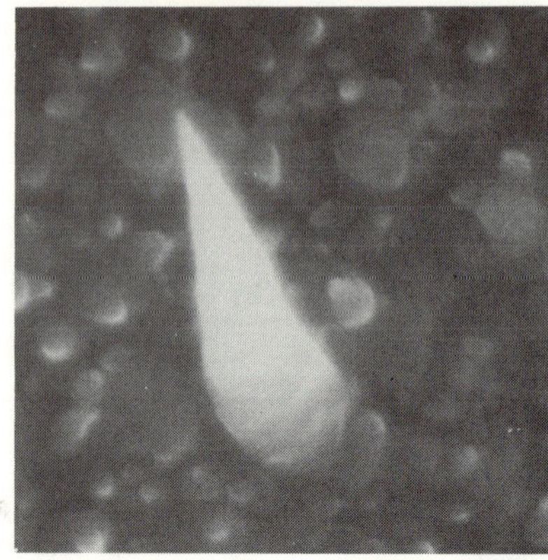

0.75 μm

22. Potassium Hydroxide

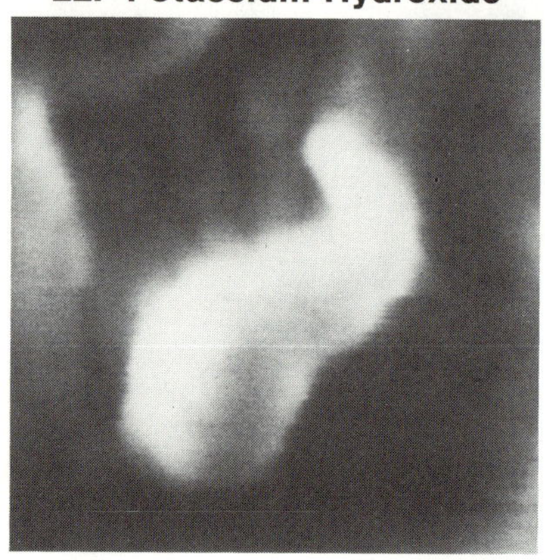

0.27 μm

Appendix B: SEM micrographs of polysilicon spikes formed when a SiO_2 wafer surface was contaminated by common alkali-metal photoresist developers.

23. Potassium Chloride

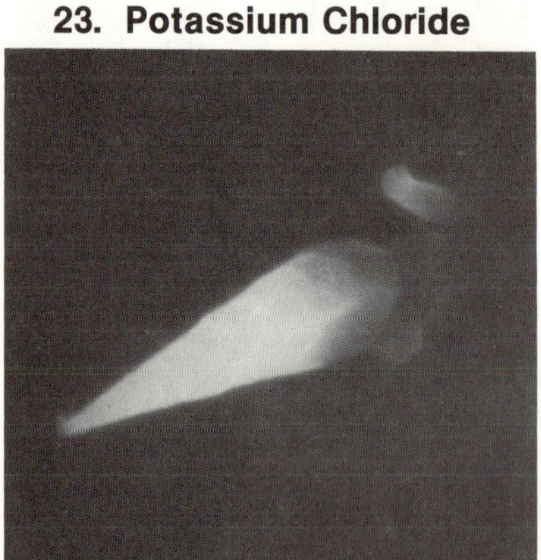

0.75 μm

24. Sodium Chloride

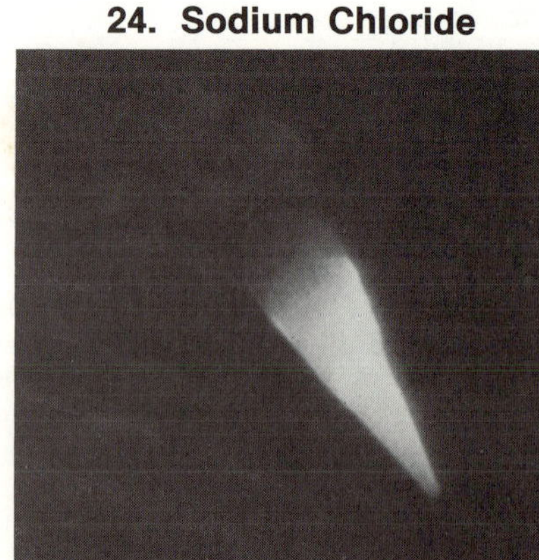

0.60 μm

Appendix C: SEM micrographs of polysilicon spikes formed when a SiO_2 wafer surface was contaminated by common salts.

DETECTION AND SUBSEQUENT REDUCTION OF SURFACE PARTICLE

INDUCED DEFECTS ON SILICON WAFERS

Betty Miller, Christopher W. Teutsch,
and Cynthia Fournier

Intel Corporation, CH2-59
5000 W. Chandler Blvd.
Chandler, AZ 85226

A serious, yield impacting defect was detected after polysilicon etch using the KLA-2028, an automated optical wafer inspection system. Visual inspection of the defect on a video monitor indicated that it was a resist particle. SEM analysis, however, showed that the defect was the result of resist pileup caused by particles. The source of these particles was traced to a wet station operation, but particle levels at the wet station were found to be at presumably acceptable levels. The product wafers were particularly susceptible to particle contamination due to the hydrophobicity following the wet operation. Several rinses to reduce particle adhesion were tested. A rinse of choline was found to decrease particle levels substantially and to increase yields on electrical defect monitor devices.

INTRODUCTION

Defect reduction is a major concern in the production of semiconductor devices. As process geometry shrinks, the minimum particle size that can cause die failure becomes smaller and smaller.

Many products are manufactured at Intel Fab 6, each one with a different defect density level. Locating the source of the defects can be very difficult and time consuming. Consequently, only one type of defect is pursued at one time to minimize confusion. The general strategy consists of, 1) selecting one defect to eliminate, 2) tracking the source of the defect, 3) verifying the source of the defect, and 4) eliminating the defect. For this study, the product of interest was a 1.5 μm NMOS, and the type of defect of interest was polysilicon extras.

On a transistor, polysilicon lines connect devices and form gates. The polysilicon is a conductive layer, and defects that cause polysilicon lines to connect unintentionally will cause short circuits, which leads to die failure. Therefore, it was crucial that the source of these defects be investigated.

Today, there are many different types of tools for defect density measurement. The KLA-2028 is an automated inspection system for

Particles on Surfaces 3, Edited by K.L. Mittal
Plenum Press, New York, 1991

detecting visible defects on patterned wafers. It has an excellent capture rate well into the submicron ranges and locates and measures defects at high speed. The time it takes for the KLA-2028 to perform inspections is dependent on the pixel size used. It can inspect wafers at the following rates.[1]

Pixel Size (μm)	Time (min/cm^2)
2.0	0.8
1.0	2.2
0.5	9.0
0.3	23.0

At Intel Fab 6, a strategy known as "Search and Destroy", using a KLA-2028 Automated Wafer Inspection system, was adopted for defect measurement and reduction. This paper will discuss how the above strategy was applied to detect and subsequently reduce surface particle induced defects on silicon wafers.

EXPERIMENTAL

1. Selecting a Defect to Eliminate

The first step in defect reduction is determining which defect to address. Electrical testing of the die at the wafer level to determine product yield indicated that the highest yield limiting defects were polysilicon defects. The next step was to determine which polysilicon defect type dominated and required attention.

```
 ---------------------------
 | POLYSILICON DEPOSITION |
 ---------------------------
             |
   ---------------------
   | PHOSPHORUS DOPING |
   ---------------------
             |
   ---------------------
   | SILICIDE PRECLEAN |
   ---------------------
             |
  -----------------------
  | SILICIDE DEPOSITION |
  -----------------------
             |
----------------------------
| SILICON OXIDE DEPOSITION |
----------------------------
             |
      ---------------
      | LITHOGRAPHY |
      ---------------
             |<-------------KLA-2028
   --------------------      INSPECTION
   | POLYSILICON ETCH |
   --------------------
             |
   ---------------------
   | PHOTORESIST STRIP |
   ---------------------
```

Figure 1. Steps Pertaining to the Polysilicon Module of the Process Flow. Step #1, Performed KLA-2028 inspections at Post Lithography.

To understand more about these polysilicon defects, a library of defects was built. This provided a pictorial representation of each defect type so that similar looking defects could be categorized together and a Pareto chart of the defects could be generated to determine which ones to address.

For the defects to cause short circuits, two polysilicon lines must connect, which meant that the pattern for the polysilicon layer must be defined. Therefore, as shown by the process flow in Figure 1, it would seem appropriate to perform the initial inspection after the lithography step where the polysilicon lines are defined.

Four wafers were selected for inspection from each of seven lots over a period of two weeks. Twenty-four die locations on each wafer, selected randomly, were inspected on the KLA-2028. A total of 319.2 cm^2 was inspected and all defects were classified by appearance. The assumption made at this point is that similar looking defects must have a common cause.[2] If there is more than one cause, one will dominate. From this information, it was found that the most frequently occurring defect was the "Blob"; which accounted for 80% of the defects.

As the name implies, a "Blob" is a shapeless splotch, as illustrated by Figure 2a. Visual inspection of the defect on the KLA-2028 indicated that the blob was a particle in the photoresist. However, as shown by Figure 2b, SEM (Scanning Electron Microscope) analysis showed that the defect was the result of photoresist pileup caused by a particle, possibly during the lithography process.

2. Tracking the Defect

The lithography step, which defines the pattern of the circuit, consists of photoresist application (spin), exposure, and development.

Figure 2a. Blob Detected by KLA-2028 at Post Photoresist Development.

Since the SEM analysis showed that the blob was the result of photoresist pile up caused by a particle, it is unlikely that the exposure or development step created it. The particle was deposited either prior to or during the photoresist spin step. Therefore, as shown by the process flow in Figure 3, the next step was to determine if photoresist spin was responsible for the creation of the blob.

KLA-2028 inspections were performed on several lots just prior to the lithography step. The results showed that the particles which created the blobs were present prior to lithography; therefore, it was concluded that the particles must have been deposited prior to the photoresist spin step.

Since the defects were polysilicon defects, it was appropriate to assume that they were present during or after the polysilicon deposition step. The fact that the particles were not induced during lithography indicated that they must have originated somewhere between polysilicon deposition and lithography. Therefore, the next step was to determine which step after polysilicon deposition but prior to lithography was responsible. This was done by stepping backwards through the process and performing KLA-2028 inspections on individual lots after these steps (Figure 4).

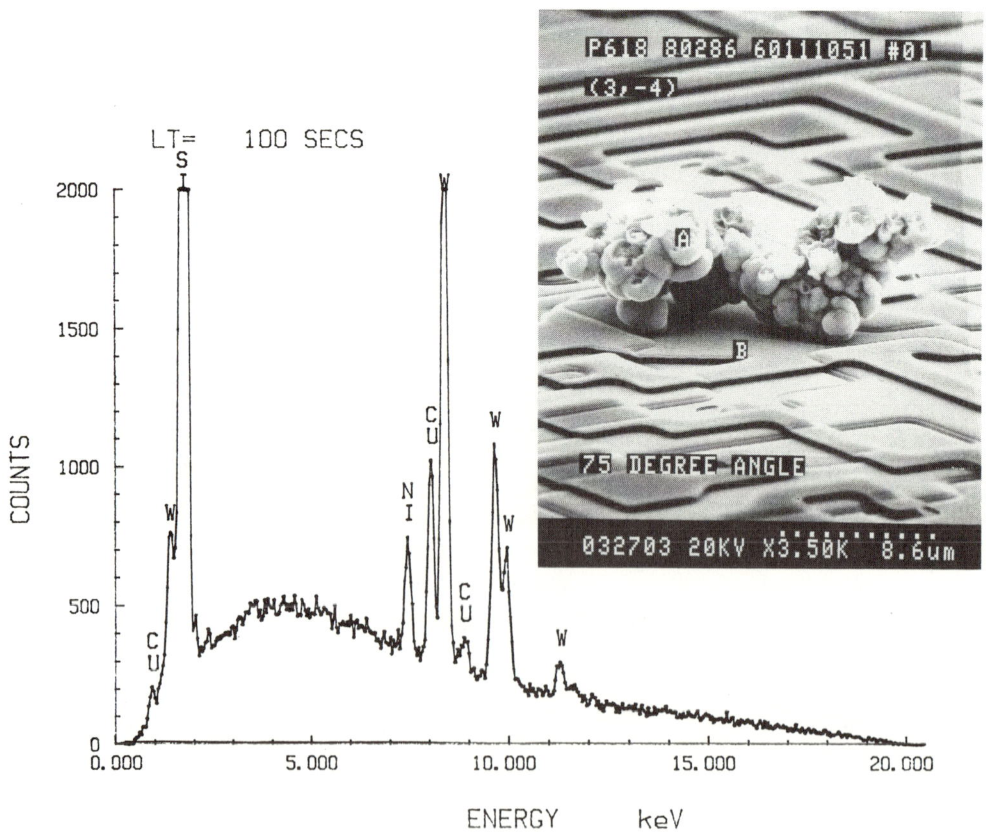

Figure 2b. Scanning Electron Micrograph (SEM) and Energy Dispersive X-ray of Blob from Figure 2a after Photoresist Removal.

The data from the above inspections showed that the preclean prior to silicide deposition generated particles which appeared similar to the ones that created blobs. In tracking the defect, the process was stepped backwards and different lots were inspected and then sent on.

```
          --------------------------
          | POLYSILICON DEPOSITION |
          --------------------------
                       |<--------------KLA-2028
            ---------------------       INSPECTION
            | PHOSPHORUS DOPING |
            ---------------------
                       |<--------------KLA-2028
            ---------------------       INSPECTION
            | SILICIDE PRECLEAN |
            ---------------------
                       |<--------------KLA-2028
           -----------------------      INSPECTION
           | SILICIDE DEPOSITION |
           -----------------------
                       |<--------------KLA-2028
       ----------------------------     INSPECTION
       | SILICON OXIDE DEPOSITION |
       ----------------------------
                       |<--------------KLA-2028
               ---------------          INSPECTION
               | LITHOGRAPHY |
               ---------------
                       |
            --------------------
            | POLYSILICON ETCH |
            --------------------
                       |
            ---------------------
            | PHOTORESIST STRIP |
            ---------------------
```

Figure 3. Process Flow, Step #2, Performed KLA-2028 Inspection at Post Silicon Oxide Deposition.

```
          --------------------------
          | POLYSILICON DEPOSITION |
          --------------------------
                       |
            ---------------------
            | PHOSPHORUS DOPING |
            ---------------------
                       |
            ---------------------
            | SILICIDE PRECLEAN |
            ---------------------
                       |
           -----------------------
           | SILICIDE DEPOSITION |
           -----------------------
                       |
       ----------------------------
       | SILICON OXIDE DEPOSITION |
       ----------------------------
                       |<--------------KLA-2028
               ---------------          INSPECTION
               | LITHOGRAPHY |
               ---------------
                       |
            --------------------
            | POLYSILICON ETCH |
            --------------------
                       |
            ---------------------
            | PHOTORESIST STRIP |
            ---------------------
```

Figure 4. Process Flow. Step #3, Stepping Backwards Through the Process and Performing KLA-2028 Inspections.

After the source of defects has been located, the next step was to follow the defects found at the preclean to verify that they did indeed transform into blobs.

3. <u>Verifying the Source of the Defect</u>

Wafers from the same lots were inspected after each process step from post polysilicon deposition to post photoresist strip (Figure 5). Suspect particles were followed from step to step. The final results showed that the silicide preclean indeed added particles which transformed into blobs. The formation of the blob is illustrated in Figures 6a,b,c,d.

4. <u>Eliminating the Defect</u>

The next step in this experiment was to further isolate the source of the particles. Since the preclean was responsible for creating the blobs, the wet station where the preclean is performed was the most likely source for introducing the particles.

This particular wet station consists of three modules; a Filtered Acid Recirculation Module (F.A.R.M.), a Quick Dump Rinser (QDR), and a Spin Rinse Dryer (SRD). The F.A.R.M. contains hydrofluoric acid (HF) and is used to etch silicon oxide on the wafer. The QDR contains deionized (DI) water and is used to rinse the acid from the wafer. The SRD uses filtered nitrogen to dry the wafer. A preclean consists of a dip in HF, a rinse in the QDR, and a dry cycle in the SRD. An experiment was designed which employed both the wet station and the KLA-2028 to determine what types of defects were added by each module.

```
      --------------------------
      | POLYSILICON DEPOSITION |
      --------------------------
                  |-------------->KLA-2028
        ---------------------      INSPECTION
        | PHOSPHORUS DOPING |<--------|
        ---------------------
                  |-------------->KLA-2028
        ---------------------      INSPECTION
        | SILICIDE PRECLEAN |<--------|
        ---------------------
                  |-------------->KLA-2028
       -----------------------     INSPECTION
       | SILICIDE DEPOSITION |<-------|
       -----------------------
                  |-------------->KLA-2028
    ----------------------------   INSPECTION
    | SILICON OXIDE DEPOSITION |<------|
    ----------------------------
                  |-------------->KLA-2028
           ---------------         INSPECTION
           | LITHOGRAPHY |<------------|
           ---------------
                  |-------------->KLA-2028
        --------------------       INSPECTION
        | POLYSILICON ETCH |<----------|
        --------------------
                  |-------------->KLA-2028
        ---------------------      INSPECTION
        | PHOTORESIST STRIP |<---------|
        ---------------------
                  |-------------->KLA-2028
                                   INSPECTION
```

Figure 5. Process Flow. Step #4, Stepped Forward from Post Polysilicon Deposition to Post Resist Strip and Performed KLA Inspections.

The simplest and quickest test would be to process wafers through each module and perform KLA-2028 inspections before and after processing to determine what defects were added by each module. However, two constraints made this simple test impossible.

First, during the phosphorus doping step, a layer of native oxide grows on the wafer. Prior to continuing on to the next deposition step, this layer of oxide must be removed. The HF in the preclean is used for etching this layer of oxide. In doing so, as shown in Figure 7, the bare silicon is exposed, creating an extremely hydrophobic surface to which particles can adhere strongly.[3] In order for the wafers in the experiment to have the same surface chemistry, all of them would require initial HF processing to strip off the oxide layer. This activates the wafer for particle attraction prior to splitting them up to test the different modules. This way, wafers that get tested in the QDR or SRD would have the same affinity for particles as the wafers that are tested in the F.A.R.M..

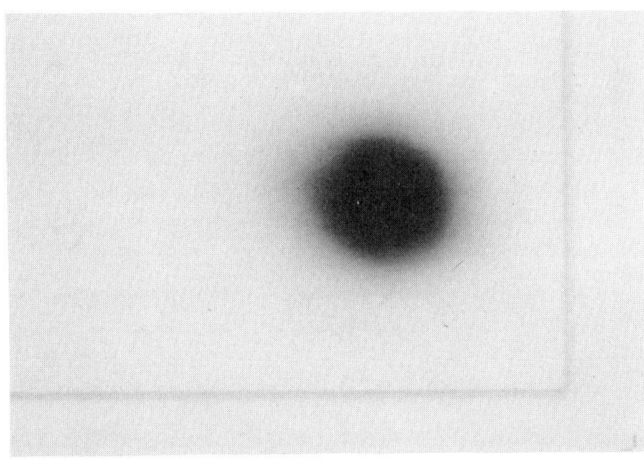

Figure 6a. Formation of the Blob. KLA-2028 Photograph of Particle at Post Silicide Deposition.

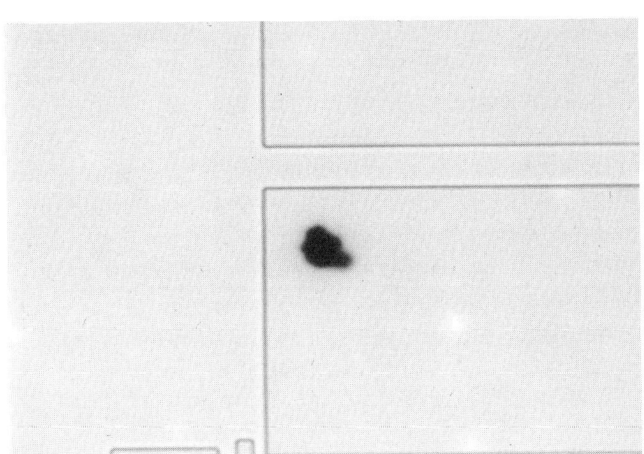

Figure 6b. Formation of the Blob. KLA-2028 Photograph of Particle at Post Silicon Oxide Deposition.

Second, it is not possible to run a wafer through HF only or deionized water only and then inspect it on the KLA. The wafer with HF must be rinsed and dried and the wafer with the DI water must be dried prior to inspecting them on the KLA. Therefore, the next experiment was designed to take these two constraints into consideration.

The experiment began with 12 product wafers that had been processed through the phosphorus doping step. All of the wafers then received an initial preclean to ensure that all of the wafers had the same surface chemistry and affinity for particles. A KLA-2028 inspection was then performed to determine the initial particle level of the wafers. The wafers were then divided into three groups of 4 wafers each. The first group was processed through the SRD. Since the wafers could not be processed through just the QDR, the second group was processed through both the QDR and SRD. The third group was processed through the F.A.R.M., QDR, and SRD. All of the wafers then received a second KLA-2028 inspection.

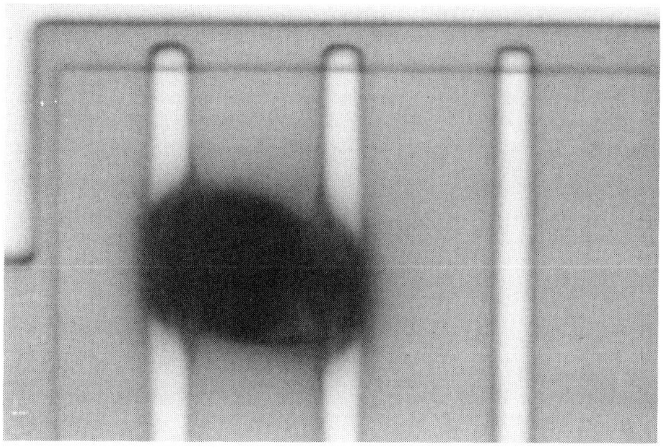

Figure 6c. Formation of the Blob. KLA-2028 Photograph Seen at Post Lithography.

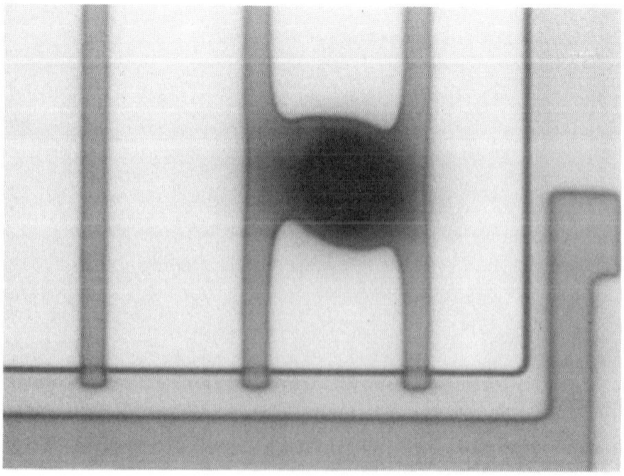

Figure 6d. Formation of the Blob. KLA-2028 Photograph of Particle Seen at Post Resist Strip.

A simplified version of the above experimental procedures is shown in flowchart form in Figure 8.

RESULTS

This experiment was repeated 12 times for a total of 144 wafers over a period of 2 months. Many different types of defects were seen, but only three occurred frequently. All of the defects were classified by appearance, and the three most common ones were: black particles, white particles and starships.

The two sets of KLA data, before and after, were recorded and analyzed. Usually, when performing before and after particle counts,

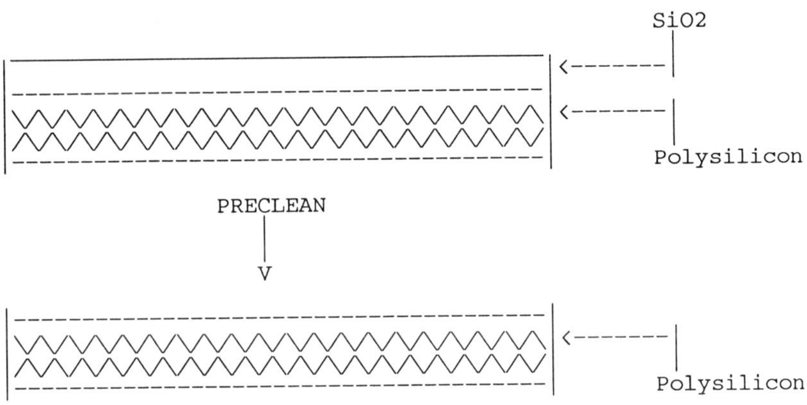

Figure 7. After HF Etches the Native Oxide from the Wafer, Polysilicon is Exposed.

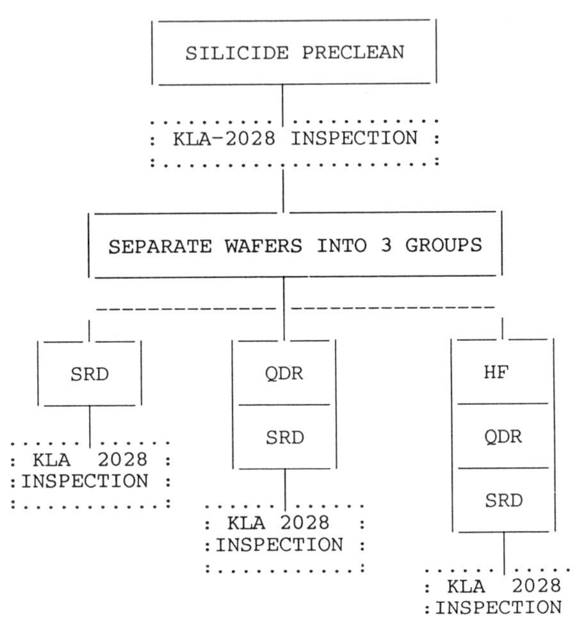

Figure 8. Flow Diagram of Wet Station Experimental Procedures.

the final results are determined simply by subtracting after counts from the before counts. However, since the KLA-2028 identifies the exact X-Y coordinates of each defect, it was possible to track each defect from one inspection to the next and determine which defects on the wafer were added, lost, or unchanged after they were processed through the individual modules. For example, if ten defects were present before the test and ten defects were present after the test, one would assume that no defects were added. However, it is possible that the initial ten defects were removed and then ten new ones were added. With the KLA-2028, it would be apparent if ten other defects had been added and ten defects were lost.

As explained earlier, the HF module had to be combined with the QDR/SRD modules and the QDR module had to be combined with SRD module. As a result, the data generated in the HF group included defects added by the HF, QDR, SRD; the data generated in the QDR group included defects added by QDR and SRD.

If:	P(SRD)	= P(SRD)
	P(SRD + QDR)	= P(SRD) + P(QDR)
	P(SRD + QDR + HF)	= P(SRD) + P(QDR) + P(HF)
Then:	P(SRD)	= P(SRD)
	P(QDR)	= P(QDR + SRD) − P(SRD)
	P(HF)	= P(HF + QDR + SRD) − P(QDR + SRD)

NOTE: P [=] PARTICLE CONTRIBUTED BY

Figure 9. Experimental Assumptions and their Mathematical Representations.

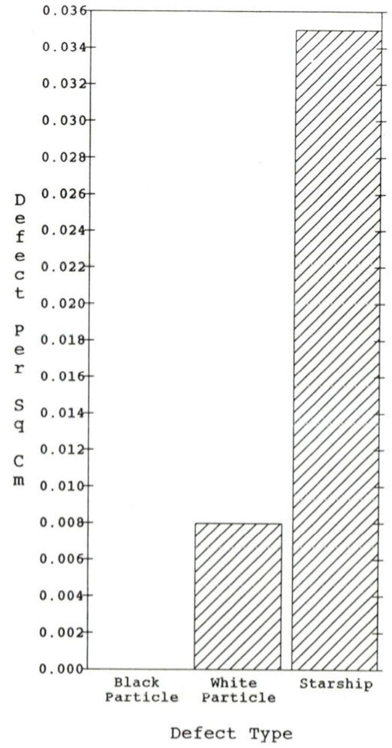

Figure 10. Defects added by HF F.A.R.M.

In order to determine the defects added by each module, two assumptions were made. First, the total defects removed by the modules are insignificant compared to the defects that are added by the modules. Second, the total defects added by the QDR/SRD combination was the sum of the defects added by the QDR and the defects added by the SRD. Second, the total defects added by the HF/QDR/SRD combination was the

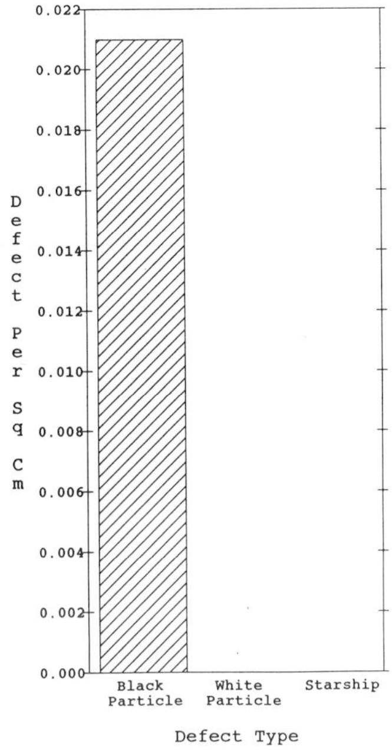

Figure 11. Defects Added by Quick Dump Rinser (QDR).

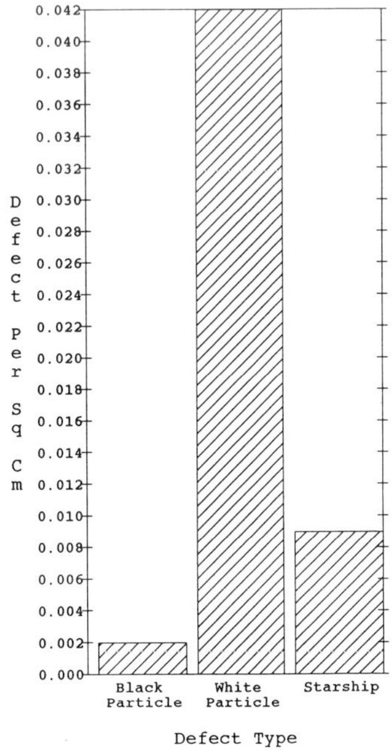

Figure 12. Defects Added by Spin Rinse Dryer (SRD).

sum of the defects added by the HF, the defects added by the QDR, and the defects added by SRD. Therefore, by subtracting the defects added by the SRD from the defects added by the QDR/SRD combination, the defects added by the QDR could be determined. Likewise, by subtracting the defects added by the QDR/SRD from the defects added by the HF/QDR/SRD, the defects added by the HF could be determined. These assumptions and their mathematical representations are shown in Figure 9.

Using the above assumptions and mathematical representations, the defects added individually by the HF, QDR, and SRD were determined. The type of defects added by each module and the corresponding defect densities are shown in Figures 10, 11, and 12. These results show that the F.A.R.M. added "starship", the QDR added "black particle" and the SRD added "white particle". An example of each of these defects as well as the SEM/EDX are shown in Figures 13, 14, and 15.

FURTHER WORK

1. Determination of the Cause of the "Blob" Excursion

While the above experiment was in progress, the defect density of the blob was also monitored. Since the results of the above experiment showed that the preclean generated particles that transformed into blobs, the data were graphed with defect density as a function of silicide preclean date. As shown in Figure 16, there was a definite excursion in blobs during month 1, followed by a dramatic drop in defect density in month 2. To further complicate the analysis, it was found that nothing had been done to the wet station to explain the improvement. The question that needed answering was "Why did the defect density of the blob drop by a factor of 10?"

Many fab operators, technicians, engineers, and facility workers were interviewed to find an answer to this question. Since the excursion occurred during month 1, when there was no historical KLA data available for comparison, the month 1 data were assumed to be the normal defect density for the blobs. It was during month 2, when there was a dramatic drop in defect density that it became apparent that there was an excursion in month 1. The interviews led to one possible explanation.

The wet station where silicide preclean was performed was a new wet station. Before this new wet station could be used for production it

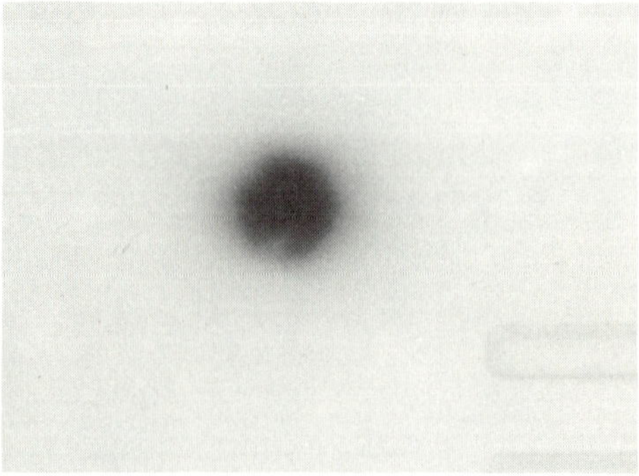

Figure 13a. KLA-2028 Photograph of Black Particle.

had to be qualified as production worthy. While this wet station was being qualified, the previous wet station was used side-by-side.

These two wet stations were both connected to the same water supply. Prior to the installation of the new wet station, the flow rate

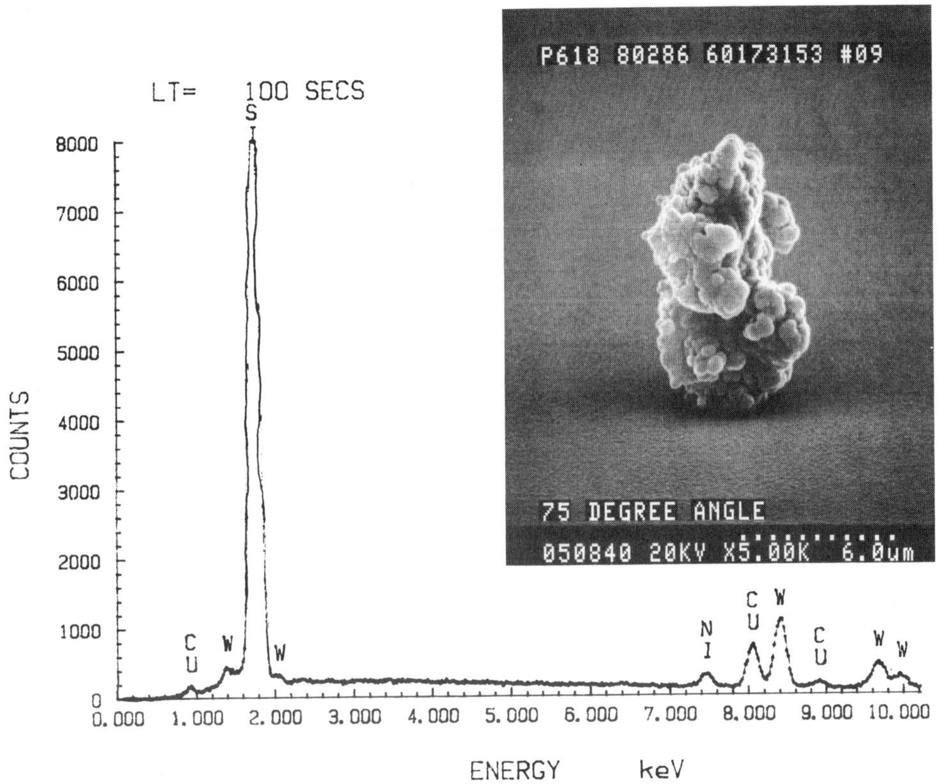

Figure 13b. Scanning Electron Micrograph and Energy Dispersive X-Ray of Black Particle.

Figure 14a. KLA-2028 Photograph of White Particle.

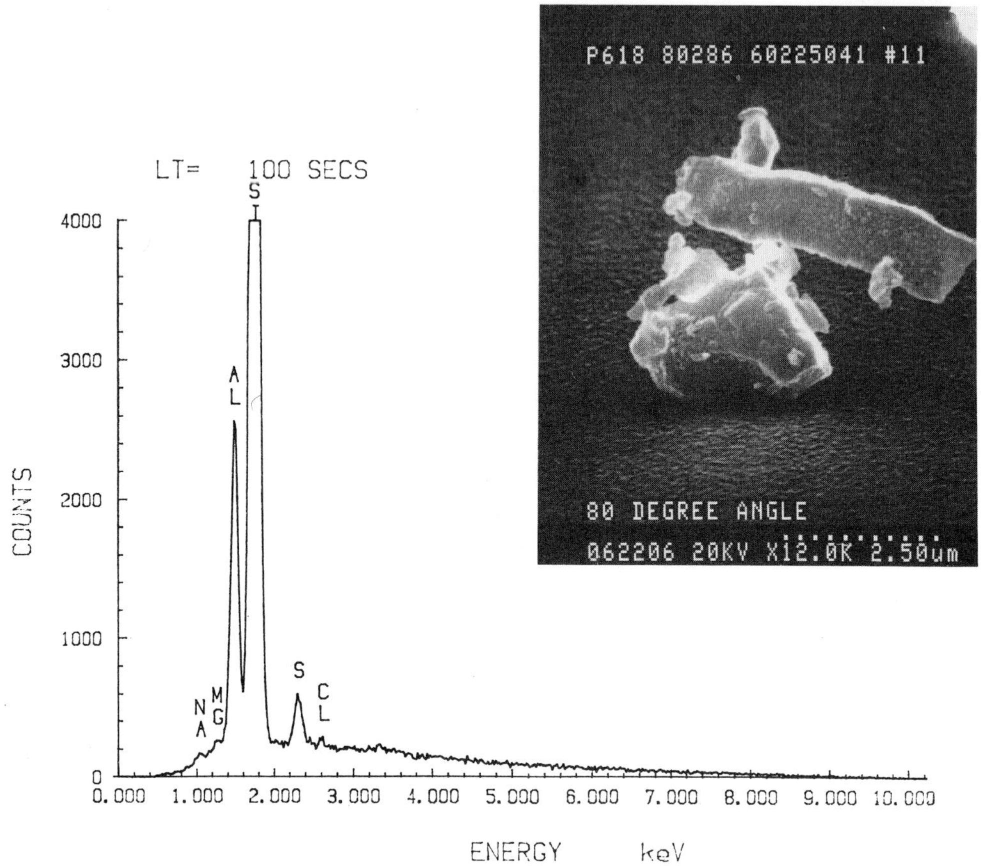

Figure 14b. Scanning Electron Micrograph and Energy Dispersive X-Ray of White Particle.

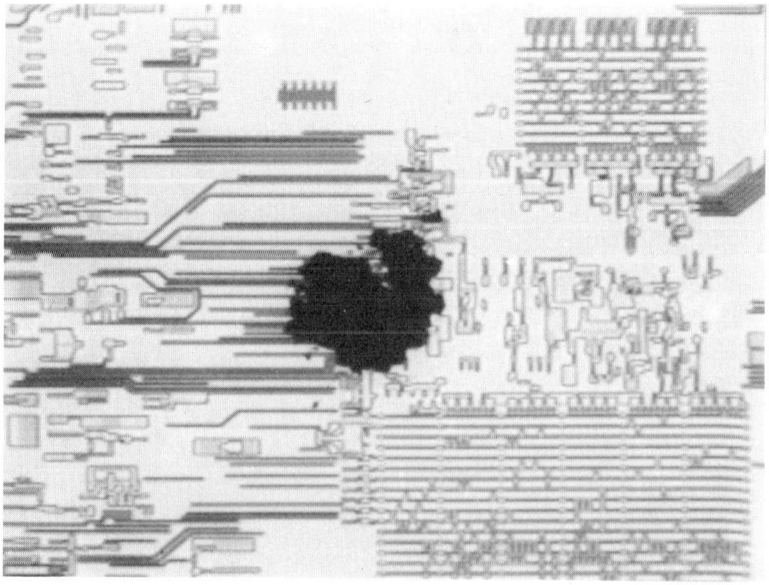

Figure 15a. KLA-2028 Photograph of Starship.

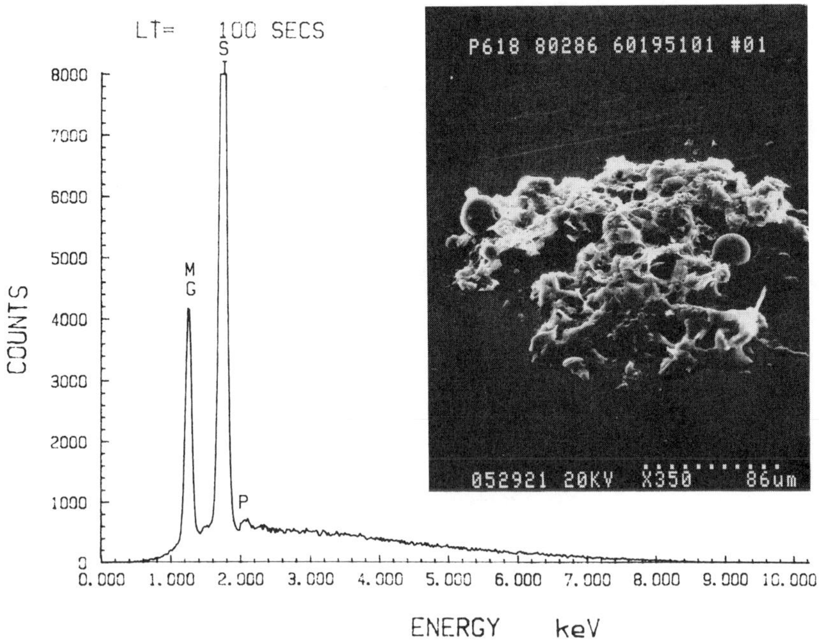

Figure 15b. Scanning Electron Micrograph and Energy Dispersive X-Ray of Starship.

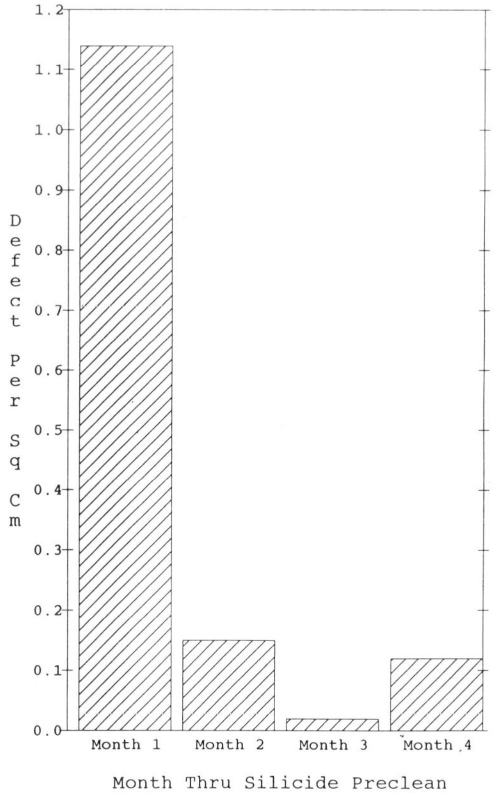

Figure 16. Blob Defect Density versus Silicide Preclean Month.

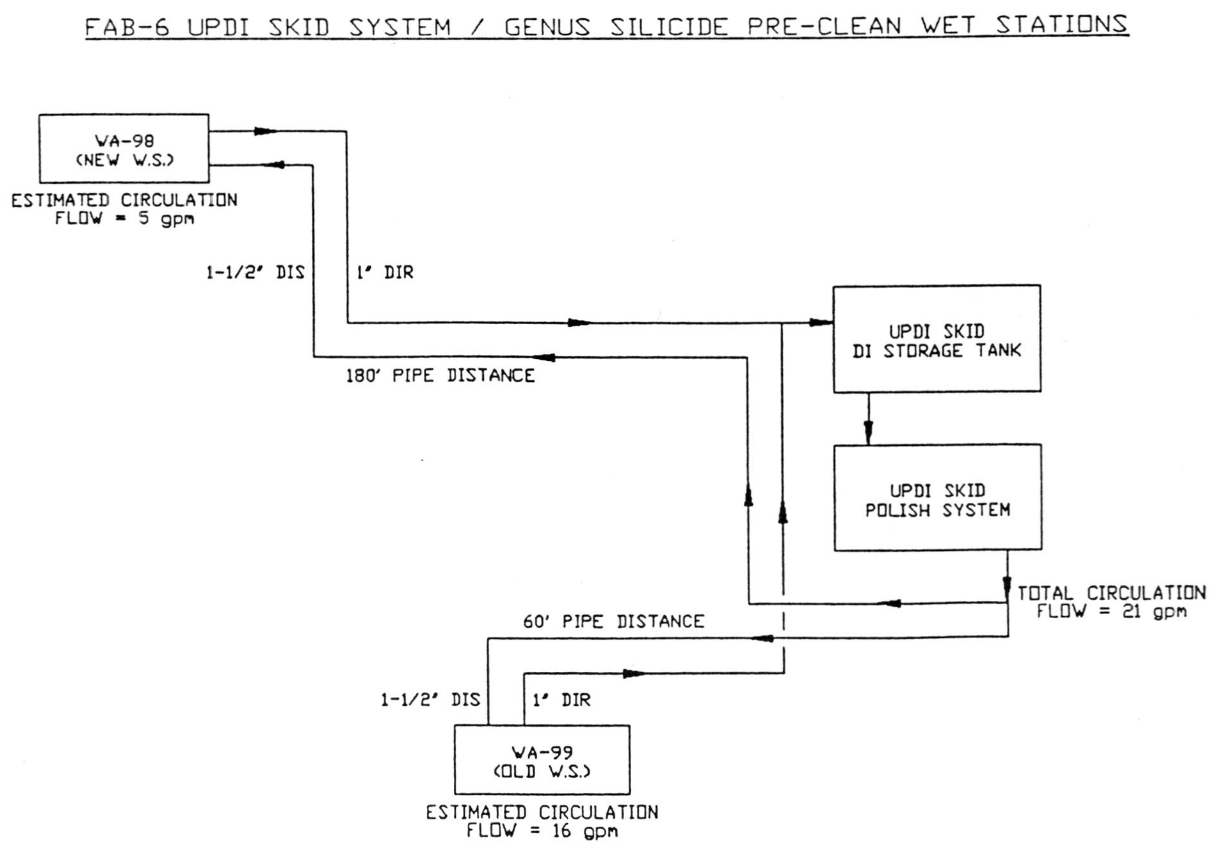

Figure 17. Ultra Pure Deionized (UPDI) Skid System / Silicide Preclean Wet Station

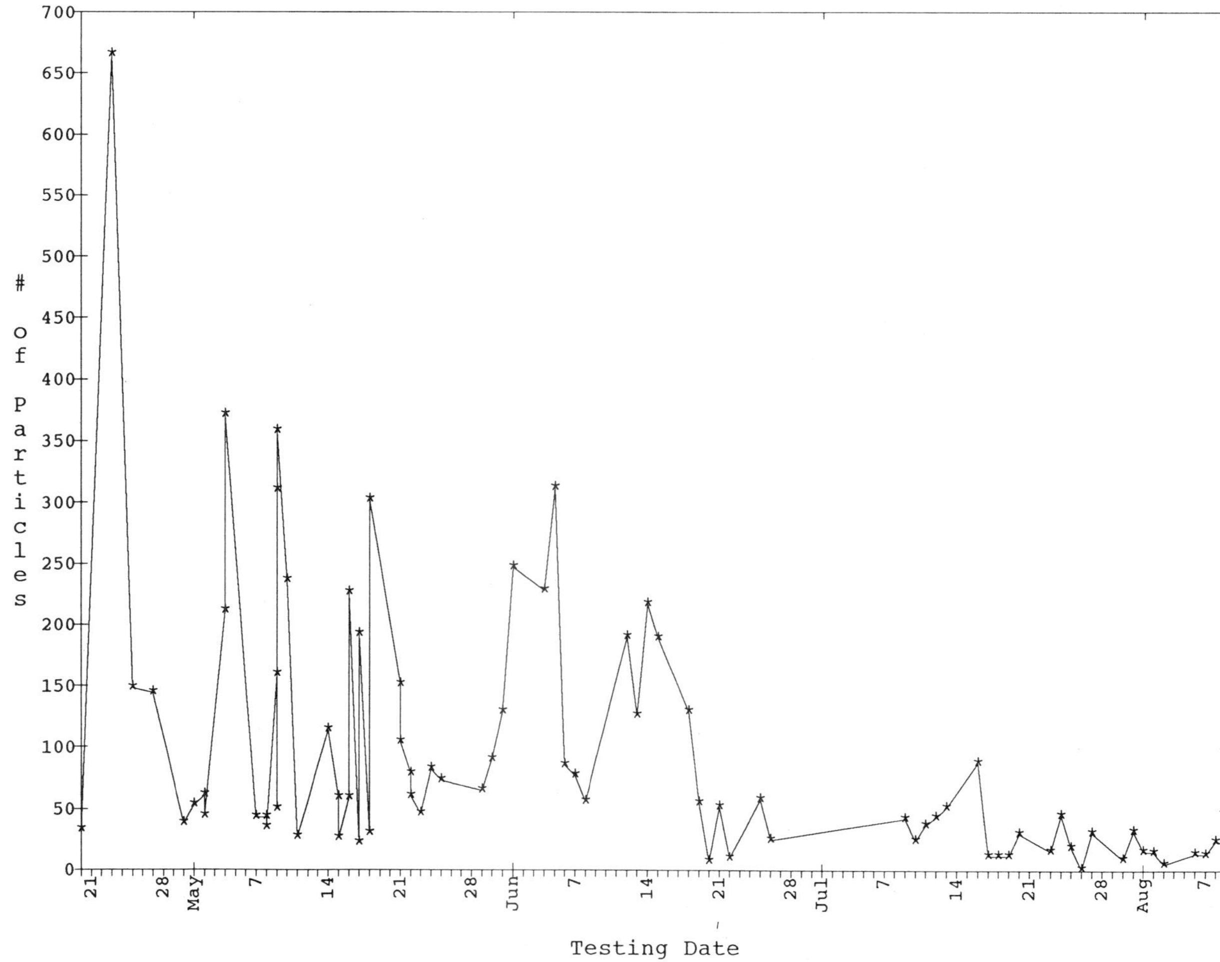

Figure 18. Engineering Particle Monitors at Silicide Preclean.

of the water from the main water supply to the old wet station was 21 gallons per minute. As shown in Figure 17, after the installation of the new wet station, most of the water continued to flow to the old wet station. The flowrate of the water to the new wet station was approximately 5 gallons per minute, which is probably not sufficient to provide a thorough rinse to remove the particles.

It was during this period, when both wet stations were being used, that the excursion occurred. After the new wet station was qualified, the old one was removed. It was on this same day that the excursion disappeared. After removal of the old wet station the water flow rate from the supply to the new wet station increased to approximately 21 gpm. Since the old wet station could not be reinstalled, it could not be proven that the addition of the new wet station caused this excursion. Also, the blobs were not being monitored prior to the installation of the new wet station; therefore, the defect density of the blobs prior to Month 1 was not available for comparison.

2. Implementation of Daily Engineering Particle Monitors on Wet Station

To characterize particle performance at the wet stations and to ensure that the above excursion did not reappear, daily engineering particle monitors on the wet station were implemented. Two bare silicon wafers were processed through the wet station and the particle level contributed by the wet station was recorded.

The Surfscan 4500, a laser based surface defect monitor for bare silicon wafers, was used to measure the particle level before and after wet station processing. All work performed on the wet station, such as filter changes to the F.A.R.M., deionized water supply sterilization, acid changes, etc. were noted. As shown in Figure 18, these data indicated causes of particle excursions. These causes were then investigated to eliminate these particle excursions. Figure 18 also shows the particle counts at this wet station are very well controlled.

3. Choline Rinse Experiment

Earlier, it was mentioned that the HF changes the surface chemistry of the wafers. Past studies have shown that an alkaline rinse after the HF dip makes the wafers less susceptible to attracting particles.[4] Electrical test structures known as "Spider Masks" were run to test whether this additional rinse would be beneficial. Spider Mask is a collection of combs and snake patterns intended to predict process defect density[5]. A detailed explanation and a figure of a Spider Mask can be found in the paper "Isolation and characterization of particle induced defects from the Lithography process using an electrical defect monitor, which follows this paper). Three lots of spider wafers were run with half of the lot receiving standard processing and the other half receiving a choline (alkaline solution) rinse after the HF/QDR/SRD step. The results showed a 6 die/wafer increased in yield.

CONCLUSIONS

A strategy known as "Search and Destroy" using the KLA-2028 was outlined. The experimental method described was used to trace the sources of surface particle induced defects on silicon wafers.

In performing the experiments, it was found that particle sensitive steps do exist in wafer processing, especially if a wet station is involved. The wafer's affinity for particles is controlled by its surface chemistry, and certain chemical processing steps could have a dramatic effect on wafer particle levels. Therefore, it is very important to control and reduce the defect level of process equipment.

Using the KLA-2028, the source of the defects was found. Further, what was perceived to be a lithography induced problem was shown to be a wet station problem. From this, it was discovered that "first seen" does not always mean "first present." The reason that lithography was suspected was that this step magnifies a particle problem. Lithography defines the pattern of the circuit; therefore, the damage to the pattern caused by the particle might not be seen until after the pattern has been defined.

Finally, DI water cleanliness and flowrate are very important on any wet station. Even though wet processing is considered a particle removal step, it is also possible that particles are introduced in such processing.

ACKNOWLEDGEMENTS

The authors would like to express special thanks to Rose Wade, Hayden Dingwall and Diane Carmichael for their work on the KLA-2028, to Cynthia Barutha and Tony Chan for their work on the SEM, to Shawn Church for his work on the wet stations, and to Kevin Pate and Milton Staudt for their work on the UPDI system.

REFERENCES

1. KLA-2028 Product Reference Manual.

2. Chris Radin, A defect reduction methodology for increased sort yield using automated defect reduction. SPIE Proc. 921 4(1988).

3. A.E. Walter and C. F. McConnell, Direct-displacement wet processing how it affects Wafer Surface Phenomena, Microcontamination 8, No.1, 35-38, 60-61(Jan 1990).

4. Charlie A. Peterson, in "Particles on Surface 1: Detection, Adhesion and Removal", K.L. Mittal, editor, pp. 37-42, Plenum Press, New York, 1988.

5. Christopher W. Teutsch and David C. Drain, Combining electrical defect monitors with automatic visual inspection systems, in "Integrated Circuit Metrology, Inspection, and Process Control," K.M. Monahan, ed., Proc. SPIE 1087, 189 (1989).

ISOLATION AND CHARACTERIZATION OF PARTICLE INDUCED DEFECTS FROM THE LITHOGRAPHY PROCESS USING AN ELECTRICAL DEFECT MONITOR

Christopher W. Teutsch, Betty Miller, and
Cynthia Fournier

Intel Corporation
M/S CH2-59, 5000 W. Chandler Blvd.
Chandler, AZ 85226

The design, processing and use of a serpentine-comb electrical defect monitor that does not require the photoresist to be stripped prior to testing is described. By use of a dual pad structure, the serpentine-comb pattern can be tested for opens and shorts while the resist remains on the pattern and on the defects. In this way, defects, (especially particulate induced ones) can be ascribed to either before or after photoresist application and in some instances to before or after resist patterning. The use of an electrical defect monitor of this type makes it possible to collect lithography related defect data from a sample area much larger than would be possible with conventional optical inspection methods.

INTRODUCTION

Yield losses in semiconductor processing are broadly attributed to either parametric defects or visual (hard) defects. Parametric defects result in changes in device performance that make the device unable to function within its specified operating range. Incorrect junction formation, out of spec critical dimensions, and incorrect dopant levels are manifestations of parametric defects. Hard defects are errors in the pattern transfer process exhibited by extra or missing patterns. Hard defects are usually particulate induced and occur when a particle mechanically interferes with lithography, film growth and/or etch process steps. Figure 1 shows a particle that has interfered with lithographic patterning on a 1.5 μm CMOS process.

Hard defects result in serious economic losses for the semiconductor industry. Large, complicated die with dense patterns are affected severely. It is not uncommon to find a product where 80% of available good die are lost to hard defects. Smaller die, due to the reduced probability of receiving a random particle, are less impacted and may only suffer a 10% yield loss due to hard defects. Since hundreds of process steps are required to fabricate integrated circuits, the actual defect density per process step must be very small in order for the total defect density to be tolerable. Thus, defect

Particles on Surfaces 3, Edited by K.L. Mittal
Plenum Press, New York, 1991

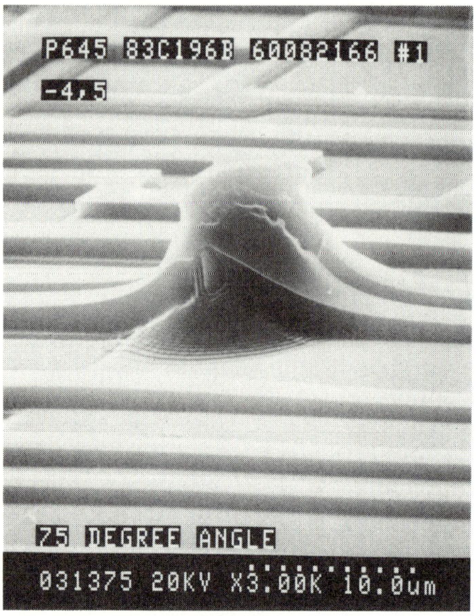

Figure 1. Particle that has interfered with lithographic patterning. Note the pileup of photoresist around the particle.

experimentation on a process step basis must frequently deal with particles as rare events, making normal statistical methods inapplicable.

As defect reduction continues to progress and defect densities are reduced, ever increasing sample sizes are required. This, combined with shrinking critical dimensions, requires that the speed and quality of defect detection and tracing methods be continually improved.

DEFECT REDUCTION ISSUES

Particle Size

Particles on the order of $^1/_{10}$th of the critical (smallest) geometry are generally considered harmful. Particles even smaller than this can cause killer defects when they contain chemical contaminants or occur in very thin layers[1]. For the purposes of this discussion, however, we will consider particles whose size is large enough to mechanically interfere with pattern transfer and thus are near the critical geometry size ($^1/_{10}$th critical geometry and larger).

It is interesting to note that although the quantity of particles increases rapidly as their size decreases, the mechanical capability of a large particle to be a killer defect is much greater than a small particle. Thus, although there are many more small particles, large particles continue to be of great concern, where large is defined as 2 to 10 times the critical geometry size. This is contrary to many defect reduction efforts within the electronics industry. Because of the greater number of small particles, many defect reduction engineers place too little emphasis on large particles. Thomas and Maly[2] showed data indicating a ratio of large to small defects of approximately 0.8 on a Dual Bridge Test Structure (DBTS) capable of determining defect size ranges as well as densities. This was a 4 μm pitch structure with a 1.5 μm space designed to report the number of adjacent lines affected by a defect. In analyzing their data, we have defined 1.5 to 5.5 μm as "small" and 5.5 μm and up as "large". Although 5.5 μm is still rather large compared to the 1.5 μm space, this is the limit of the test device's resolution.

Work by Dean[3] also showed the effect of particle size on the particle-defect ratio. Using sized latex spheres and E-beam lithography with a 1 μm spot size, 2 μm spheres deposited prior to exposure resulted in no defects while 4 μm spheres consistently produced extra pattern defects. In any event, it is evident that large particles are a major portion of the yield problem and this agrees with our daily observations in the factory. In practice, we find that many serious yield "crashes" continue to be related to out of control situations for large and even very large particles.

Defect Reduction Methods

Reduction of particulate induced defects requires methods of measuring defect levels and tracking down the sources of these defects. The most common method of reducing particulate caused defects is to reduce the general level of particles at every process step. This method assumes that all particles are equally harmful. Another method is failure analysis where actual defects on product are studied and analyzed after fabrication to determine the source of the destructive particle. Unfortunately, the destructive particle is frequently no longer present on the silicon wafer surface, making identification of the particle source virtually impossible, see Figure 2. It is then necessary to design methods for observing and analyzing the interactions of particles and product within the factory and to characterize the defect formation process.

All defect tools make tradeoffs between the quality and quantity of the data provided[4]. Knowledge of these capabilities is critical in deciding which tool to use. The automated optical inspection system is one tool that is used to directly inspect product wafers at various process steps and to trace defects to their sources. Although extremely effective, it suffers from being very expensive and many times is too slow to provide a large enough sample size for statistically valid decisions. Electrical defect monitors are another defect detection tool used to analyze defects within process modules. They have the advantage of detecting only killer defects and allowing large areas to be sampled for statistically valid decision making. This paper will present a method for improving the information quality of defect data gathered by electrical defect monitors with a corresponding increase in complexity, see Figure 3.

ELECTRICAL DEFECT MONITORS

Electrical defect monitors are typically single layer structures consisting of a dense, serpentine-comb pattern designed to capture open and short defects. The width of the lines and spaces is dependent upon the process and the defect size of interest. The pitch of the device (line width + space width) is the most common sensitivity indicator. In order for electrical testing to be possible, the electrical defect monitor (spider device) must be fabricated using a conductive film. This is one of its major limitations since not all films used in semiconductor processing are conductive, see Figure 4.

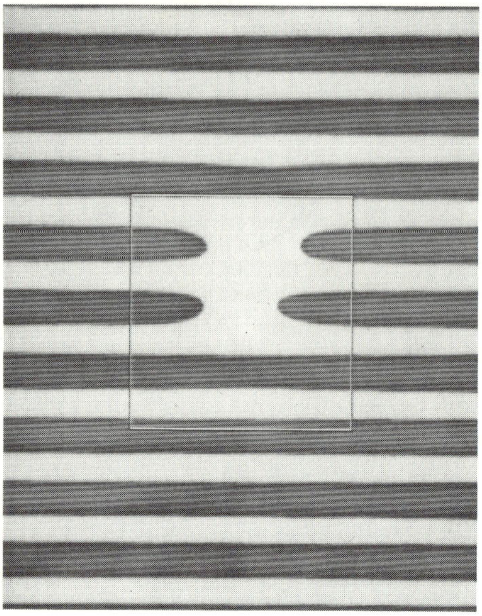

Figure 2. Defective area with particle no longer present. Three polysilicon lines are shorted together.

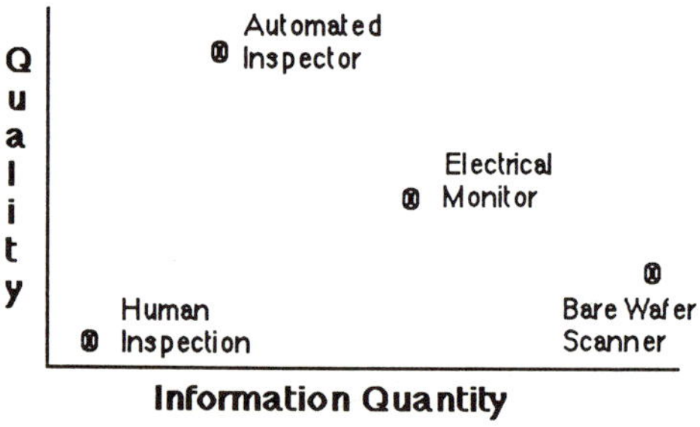

Figure 3. Comparison of various tools showing degree of quality of information vs. quantity of information.

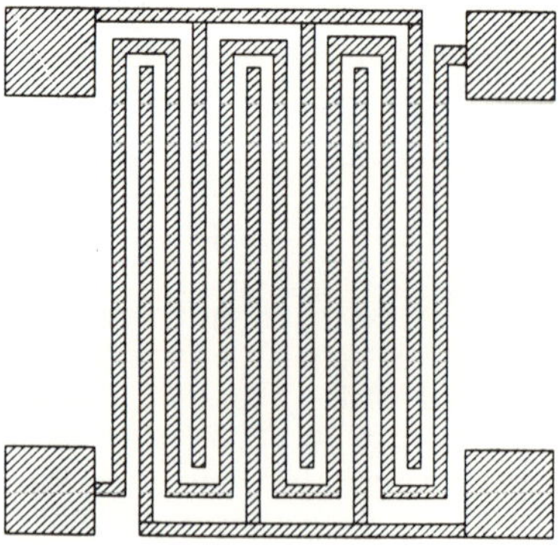

Figure 4. Serpentine-comb spider device.

Advantages of Normal Electrical Defect Monitors

Spider devices are excellent tools for monitoring the level of process defects. They require minimal (single layer) processing and can be tested quickly and easily. Processing can be completed in 1-2 days and testing requires 5-10 minutes per 6″ wafer. Since approximately 40 percent of a spider wafer is active circuit area, spider devices are one of the few tools capable of sampling a large enough area to be statistically significant when testing for defect reduction during experiments. Another advantage is that spider devices require no special equipment for fabrication or testing. Most automated defect detection tools are quite expensive, ranging to well over $1 million.

Disadvantages of Normal Defect Monitors

Limitations of the spider device include the requirement for a conductive film, mentioned previously, and the cost of test wafer processing. Also, only electrically active defects are detected and, without microscope inspection, nothing more is known about the nature of the defect other than whether it is a short or an open and possibly the general size of the defect in the case of the previously mentioned DBTS.

A serious drawback of the spider device is that it is an after-the-fact monitor. Since the photoresist and non-conductive surface films must be removed prior to probing, the particle causing the defect is frequently lost. We have observed this at the polysilicon layer where particles seem to be tightly bonded prior to resist application but to release their bond after resist removal. Although the artifact is detected as a short in the polysilicon pattern, the particle is gone. In the few instances where the particle still remained, attempts to study it with an SEM (Scanning Electron Microscope) were generally unsuccessful. The particle was almost always repelled from the surface as soon as it received a slight charge from the electron beam. Sometimes it is possible to deduce the process step at which the particle causing the artifact defect originated from the appearance of the artifact defect but this can result in incorrect conclusions. In any event, even if the guilty process step can be deduced but the particle has been lost, it cannot be analyzed. There are, of course, some particles that remain on the surface, even after resist removal. A good example is a particle that is imbedded in or under the conductive film.

RESIST-TESTABLE ELECTRICAL DEFECT MONITORS

It is obvious that if a spider device could be tested without removing the resist film, a great deal more information would be available. It would be much easier to determine the process step which caused the defect and, frequently, the particle responsible for the defect would still be present for analysis. Defects caused by particles prior to lithography would have particles under or in the resist, see Figure 5. Particles from etch would have an undisrupted resist pattern under them, see Figure 6. Although particles from lithography that blocked exposure might be lost during pattern development, it would still be clear that it was a lithography and not etch related defect due to the presence of the photoresist above the artifact, see Figure 7. It is our experience that defects due to blocked exposure are a small portion of the defect problem and that most particles that caused defects are still present prior to resist strip.

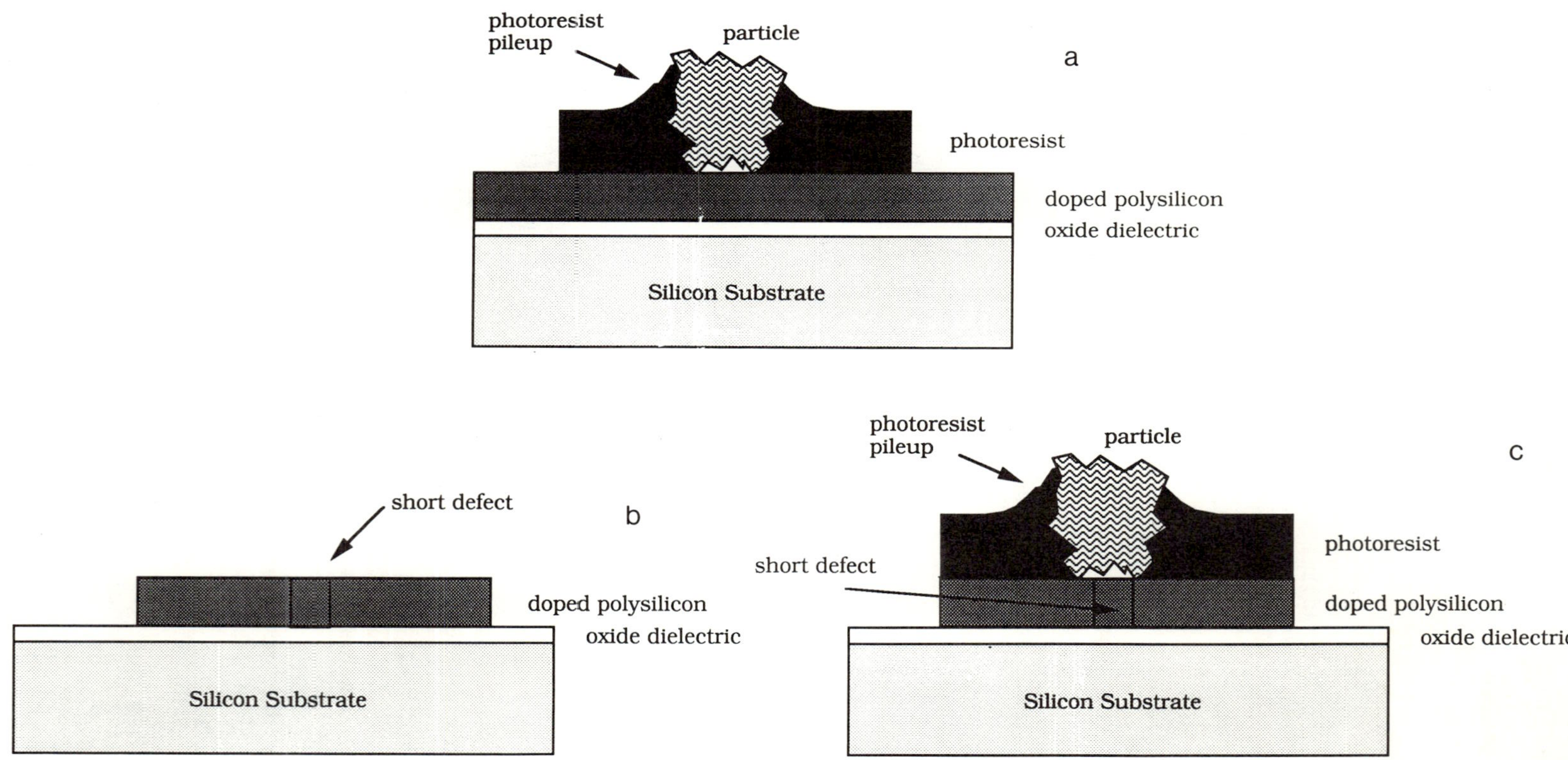

Figure 5. Particle-induced lithography defect. Particle increases resist thickness beyond exposure limit preventing complete UV exposure (a). Resulting defect on normal spider device (b). Resulting defect on resist-testable spider device (c).

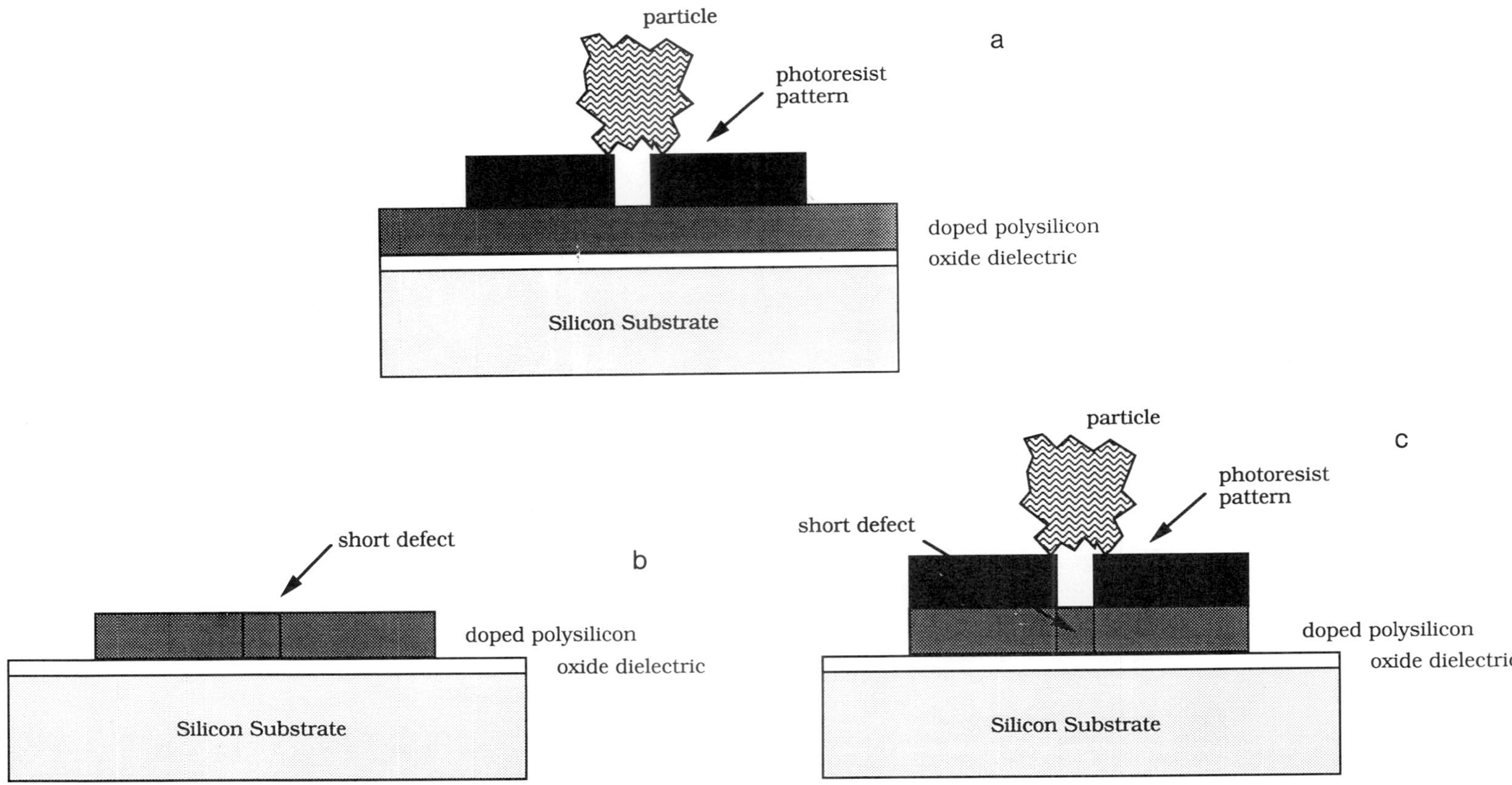

Figure 6. Particle-induced etch defect. Particle blocks oxide/polysilicon etch (a). Resulting defect on normal spider device (b). Resulting defect on resist-testable spider device (c).

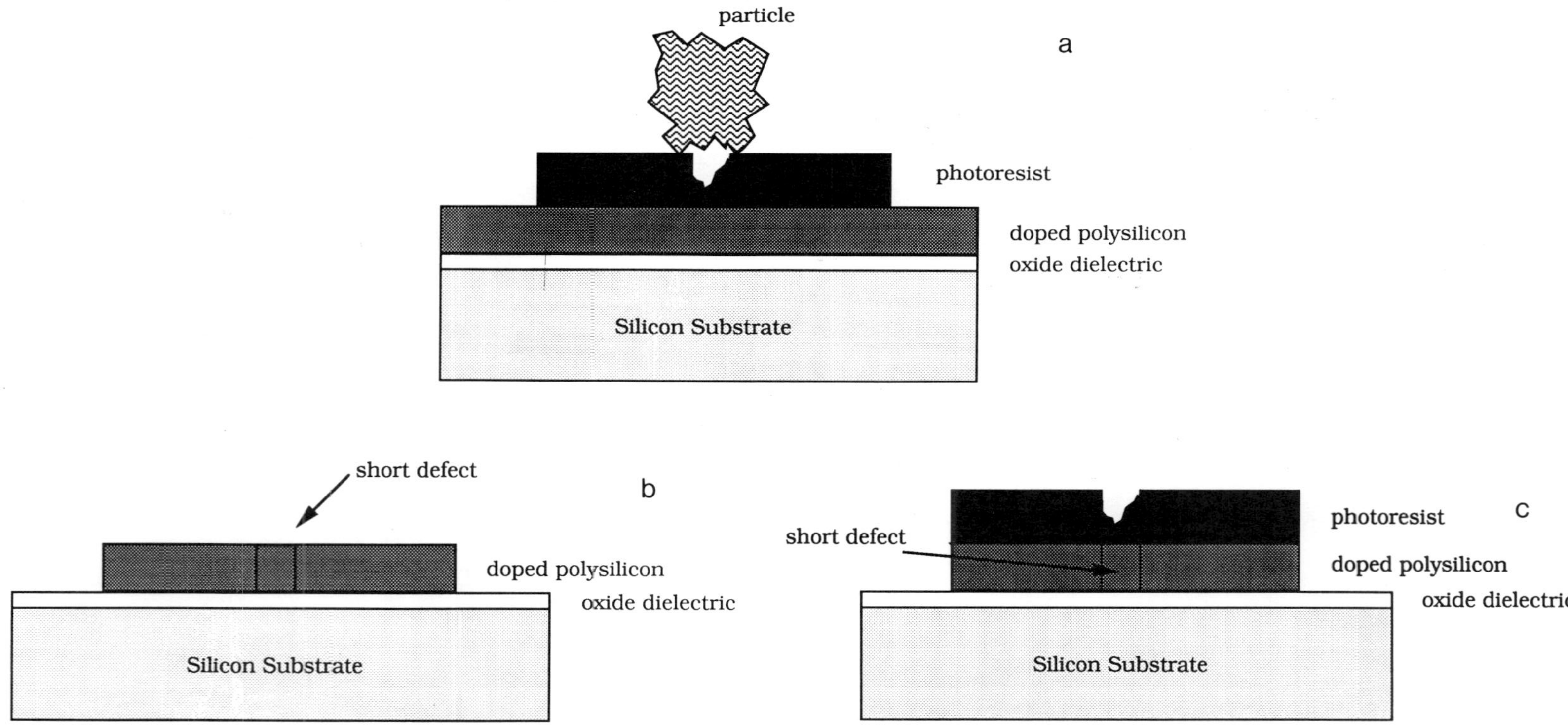

Figure 7. Particle-induced lithography defect. Particle blocks UV exposure (a). Resulting defect on normal spider device (b). Resulting defect on resist-testable spider device with particle lost during pattern development (c).

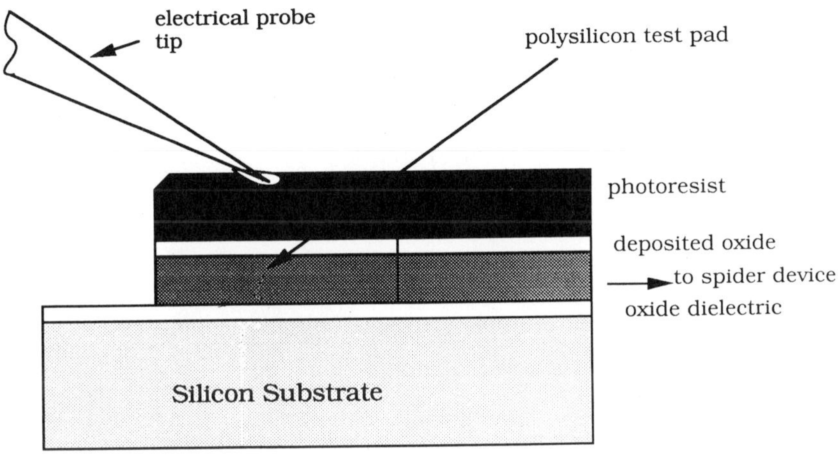

Figure 8. Electrical probe tip is prevented from making contact with the polysilicon test pad due to photoresist and an oxide film.

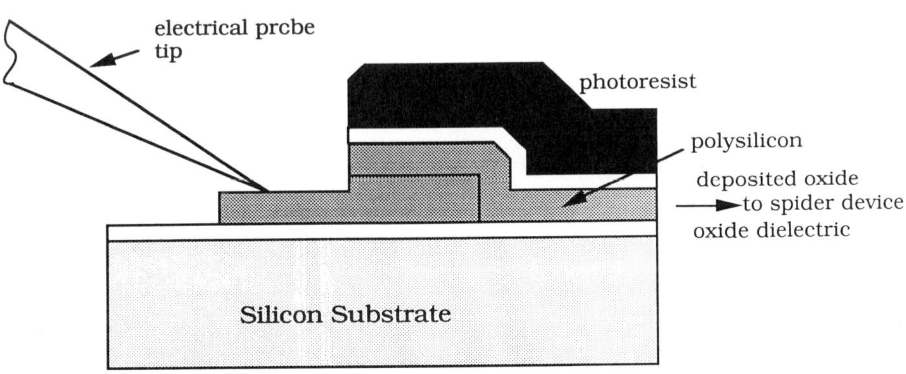

Figure 9. Special probe pad structure allows the device to be tested without stripping the photoresist and oxide.

Unfortunately, in order to form probing pads, a resist pattern is required to protect the pad structure during etch. This resist film, sitting on top of the pad, prevents the probe needle from making electrical contact with the pad, see Figure 8. Attempts to probe through the resist have been unsuccessful. In addition, for the process in question, an additional layer of nonconductive oxide covers the pad, preventing prober contact to the polysilicon below. An Intel-designed two-layer spider device resolves this problem with a special probe pad structure that allows the device to be tested without stripping the photoresist and oxide, see Figure 9.

The process flow for the resist-testable spider device is shown below. The process cross section sequence for fabrication of the dual pad structure is shown in Figures 10 through 15. Electrical contact between the double wide first layer (bottom) pad and the second layer (top) pad provides an electrical pathway under the photoresist. This is accomplished by making the combined probe pad so thick that even though part of it is unprotected by photoresist during etch, it is not completely removed. The spider device can also be used as a single layer device which would require only the steps shown in Figures 10 through 12 (deposition, patterning, etch and resist strip). For resist testability, a second deposition, patterning with alignment, and a second etch are required.

Table 1

Normal Spider Device Process Flow	**Resist-testable Spider Device Process Flow**
Oxide Growth	Oxide Growth
Polysilicon Deposition	Polysilicon Deposition
Polysilicon Doping	Polysilicon Doping
Oxide Deposition	Dual Pad Patterning
Spider Device Patterning	Polysilicon Etch
Oxide/polysilicon Etch	Photoresist Strip
Photoresist Strip	Polysilicon Preclean
Oxide Strip	Polysilicon Deposition
Electrical Test	Polysilicon Doping
	Oxide Deposition
	Spider Device Patterning
	Oxide/polysilicon Etch
	Electrical Test

Figure 10. Process cross section sequence for fabrication of the dual pad structure. The starting material is a silicon wafer with a layer of thermally grown silicon oxide and a layer of CVD (Chemical Vapor Deposition) polysilicon. The polysilicon has been doped with phosphorous to make it conductive.

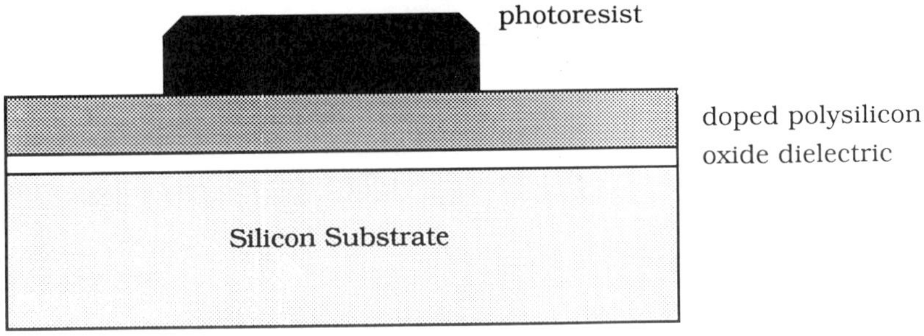

Figure 11. Photoresist application and patterning of dual pad structure. This will form the bottom pad.

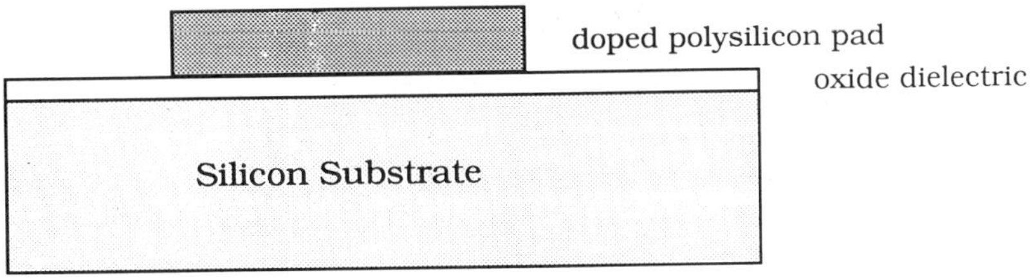

Figure 12. Polysilicon etch and photoresist strip. This step transfers the photoresist pattern to the polysilicon.

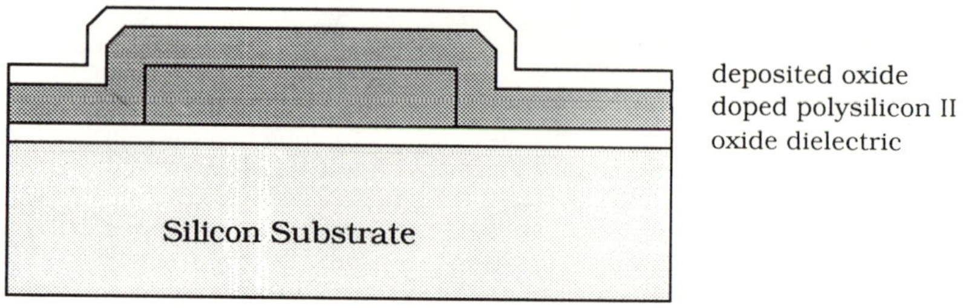

Figure 13. 2nd doped polysilicon and oxide deposition. The two polysilicon layers are in electrical contact.

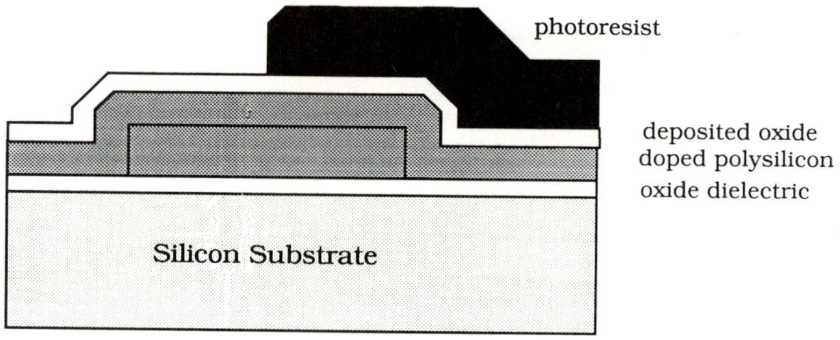

Figure 14. Photoresist application and patterning of spider device.

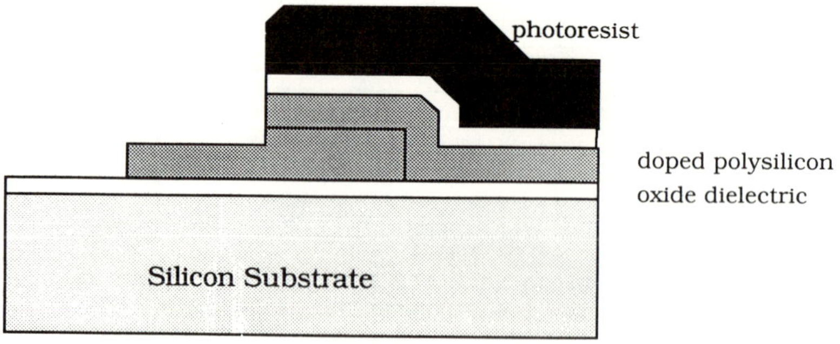

Figure 15. Oxide/2nd polysilicon etch.

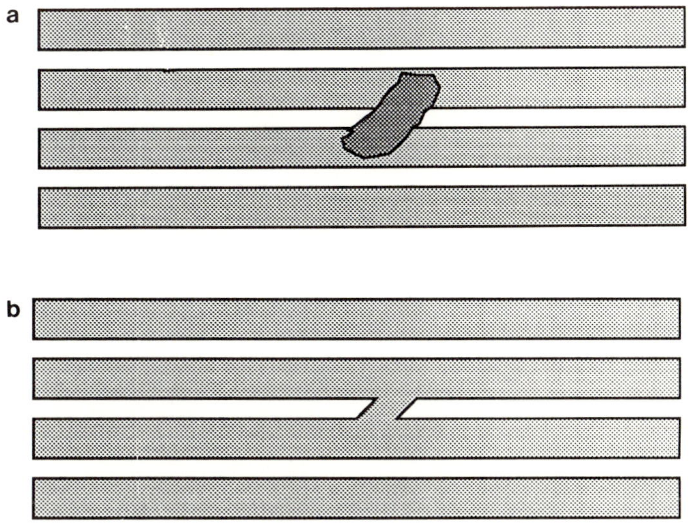

Figure 16. Resist defects are much larger than their defect artifacts and tend to be dark and stand out from the resist pattern. Bridge defect prior to resist strip (a). Bridge defect after resist strip (b).

Location of the Defect for Observation

In order for the particle to be studied, it must of course be located. Consequently, it is important to provide many test points on electrical defect monitors. This allows electrical testing to isolate the defect to a small area before the tedious microscopic search is begun. We have designed our structure to cover approximately 40% of the silicon wafer surface with active circuit area and still enable pinpointing of the defect to within a 0.0024 cm^2 area. Particles and resist defects are much larger than their defect artifacts so defects on the resist-testable device are much easier to locate than those on a standard monitor. The resist defects also tend to be dark and to stand out from the resist pattern which also makes inspection easier, see Figure 16.

Drawbacks of the Resist-testable Device

As shown by the process flow, the resist-testable spider device requires extra process steps, adding to its complexity. This is the main drawback of the device since one of the advantages of normal spider devices is their short processing time. Also, by adding a second layer, the interference of PLDs (Previous Layer Defects) is added. Fortunately, the first layer contains only large geometries (pads) and can be overexposed to reduce defect levels, see Figure 17.

Detection of Particulate Induced Lithographic Defects

For defect engineers lacking expensive automated inspection equipment, the resist-testable spider device may be the only way of viewing a large number of particle defects with the photoresist still intact. Human microscopic inspection is simply too slow and error prone to dependably capture defects at today's extremely low defect densities[5]. McCarthy et. al.[6] described a technique for detecting and

Figure 17. The first layer of the resist-testable spider device consists of very large, uncomplicated geometries.

discriminating photolithographic defects but removal of the resist layer was still necessary, leaving only artifacts for analysis. With their technique, a resist pattern is transferred to an aluminum conductive film using accelerated electrolytic etching of the metal with the developer solution. The etching equipment and its defect contribution are eliminated but the particulate evidence is lost.

EXPERIMENTAL SECTION

In order to verify the resist testable technique, twelve 6″ silicon wafers were processed using a 7 μm pitch spider device. Line and space sizes were 4.5 and 2.5 μm, respectively. The process flow shown previously was followed, resulting in two 4500 Å polysilicon depositions, a 1000 Å oxide deposition, two lithographic steps and two etch steps. An Ultratech 1100 stepper was used for resist patterning, followed by a metal-ion-free spray develop. Etching was done with a Lam Research Corporation plasma etcher. Following the second polysilicon etch, the resist patterned wafers were sent to electrical test and probed successfully.

After electrical test, 10 failed die/wafer on 3 of the wafers were visually inspected under a microscope. The defects seen were then cataloged according to type (i.e., - bridge with particle, extra pattern, scratch, etc.). The resist was removed using an oxygen plasma only, from the three wafers. The same defects were then re-inspected under a microscope and re-cataloged as before. Optical photographs were taken of each type of defect before and after resist removal and several SEMs were taken of defects after resist removal. SEMs were taken of defects on a fourth wafer with the photoresist still remaining.

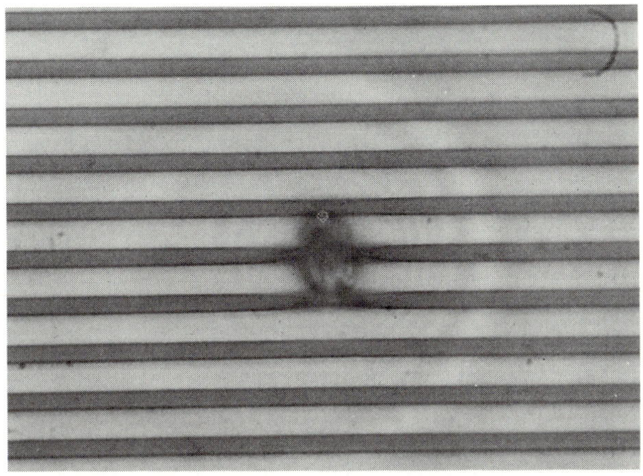

Figure 18. Particle induced electrically detected defect. The rings are due to resist pileup.

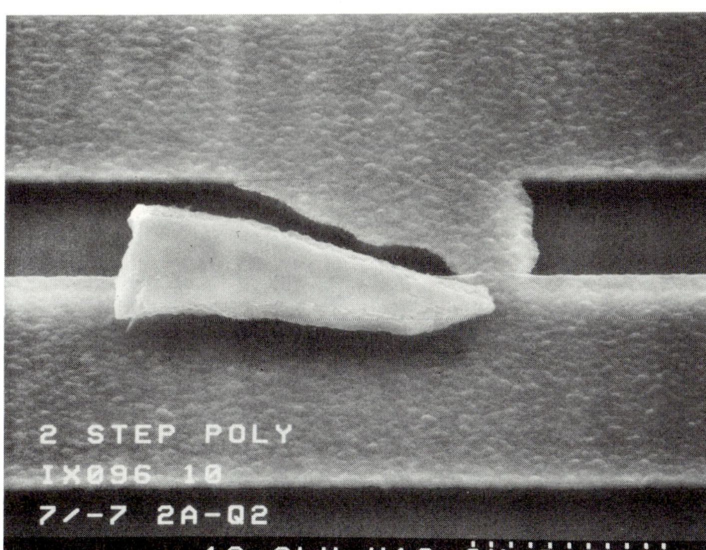

Figure 19. Scanning electron micrograph of a defect after photoresist removal using an oxygen plasma strip.

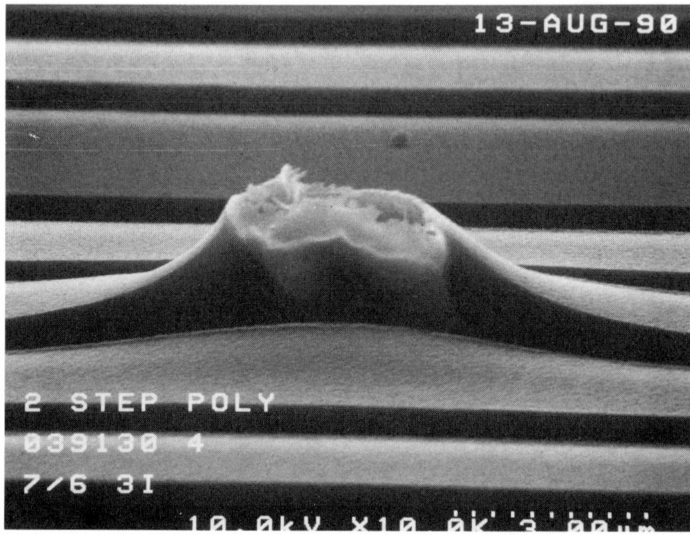

Figure 20. Scanning electron micrograph of defect. Resist has piled up around particle, resulting in a thick resist layer that couldn't be completely exposed and developed.

Figure 21. Scanning electron micrograph of a post-lithography defect.

RESULTS

All of the 30 electrically failing die that were inspected had visually apparent pattern defects prior to resist strip. Particles were present on 17 of the sites. After the plasma strip, 2 defects could no longer be located and 12 particles remained. Thus 5 particles were lost even with the most gentle resist strip possible and 2 of the defects were too small to be seen without the resist artifact. Figure 18 shows a particle induced defect that has been detected electrically. The rings due to resist pileup are clearly visible. Figure 19 shows a bridge defect after plasma resist strip with the suspect particle sitting next to it. Apparently the particle moved slightly and settled next to the artifact.

Figure 20 shows a particle induced lithography defect. Resist has piled up around the particle, resulting in a thick resist layer that could not be completely exposed and developed. Figure 21 is a post-lithography defect. The photoresist pattern is clearly delineated.

One interesting observation was that a defect we refer to as a microbridge appeared after resist strip from beneath a previous particle defect. Prior to this experiment, we had thought that microbridges were caused by very thin resist bridges, left over from poor exposure or develop. Figure 22 shows a particle induced lithography defect after resist strip. Not all of the resist surrounding the particle has been removed due to the greater thickness. The shape of the artifact defect is clearly visible and is similar to the artifact shown previously in Figure 2.

The defect density of the resist-testable spider was high compared to normal spider lots. This is probably due to the long throughput encountered while the process was debugged and to the additional process steps. Difficulties were encountered at the second polysilicon etch which etches partially into the dual pad structure. A confusing endpoint signature was created which had to be carefully controlled.

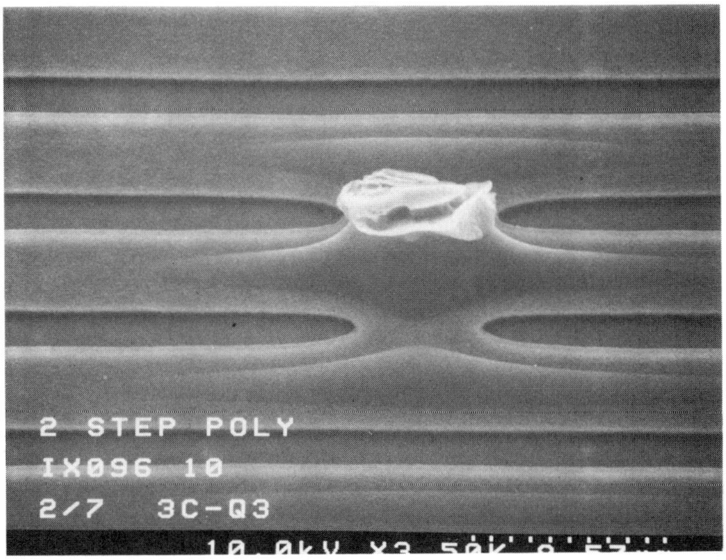

Figure 22. Particle induced lithography defect after resist strip.

CONCLUSIONS

A technique to enable testing of electrical defect monitors prior to resist strip has been demonstrated. Particulate induced lithographic defects have been electrically tested with the causative particle still present in the photoresist. The technique allows for the discrimination of defects to before, during or after photolithography. Particles that induced lithographic defects have been preserved for further analysis and study. The technique also allows for monitoring of electrical defects in other nonconductive films, such as deposited oxide, by transferring the defect to a conductive film below. Additional process steps are required for resist testability including a second polysilicon etch that requires careful endpoint control to preserve the test pad structure.

ACKNOWLEDGEMENTS

The authors wish to acknowledge the technical advice of Edward J. Bawolek, the SEM photography work done by Tony Chan and Cynthia Barutha, test programs written by Brian Shinseki, and the work of the Intel Defect PMT group for designing this test structure.

REFERENCES

1. Wayne G. Fisher, Particle interaction with integrated circuits , in "Particle Control for Semiconductor Manufacturing," Robert P. Donovan, et al, , SRC Technical Report T88105, Semiconductor Research Corp., P.O. Box 12053, Research Triangle Park, NC 27709, 1-3 (1988).

2. Michael E. Thomas and Wojciech Maly, Multilevel interconnect yield estimates using the double bridge test structure, V-MIC Conf., IEEE CH-2624-5/88, 232-233 (1988).

3. R.L. Dean, Implications of particulate contamination in E-beam lithography, in "Particles on Surfaces 2: Detection, Adhesion, and Removal," K.L. Mittal, editor, pp 253-265, Plenum Press, New York, (1989).

4. Christopher W. Teutsch and David C. Drain, Combining electrical defect monitors with automatic visual inspection systems, in "Integrated Circuit Metrology, Inspection, and Process Control," K.M. Monahan, ed., Proc. SPIE 1087, 189 (1989).

5. Christopher Radin, A defect reduction methodology for increased sort yield using automated defect inspection, in "Integrated Circuit Metrology, Inspection, and Process Control," K.M. Monahan, ed., Proc. SPIE 921, 224-233 (1988).

6. Anthony M. McCarthy, Wes Lukaszek, Chon-Cheung Fu, David H. Dameron and James D. Meindl, A Novel Technique for Detecting Lithographic Defects, IEEE Trans. Semiconductor Manuf., 1, No. 1, 10-14 (1988).

PART III. PARTICLE REDUCTION AND REMOVAL

REDUCING UNCERTAINTIES IN PARTICLE ADHESION AND REMOVAL MEASUREMENTS

Hwa-Chi Wang

American Air Liquide
Chicago Research Center
5230 S. East Avenue
Countryside, IL 60525

The sources of uncertainties in particle adhesion and removal studies are identified. Uncertainties can be reduced by proper selection of particle material, consistent cleaning of substrate surfaces, uniform deposition technique, and proper storage environment and duration. In addition, the method of applying removal forces has to be considered for data comparison. A standard set of experimental conditions is proposed to minimize these uncertainties.

INTRODUCTION

Particle adhesion to a surface and/or the removal from it by external forces are widely encountered in many industrial processes, such as spray coating, surface cleaning, microelectronic manufacturing, etc. Successful control of these industrial processes requires understanding the interfacial interactions between the particle and the surface at the area of contact. Several theoretical models have been proposed[1-4] in the literature to describe the fundamental contact phenomena. Modifications of the basic adhesion models have been made to include plastic deformation[5,6]. Extensions of the basic adhesion models have also been reported to deal with various issues such as particle bounce[7], particle reentrainment[8], contact charge transfer[9], and filtration efficiency in the thermal rebound regime[10].

In spite of the active research on the theoretical aspect of adhesion, experimental verifications of these proposed models are in general not conclusive. For example, the slope of critical velocity vs. particle diameter in the log-log plot can be used to indirectly verify the basic adhesion models. A compilation of the literature results[11], however, shows a range from -0.39 to -4.79. This wide variation may be a result of various mechanisms involved in adhesion; most of them are specific to the particular experimental conditions used. These results

Particles on Surfaces 3, Edited by K.L. Mittal
Plenum Press, New York, 1991

do not offer much insight to distinguish between various adhesion models but they underline the necessity of a specified experimental procedure under well-controlled conditions to tackle the complex adhesion problems.

We encountered the similar need for a standard set of experimental conditions when investigating the behavior of particle reentrainment under turbulent flow conditions. Extensive efforts were made in our laboratory to identify and reduce the sources of uncertainties. In this article, we summarize our efforts and propose a set of experimental conditions that can yield consistent and reproducible results. The behavior of particle reentrainment observed under the proposed conditions is reported elsewhere[12].

SOURCES OF EXPERIMENTAL UNCERTAINTIES

Quantitative measurements of particle-surface interactions at the interface involve (1) depositing known particles on a given surface, (2) removing particles from the surface by controlled external force and (3) correlating the removal results with the applied force. It is obvious that a well-defined **particle generation** technique is required to ensure consistent results. Equally important is the uniformity of **surface cleanliness** of the substrate. **Deposition technique**, often chosen arbitrarily among researchers, not only plays an important role in obtaining meaningful data but also in interpreting and comparing results. For convenience, it is a common practice to prepare a batch of particle-loaded substrates at the same time and to store them for subsequent removal experiments. The storage environment, especially the **moisture level** and the **storage duration**, can affect the results dramatically. Finally, the **method of applying removal force** determines how the particle is removed. Different **inceptive motions** such as lift-off, slide, or rotation have different onset removal forces, which must be accounted for when correlating experimental results.

The potential sources of uncertainties mentioned above, namely particle generation, surface cleanliness, deposition technique, storage environment and duration, and inceptive motions, will be discussed.

EXPERIMENTAL

Direct observation of particle removal in a turbulent channel flow is used to quantify the effects of various experimental parameters.

Figure 1 shows the experimental setup. Key components include a rectangular flow cell mounted on an optical microscope stage, a gas introduction system capable of reaching a desired flow rate within 0.001 seconds, and a camera / video recorder system to monitor particle removal with a time resolution of 0.01 seconds.

The flow introduction system consists of a needle valve, an air-actuated bellows valve, a solenoid valve, a flow meter, and an absolute filter. The needle valve and the flow meter are used to preset a desired flow rate within the range of 0-200 l/min. The solenoid valve is used to control the air actuated bellows valve, which provides a consistent and instantaneous means to start the flow. A cartridge filter was placed downstream of the flow meter and immediately upstream of the flow cell to provide particle-free gases.

The flow cell, as shown in Figure 1, is constructed of highly polished brass with gradually reducing entrance from a 2.5 cm ID tube to a 2x0.3 cm rectangular channel. A cleaned glass slide on the top of the flow cell provides the substrate for particle deposition and a port for viewing particle movement. Mounting the flow cell on the stage of the microscope allows precise positioning of the viewing area and gains a range of 80-500x magnification. A video camera / VCR / monitor system was connected to the microscope for viewing and recording in-situ particle release.

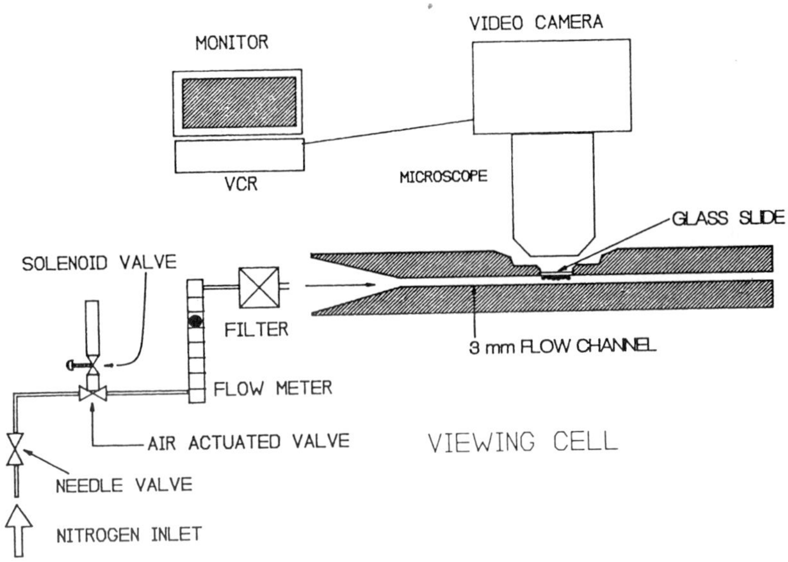

Figure 1. Schematic diagram of the experimental setup.

Experimental parameters include substrate cleaning, particle generation, particle deposition, sample storage duration, and the moisture level of the storage environment. Each of the above parameters is a complicated function of several independent variables. Trying to obtain a unique set of optimum conditions is not an easy task and may not be necessary. Instead, we tried to find a solution that yielded results within a 10% variation. Therefore, our approach consisted of (1) judicious selection of a set of experimental conditions, usually guided by experience and availability of materials and instruments, (2) preparation of 5 duplicates using the selected conditions, (3) measurement of particle removal in the turbulent channel flow at a fixed flow rate roughly corresponding to 50% removal, and (4) comparison of the percentage of removal among these 5 replicas. Most particle removal occurred in the first 2 minutes; no further particle removal was observed after 5 minutes in all experimental runs. Thus, particle removal was obtained by counting particles in a given view field before and 5 minutes after the flow was started. Four particle sizes were used: 3, 5.5, 6.5, and 9.5 μm.

REDUCING EXPERIMENTAL UNCERTAINTIES

Particle Material and Generation

Monodisperse spheres having uniform surfaces are preferred for the adhesion and removal experiments. Polystyrene latex (PSL) spheres, although popular and convenient to use, did not yield consistent results in our experiment. The particle surfaces may be contaminated by the stabilizing agents coexisting with PSL particles[13]. Particles generated by spray-drying of fluorescein / ammonium hydroxide solution form a consistent, featureless surface of ammonium fluorescein[14], and are thus a good candidate as the model particles. These particles have been successfully used for instrument calibration[15,16] and adhesion measurements[17,18]. To produce monodisperse particles of desired sizes from a solution, a vibrating orifice aerosol generator (TSI 3450) can be used, which offers a geometric standard deviation of the size distribution better than 1.05[19].

Substrate Surface Cleanliness

Microscope glass slides used as the substrate in our study were often contaminated in the manufacturing and packaging process. The percentage of particle removal from the same batch of uncleaned glass slides could vary from 0 to 80%.

Even more interesting was the particle motion on a contaminated glass slide. We deposited the model particles on an uncleaned glass slide and observed the behavior of particle removal in the turbulent channel flow. Some particles appeared to jump straight off the slide while others slid along the surface of the slide. It was found that grease-based contaminants were the cause of particle sliding. After removing the contaminants, no particle sliding was observed. All particles removed appeared to jump straight off the surface. (The use of spherical particles does not allow us to distinguish between rotation and lift-off. However, experiment with irregular particles shows evidence of rotation.)

The commonly used methods of ultrasonic cleaning of the substrates in clean DI water and in clean solvent such as IPA or Freon can effectively remove the surface contaminants. It was found, however, that the procedure of cleaning and the handling of the cleaned substrate glass are critical. Residues from the cleaning solution and contamination during handling must be avoided in the cleaning process.

Deposition Technique

Inertial impaction is a commonly used technique to collect particles on a substrate surface for later analysis. The impactor used in our study had a D50, the particle diameter corresponding to 50% impaction efficiency, of 3 μm. Particles used in our study were mostly greater than 5 μm and had practically 100% efficiency. Removal experiments conducted on the impaction deposits showed a dependence upon the particle's location. Particles closer to the center of the impaction deposits were less likely to be removed. In certain cases, we observed a complete particle removal except those immediately under the nozzle opening. This can be explained by the differences in contact area, which depends on the local impact velocity during deposition. Figure 2 shows that particles closer to the center of impaction deposits have larger contact areas and thus are more difficult to remove. Contact area as a function of impact velocity was previously treated by Wang and John[9]. It can be easily shown that the elastic limit is exceeded in our situation and thus the contact area increases linearly with the impact velocity.

Uniform deposition velocity is essential to ensure uniform contact area and thus uniform adhesion force for each particle. This can be achieved by gravitational settling of particles through stagnant air onto the substrate surfaces or by electrostatic deposition of uniformly charged particles in a uniform electrical field. We adopted the electrostatic deposition method because of the availability of a commercial instrument (TSI 3100), which provides particle charging and subsequent precipitation in a pulsed electric field. Particle removal results were consistent (within 5%) for glass slides prepared by electrostatic deposition.

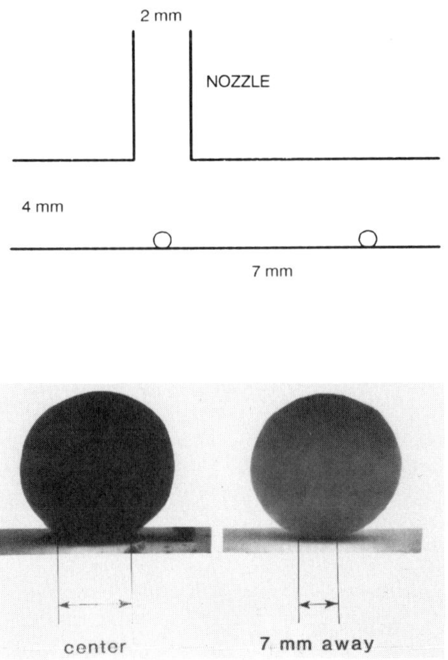

Figure 2. SEM photographs of impacted particles showing different contact areas at different impact locations.

Storage Duration

Continuous deformation due to material creep may occur for soft PSL particles sitting on substrate surfaces[2]. As a result, the contact area and thus the adhesion force may increase with storage duration. Besides material creep, contaminants in the storage environment may build up around the particle-surface interface. Therefore the allowable storage duration has to be specified for a given storage condition. A clear plastic box (Figure 3) continuously purged with dry, filtered, high-purity nitrogen was used for storage in our study.

Experiments were performed under identical conditions except for different storage duration. Three particle sizes were tested. To show the effect of storage duration, the results for each particle size were normalized by the corresponding removal after one day's storage. Figure 4 shows the normalized fraction of removal versus storage duration. For ammonium fluorescein particles, less than 10% variation was obtained within two weeks. For longer storage duration, significantly lower removal was observed, i.e., below 30% of its original value after 40 days. Similar experiments[20] for PSL particles stored in a less stringent environment were reported to have a faster decay in removal versus storage duration. This confirms the necessity of a well-controlled storage environment.

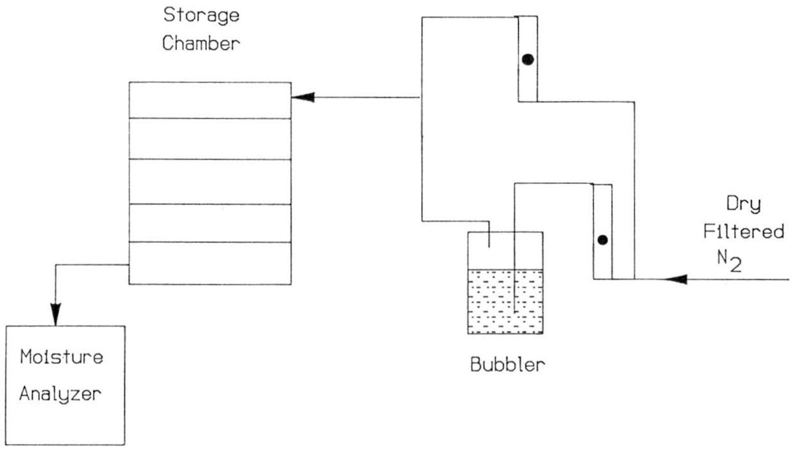

Figure 3. Experimental setup for testing storage environment.

Moisture Level

It is known that moisture can modify the surface condition and alter the adhesion force for a given particle-surface system. As shown in Figure 3, the moisture level in the storage environment is controlled by mixing dry nitrogen with various amounts of a wet N_2 stream produced by a water bubbler. The moisture level is continuously monitored by a moisture analyzer. The range of moisture studied is from 0.2 ppm to 20,000 ppm. Figure 5 shows particle removal versus moisture concentration for a 24-hour storage period. A sharp drop in particle removal is observed at around 1000 ppm which corresponds to 4% RH at 20° C. For moisture level below 500 ppm, less than 10% variation is found.

Another interesting observation indicated that once the particle-loaded substrate was exposed to high moisture condition, subsequent lowering of moisture has no effect on particle removal. It leads to a speculation that a ring of water just outside the particle-surface contact must be formed at high moisture level by condensation. Subsequent drying of this ring of water form a strong bridge between the particle and surface.

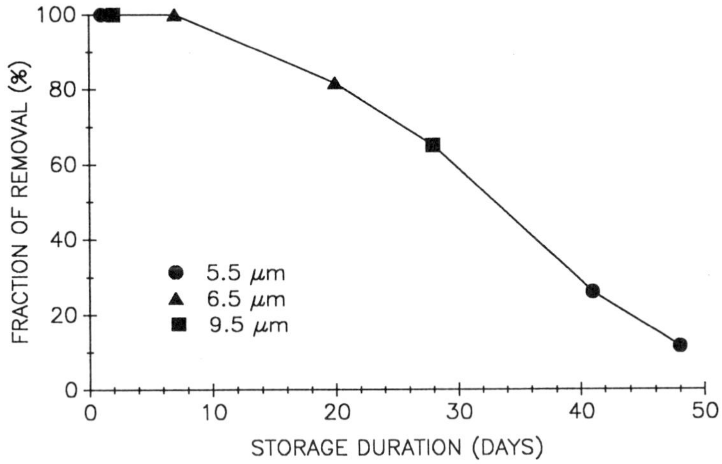

Figure 4. Effect of storage duration on particle removal.

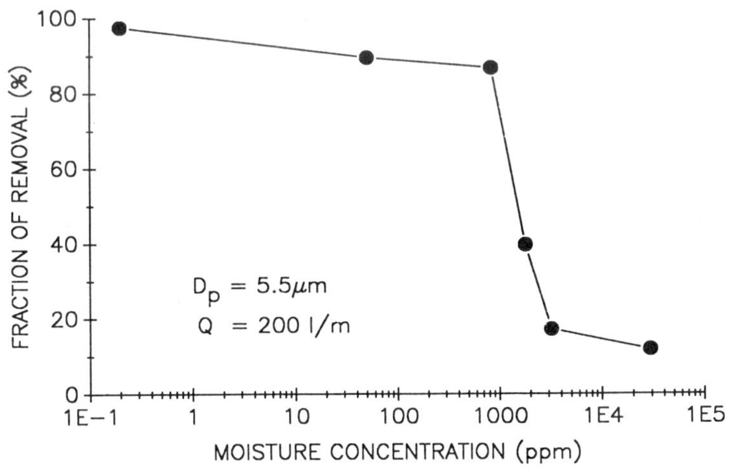

Figure 5. Effect of moisture level on particle removal.

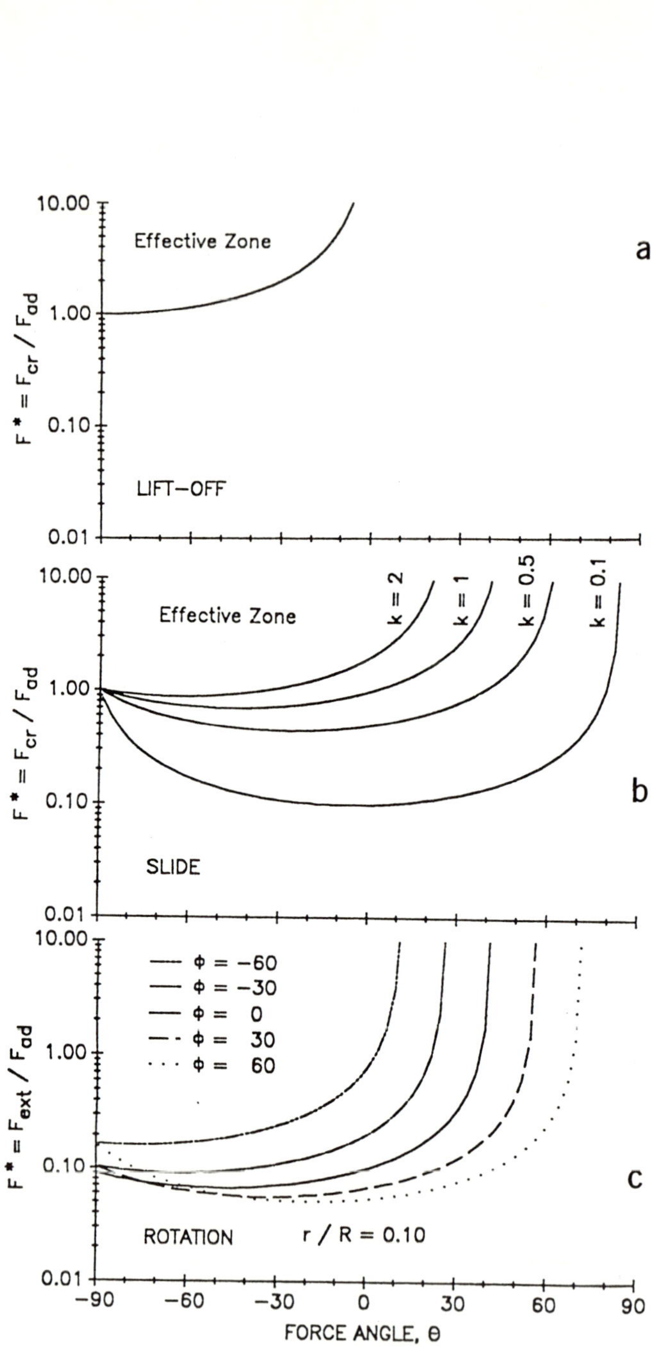

Figure 6. External force required for particle removal as a function of the direction of the applied external force for (a) lift-off, (b) slide, and (c) rotation.

Effects of Inceptive Motion

Removal of particles from a surface can be initiated by direct lift-off, slide, or rotation. The corresponding critical removal force F_{cr}, defined as the applied external force at the onset of particle movement, would change accordingly. Wang[21] proposed a simple model to calculate the ratio of the critical removal force to adhesion force, F_{cr}/F_{ad}, as a function of the removal force angle, being -90° when the particle is pulled perpendicularly off the surface and 90° when the particle is pushed toward the surface.

Typical results are shown in Figure 6a for lift-off, 6b for slide at various coefficients of friction k, and 6c for rotation. For detailed description, the reader should refer to Wang[21]. In this example, it can be seen that an external force twice the adhesion force is necessary to pull the particle off the surface when the removal force angle is 30°. Under the same conditions, the external force is approximately half of the adhesion force for slide (k=0.5) and is only 7% of the adhesion force for rotation. How the external force is applied to remove the particles is therefore extremely critical and must be taken into account when comparing results obtained from different sources. For example, the adhesion force measured for particles deposited on the top of a spinning disc is definitely smaller than that on the edge of the disc[22].

SUMMARY

Sources of uncertainties in particle adhesion and removal measurements were identified. Methods of reducing these uncertainties were discussed. A set of experimental conditions that produces results within 10% variation is given as follows:

Particle Material:	Ammonium Fluorescein
Particle Generation:	Vibrating Orifice Generator
Surface Cleaning:	DI Water + IPA or Freon in Ultrasonic Bath
Deposition Technique:	Sedimentation or Electrostatic Deposition with Uniform Velocity
Storage Duration:	Less Than 2 Weeks
Moisture Level:	Less than 500 ppm

For inter-comparison of experimental data, method of removal must be considered. A simple model of Wang[21] can be used to extrapolate the data and compare at the same basis.

ACKNOWLEDGMENT

The author wishes to thank Dr. T. Niida for designing the experiment, Dr. H. Y. Wen for involvement in the initial stage, and Mr. R. Udischas for performing the experiments.

REFERENCES

1. R. S. Bradley, Phil. Mag. 13, 853 (1932).
2. B. Dahneke, J. Colloid Interface Sci., 40, 1-13 (1972).
3. K. L. Johnson, K. Kendall, and A. D. Roberts, Proc. Royal Soc. A, 324, 301-313 (1971).

4. B. V. Derjaguin, V. M. Muller, and Y. P. Toporov, J. Colloid Interface Sci., 53, 314-326 (1975).
5. M. Fichman and D. Pnueli, J. Appl. Mechanics, 52, 105-108 (1985).
6. D. Maugis and H. M. Pollock, Acta Metall., 32, 1323-1334 (1984).
7. L. N. Rogers and J. Reed, J. Phys. D, 17, 677-689 (1984).
8. M. W. Reeks, J. Reed, and D. Hall, J. Phys. D, 21, 574-589 (1988).
9. H. C. Wang and W. John, J. Aerosol Sci., 19, 399-411 (1988).
10 H. C. Wang and G. Kasper, J. Aerosol Sci., 22, 31-41 (1991).
11. S. Wall, W. John, H. C. Wang, and S. L. Goren, Aerosol Sci. Technol., 12, 926-946 (1990).
12. H. C. Wang, B. Jurcik, R. Udischas, and G. Kasper, Time dependence of particle reentrainment in turbulent flow, J. Aerosol Sci., in press.
13. B. Dahneke, J. Colloid Interface Sci., 51, 58-65 (1975).
14. R. W. Vanderpool and K. L. Rubow, TSI Quarterly, 3-6 (1984).
15. H. C. Wang and W. John, Aerosol Sci. Technol., 7, 285-299 (1987).
16. H. C. Wang and W. John, Aerosol Sci. Technol., 8, 157-172 (1988).
17. H. C. Wang and W. John, in "Particles on Surfaces 1:Detection, Adhesion, and Removal" K. L. Mittal, Editor, pp. 211-224, Plenum Press, New York, 1988.
18. S. Wall, W. John, and S. L. Goren, in "Particle on Surface 2: Detection, Adhesion, and Removal" K. L. Mittal, Editor, pp. 19-34, Plenum press, New York, 1989.
19. R. N. Berglund and B. Y. H. Liu, Environ. Sci. Technol., 7, 147-153 (1973).
20. K. C. Hsieh and B. Y. H. Liu, AAAR Annual Meeting Abstract, 66 (1990).
21. H. C. Wang, Aerosol Sci. Technol., 13, 386-393 (1990).
22. C. Kanaoka, H. Emi, and N. Kikukawa, Japan J. Powder Technol., 24, 233-240 (1987).

PARTICULATE AND DEFECT REDUCTION STRATEGIES FOR SEMICONDUCTOR DEVICES: TOOLS AND METHODOLOGIES

Robyn Coleman

KLA Instruments Corporation
3530 Bassett Street
Santa Clara, California 95052

The reduction and control of defect densities is critical to the successful manufacture of semiconductor devices. Automated wafer inspection tools, detecting smaller defects and providing more consistent results than previous techniques, are being used more and more in this effort. In order to use these tools effectively, a structured approach must be taken. In addition, understanding the performance of an automated inspection tool is key to correctly interpreting the information it provides.

This paper describes a defect reduction methodology employing an automated wafer inspection tool which has been successfully applied by several semiconductor manufacturers. The critical aspects of applying such a program are discussed in a case study format.

The second portion of this paper presents possible barriers to a successful defect reduction program. The organizational elements of a successful program are discussed. A methodology for evaluating the performance of an automated wafer inspection tool is outlined. The design of a test vehicle useful for tool characterization is described. Parameters which have been shown to have significant impact on the performance of wafer inspection tools are identified, and data illustrating the methodology are presented.

INTRODUCTION

As design rules decrease, devices are impacted by smaller defects. In addition, lower defect densities must be achieved to obtain comparable yields as illustrated in figure 1. Smaller defects with lower densities are increasingly difficult to detect with manual operator inspection[1]. These factors have driven an increase in the use of automated inspection tools.

The first automated inspection tools used light scattered off the surface of bare wafers to detect large surface particles. Light scattering technology has also been extended to detect particles on patterned wafers. Automated tools using digital image processing are used to

Particles on Surfaces 3, Edited by K.L. Mittal
Plenum Press, New York, 1991

detect particulates as well as process defects such as film residue, lifting, or bridging. Papers have been published describing the applications and benefits of these different types of tools[2,3,4]. Automated inspection tools, when used in a structured and well planned defect reduction effort, have been instrumental in assisting engineers to improve device yield.

DEFECT REDUCTION METHODOLOGY

A yield curve for a typical semiconductor device is shown in figure 2. During the early stages of device production, emphasis is placed on identifying and removing baseline defects so that yield can be increased. Once yield is at an acceptable level, the process is often monitored for excursions in defect density. Monitoring defect levels in-line rather than relying solely on end-of-line data minimizes the effects of excursions by increasing the timeliness and accuracy of feedback[6].

A methodology which has been used successfully by several firms to reduce both baseline defect density as well as control process excursions is depicted in figure 3. The basic steps are to identify and prioritize existing problems, quantify and understand a problem, isolate its source, eliminate its cause, verify that it is fixed, and finally, document the fix. The process of controlling and reducing defect levels is cyclical. Omitting a step in the cycle may jeopardize results.

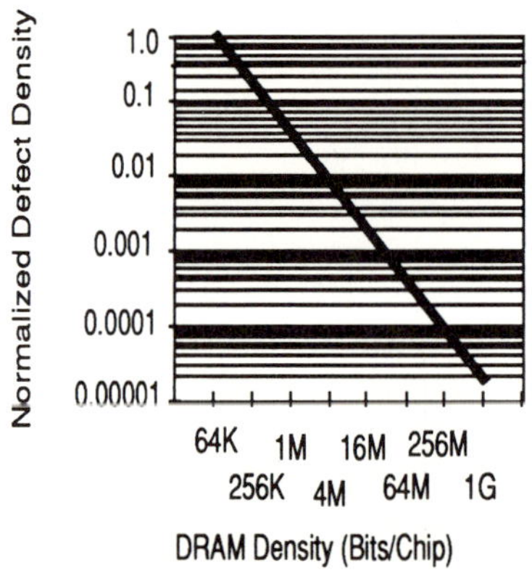

Figure 1. Defect density versus device density[5].

Identify and Prioritize Defect Problems

To begin a defect reduction effort, problems must be identified and prioritized. Several means useful for identifying defect problems may already be implemented in the wafer fabrication area (fab). For example, VLSI Inc. used data collected in a manual defect analysis program which had been running for several years to identify the contact layer as their largest source of yield limiting defects[7]. A fab at AT&T used results from an electrical test structure designed to detect opens and shorts in meander and comb patterns to identify the silicide gate layer as a source of significant yield loss[8]. HP used data derived from analysis of Random Access Memory cells along with in-process monitors to identify yield impacting layers. Low yield analysis[1], where the average number of failures at sort is compared to a defect mechanism, may also be used.

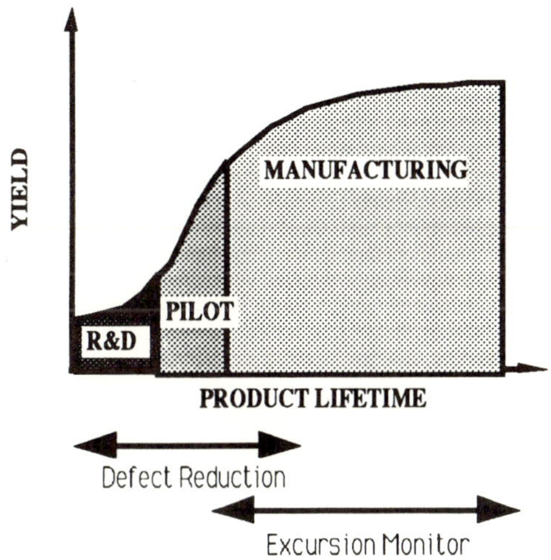

Figure 2. Product lifecycle of a typical semiconductor device.

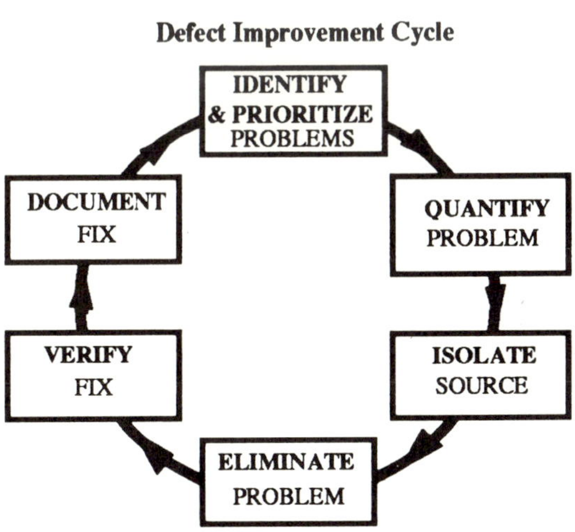

Figure 3. Critical steps of defect reduction methodology.

Quantify the Problem

Once a process level with significant yield impact has been identified, the next step is to quantify the defect problem by inspecting and reviewing random lots processed through that level. These inspections reveal what the defects look like, how many and what types of defects are present, and how the defect types are distributed from lot-to-lot, wafer-to-wafer, and die-to-die.

Understanding the defect type distribution allows the development of an effective sampling plan. For example, suppose inspection revealed that 25% of the wafers and 100% of the die on those wafers contained the defect of interest. A good sampling plan in this case would include several lots, a large number of wafers per lot, and a small area of each wafer. Noting if a defect tends to cluster around the edges of the wafer or is distributed randomly across the wafer would indicate where to focus the inspection within the wafer.

Data analysis tools are extremely helpful in compiling defect information into useful formats. Figure 4 shows a trend chart illustrating the lot-to-lot variation of defect density found by VLSI while quantifying a problem[7]. The Pareto chart in figure 5 shows the relative density of defects for all lots inspected[7]. Other types of analysis include various graphic representations of the spatial distribution of defects across the wafer, and the frequency of defects of various sizes and types.

In this type of defect reduction application, many defects should be reviewed and classified. Establishing a baseline of densities based on defect types allows the engineer to focus on a specific defect problem. The classification scheme used, however, is important to the integrity of results. It should be logical, production worthy, flexible, and accurate. One example of a flexible classification scheme is shown in Table I. Sometimes a different scheme is appropriate. For example, Fujitsu classified defects by appearance as 'killer' and 'non-killer'[9]. Killer defects were those causing circuit failure and non-killers were all others. Generally, defects should be classified by appearance not by cause. Assigning a cause to a defect requires experience and process knowledge and tends to be subjective, jeopardizing the accuracy of results.

Effective training and procedures are critical to the accuracy of defect classification. A group in Motorola had operators record classifications on a VCR with a microphone[10]. The engineer was then able to review sixteen hours of inspections in thirty minutes. Any errors in classification were discussed with the operators immediately, resulting in a high degree of classification accuracy. Also, the engineer's familiarity with the defect problems aided him/her in providing timely feedback on the problems being found.

Isolate the Source

Once a defect problem has been quantified and baseline data have been accumulated, further analysis is often necessary to isolate the defect source to a specific process step within the layer. Some people find this the most difficult part of defect elimination[9]. A Defect Source Analysis (DSA) experiment, also called Partitioning or Sequential Analysis, can be an effective way of identifying a defect source. In a DSA experiment the same area on the same wafer is inspected at several steps in the process. After inspection, defects are reviewed and classified according to type. The inspection results are analyzed by overlaying the location of defects found at each step. The density of unique and common defects for each process step can then be determined, indicating how each process step contributes

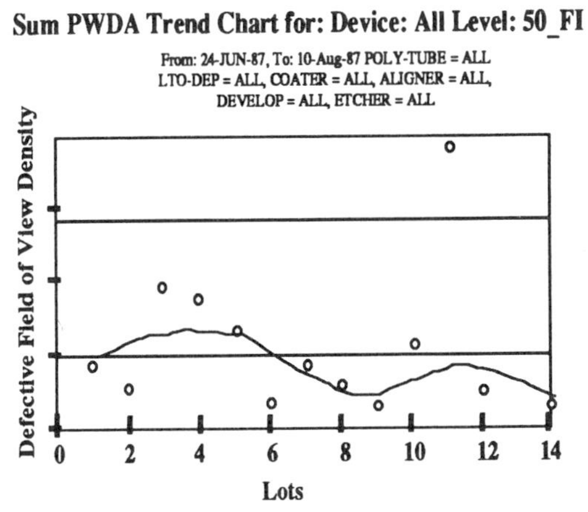

Figure 4. Trend chart of defects detected with an automated inspection tool while quantifying a problem. Data taken by VLSI[7].

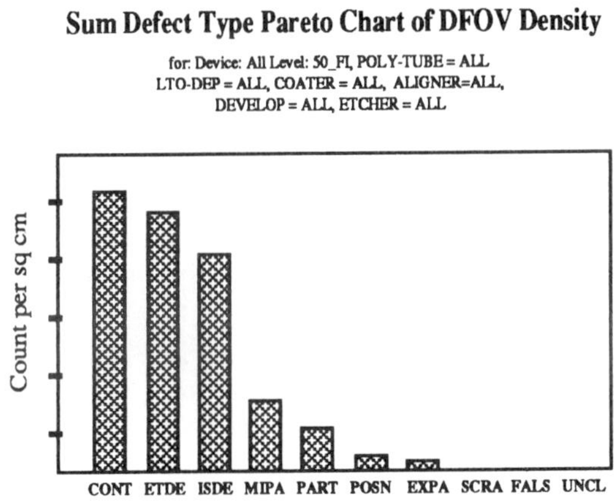

Figure 5. Pareto chart showing relative density of defects detected by an automated inspection system while quantifying a problem. Data taken by VLSI[7].

Table I. Sample Classification Scheme.

CLASS	DEFECT	
10	Particle :	small
11		medium
12		large
20	Scratch	
30	Extra Pattern:	bridging
31		extension
32		isolated
40	Missing Pattern:	line break
41		edge intrusion
42		isolated

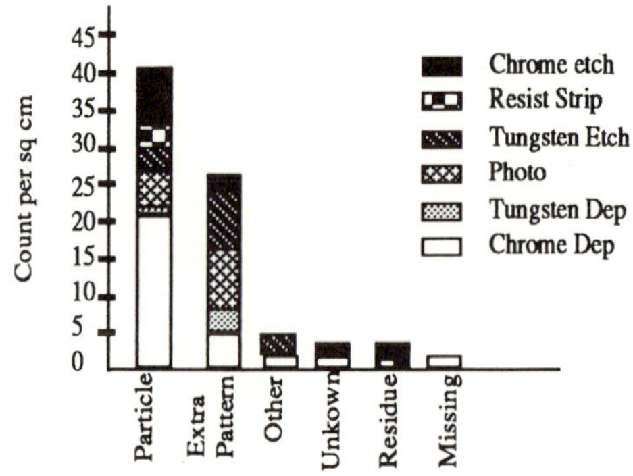

Figure 6. 'Type by Source' Pareto Chart showing sources of defects contributed by metal 1 processing. Data taken by HP[9].

to the defect density. The positional accuracy with which defects are located is critical. Using an automated inspection tool and analysis software simplifies this procedure.

DSA has been used by several companies to isolate defect problems. For example, after HP quantified a defect problem on a critical metal 1 layer, a DSA experiment was set up to isolate the source of the defects[6]. A specific group of wafers was inspected at seven steps in the metal 1 process. Defects were classified after each inspection. A software package called Defect Source Analysis (DSA) available on the KLA Analysis Station was used to analyze the data. Analysis revealed that the most common defect (particles) was generated at chromium underlayer deposition as shown in the 'Type by Source' Pareto Chart in figure 6. Photo and tungsten etch were also found to contribute greatly to extra pattern defects which were the next most common defect at metal 1. With this information, the group could focus efforts on reducing these defects at their sources.

Fujitsu also used a partitioning technique to track down a targeted defect type. A group of wafers was inspected before and after four key process steps. The Added Defect Pareto chart in figure 7 illustrates the contribution of each process step to total defects. Figure 8 shows the contribution of each process step to 'killer' defects[9]. By taking pictures of the defects during this experiment, engineers were able to observe the changing appearance of defects as they moved through the process. They discovered that a particle added before process 1 (aluminum deposition) caused a contact failure after process 4 (aluminum etch/strip). This defect was later eliminated by using EPMA analysis to identify the particles as stainless steel from their etch chambers.

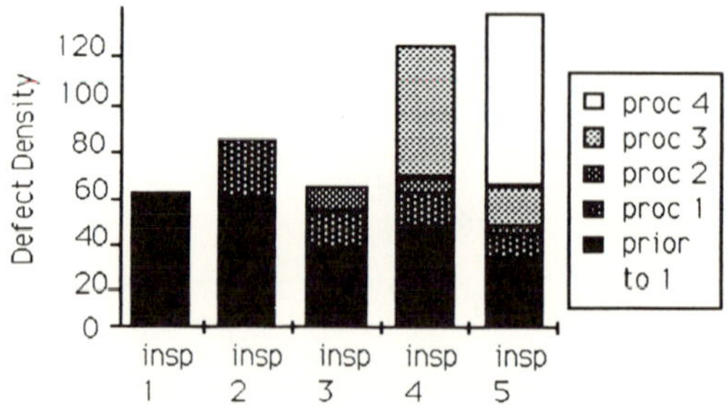

Figure 7. Added Defect Pareto Chart showing defects added at each process step. Data taken by Fujitsu[10].

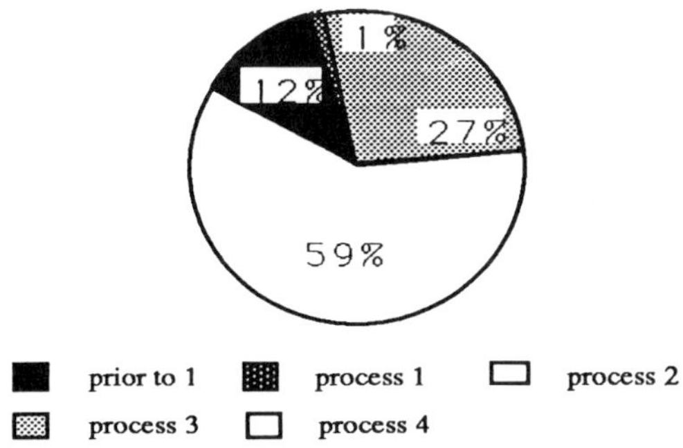

Figure 8. Pie chart showing proportion of killer defects added at each process step. Data taken by Fujitsu[10].

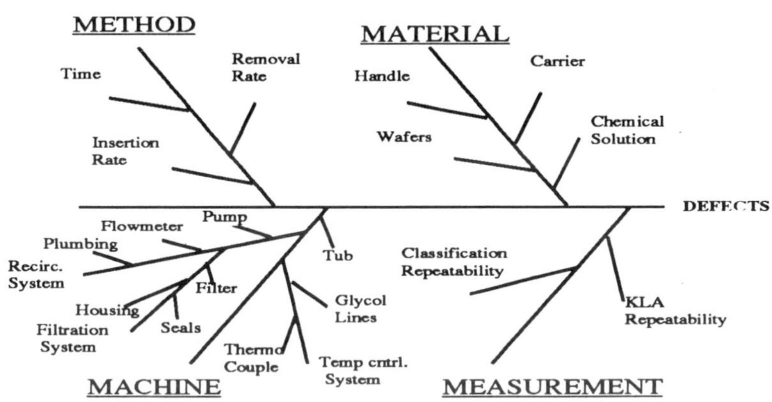

Figure 9. Ishikawa diagram of probable causes drawn up by AT&T to eliminate source of Teflon particle.

Eliminate the Source

Once the process step causing a defect has been identified, the cause must be determined and eliminated. The basic steps in eliminating the source of a defect include characterizing the problem, generating a list of probable causes, identifying and evaluating the validity of the solution (Does statistics support conclusions?) and implementing a fix (a change in process equipment, procedures, or materials).

AT&T followed this procedure to eliminate a residual polysilicon defect which had been isolated to pre-silicide cleaning[8]. The problem was characterized by using a Scanning Auger Microprobe to identify the particle of interest as some type of Teflon. With this information and their understanding of the process, they generated an Ishikawa[11] diagram as shown in figure 9. This diagram outlines possible sources of the defect. Next, they sought to identify the actual cause of the Teflon particle. Multi-variable Taguchi style experiments and single factor experiments may be useful for finding the root cause of a defect problem. AT&T chose to use single factor split lot experiments which were considered simpler to execute in a production environment. Different variables were changed to find a statistically significant change in the variation of the mean defect density from wafer to wafer within a lot as observed during the silicide process audit where the problem was quantified. After several attempts to isolate the cause in the Machine branch of the diagram, the problem was attributed to Teflon carriers and cleaning procedures. A split lot experiment replicated three times showed that cleaning carriers in a sulfuric acid solution reduced the level of particles at the pre-silicide cleaning step. Teflon particles from carriers were being redeposited from the edges of wafers to other parts of the wafer during subsequent cleaning steps. The proposed solution was replacement of Tefloncassettes with polypropylene carriers and a new carrier cleaning system.

An important aspect of eliminating a defect is capitalizing on the knowledge of people familiar with the process and the problem. By assembling a group of experts to brainstorm on a list of probable causes, a defect problem can often be narrowed down to a few possibilities. Ranking the remaining possibilities in order of likelihood may help lead to the quickest solution.

Verify and Document the Fix

Once a problem has been identified and fixed, verification is needed. A point of source monitor is often used as a quick and simple method to ensure that a problem has not returned. This type of monitor may be a manual inspection, a bare wafer monitor, or a quick automated inspection. Although a point of use monitor may not always be necessary, it is worthwhile for those defect problems which could have a large impact on yield, should they return.

An example of the importance of verifying a fix is illustrated by the residual polysilicon problem tracked down at AT&T. Through experimentation, the engineers thought they had traced the defect problem to the type of filter being used in their recirculating chemical sink[11]. Replacing this filter showed a significant but temporary decrease in defect density. Over time, high defect densities returned, indicating the problem had not been fixed. Had they not continued to monitor for the residual polysilicon defect, their yield would have been at risk.

Once implemented and verified, the solution to a defect problem should be documented. This information can be used to solve future problems. The data can also be useful for training people on the defect reduction program on-site or at other fab sites.

PROGRAM ORGANIZATION AND TOOL CHARACTERIZATION

Risks and Limitations exist with both the tools and administration of a defect reduction program. An appropriate organizational structure and a good understanding of the tools being used for defect reduction are instrumental in overcoming these barriers to a successful program. The following sections discuss technical and organizational concerns related to running a defect reduction program. An example of a successful organizational structure is presented and discussed. A methodology for characterizing the performance of automated inspection tools is presented along with characterization data from a KLA inspection tool.

Program Organization

There are many ways to structure an organization for effective defect reduction and control; however, many successful programs have certain elements in common. Cooperation and communication among the process groups is critical along with a sense of responsibility and priority for defect reduction at all levels of the organization.

Figure 10 displays the organizational chart of a company with a highly successful defect reduction program. This company maintains a separate Defect Reduction Group which is responsible for maintaining and reducing defect levels. They manage all automated, visual, and particle monitor inspections for the fab and have expertise with defect reduction techniques and methodologies. By having the Defect Reduction Group report in to the Engineering Manager, a common goal is maintained among all individuals.

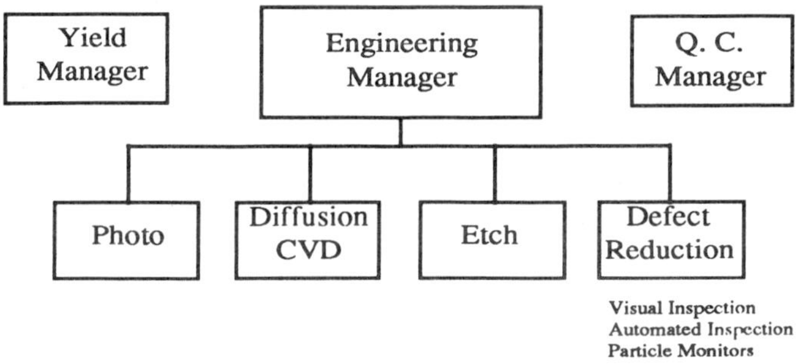

Figure 10. Organizational chart of a company with a successful defect reduction program.

Supervisors from each group, including Yield, attend a weekly meeting held by the Defect Reduction Group. Here, defect results from the week are discussed. The group, representing experience from all areas of the fab, brainstorms on possible sources of and solutions to defect problems they are seeing. Issues are prioritized and responsibilities assigned. A member of the Defect Reduction Group is teamed with members of the selected process groups to tackle each defect problem in order of priority. Both the process group and the defect reduction group are accountable for results.

This type of organization works well because there is cooperation between the functional groups and it has been communicated from management that defect reduction is an important priority. Resources have been allotted in terms of equipment, personnel, and priority to carry out defect reduction efforts.

Tool Characterization

Organizational structure and support is a significant factor in the success of a defect reduction program. Also important is assessing the performance characteristics of the tools being used. As with any type of measurement tool, it is risky to interpret results without understanding performance characteristics.

Defect capture rate, detection sensitivity, false event rate, and tool stability are parameters which have been found useful for quantifying the performance of automated inspection tools. Defect capture rate is the probability with which a defect of a specific type and size will be captured. Detection sensitivity is defined as the smallest defect which can be captured with a specified probability or capture rate. False event rate consists of two categories, nuisance and false detections. Nuisance defects refer to the capture of defects which are not of interest such as metal or polysilicon grains. False defects are those which are not visible upon review. Stability characterizes the consistency with which the tool performs.

Together, capture rate, sensitivity, false event rate, and stability quantify the performance of an automated inspection system. These parameters are interrelated and therefore should be considered simultaneously. For example, many tools use a threshold adjustment to increase sensitivity, thereby increasing the capture rate of small defects. Increasing sensitivity, however, may also increase the detection of artifacts such as grain, thereby increasing the false event rate. Depending on the user's tolerance to a higher false event rate, this may or may not be acceptable. Sensitivity, therefore, should be defined in conjunction with an acceptable false event rate.

Capture rate, sensitivity, false event rate, and stability may be evaluated with a standard test vehicle which is described in this section. The considerations involved in designing such a wafer are presented along with a methodology for using it (or any other test vehicle) for evaluating the performance of an automated inspection tool.

Test Vehicle. When characterizing the performance of a wafer inspection tool it is important to use a test vehicle which provides a thorough and meaningful basis for evaluation. Several companies have designed patterns expressly for this application. The DSW75 test pattern described here was designed by KLA for qualifying advanced inspections tools prior to and after installation in the wafer fabrication area. This test pattern is also used at KLA as a benchmark for evaluating the impact of new engineering developments on tool performance.

The DSW75 reticle layout consists of two dice, a defective die and a non-defective die. Each die is identical except for the defects programmed into the defective die. Within each die are four sections of repeating geometry representing 1 µm, 0.75 µm, 0.6 µm, and 0.5 µm design rules respectively. These geometries represent those characteristic of advanced 4M and 16M DRAM devices. Figure 11 illustrates the 12 different types of defects which are programmed into the test pattern in each defective die. Defect design rules range from 0.1 µm to 1.5 µm. The DSW75 test pattern is typically etched into 1,000 Å of thermally grown oxide although other materials may be used.

This test pattern was designed to represent advanced products being inspected by automated wafer inspection tools. Defects of several types and sizes were programmed into the test pattern to facilitate thorough evaluation of capture rate performance. Defects were distributed throughout the test pattern in known, easily located sites so that their existence could be quickly verified by visual inspection. If defect capture is evaluated without this verification, it is impossible to know whether defects were not resolved during printing or whether the inspection tool was not sensitive enough to capture them.

Actual product wafers may also be used to characterize system performance. However, these wafers often contain a poor distribution of defect types and sizes for thorough evaluation. Also, because defects are not in known locations, it is difficult to discover what types of defects the system has trouble detecting.

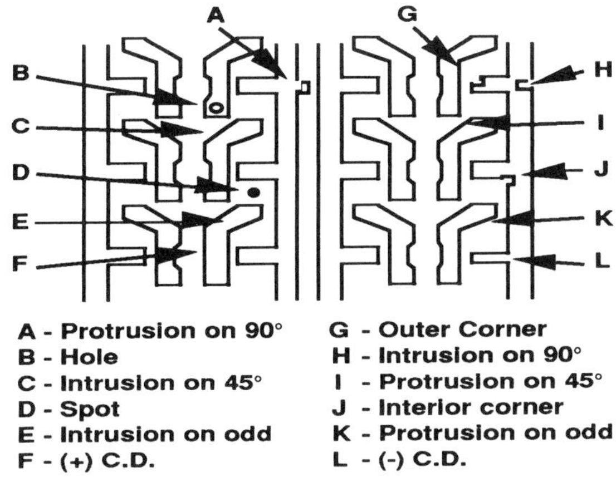

Figure 11. Defect types programmed into DSW75 test pattern.

Ideally, a multilayer test vehicle with programmed defects would be employed to better represent the complexities presented by multiple process levels. Such a wafer is currently under evaluation at KLA.

Capture Rate and Sensitivity. Capture rate, expressed as a percentage, is the probability with which defects of a certain type or size will be detected by an inspection tool. To evaluate capture rate, an area containing defects of known size, type, and location is inspected multiple times. A minimum of 20 cycles is suggested to improve the statistical validity of the data[2]. Capture rate is computed for each defect by dividing the total number of times it is captured by the total number of inspection cycles performed. To attain an overall understanding of the system's capture rate performance, the capture rate of each defect may then be plotted against its minimum dimension as shown in figure 12a. These data were generated using the KLA2029 wafer inspection system to inspect 0.75 μm design rule geometries on the DSW75. False event rate was < 1%. Because several defects in each size range were inspected, each point on the scatter plot represents many defects. Therefore, it is useful to draw a mean capture rate curve as shown in figure 12b.

Sensitivity, defined as the minimum defect size detected with a given mean capture rate, can be determined from the mean capture rate curve. For example, figure 12b indicates 0.2 μm sensitivity at a 90% capture. In other words, 90% of the 0.2 μm defects were captured. This capture rate experiment represents an average of all defect types available on the DSW standard wafer. Sensitivity can also be determined in a similar manner for a specific defect type.

Understanding the sensitivity and capture rate characteristics of a tool aids the engineer in evaluating its utility for different applications. For example, suppose a tool is being used in defect source analysis where the defects range in size from 0.2 to 0.5 μm at their source. If the inspection tool's mean capture rate for such defects is ≤ 50%, then identification of the source may be very difficult.

False Event Rate. False event rate may be divided into two categories, nuisance detections and false detections. Nuisance detections may be caused by occurrences such as grain or contrast variation which exist but are not considered defects. False detections are detections of defects where no visible defect is apparent. False event rate is calculated for a KLA

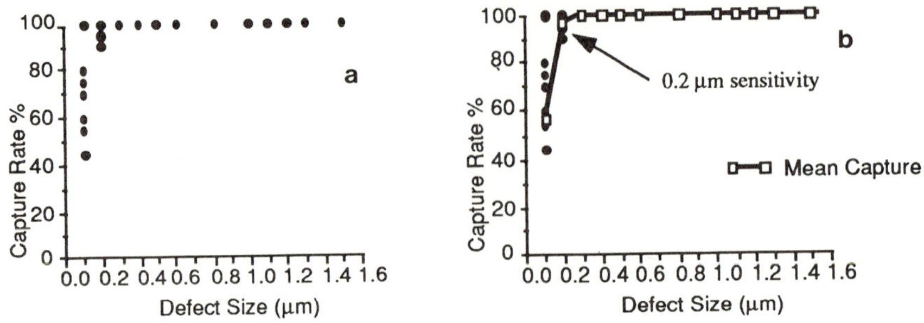

Figure 12. Plot of defect capture rate versus defect size. Inspection done on 0.75µm design rule section of DSW75 using KLA2029 inspection system (false event rate < 1%). a) Scatter plot of capture rates. b) Mean capture rate curve overlaid on scatter plot.

inspection station by dividing the number of Fields of View (FOV) containing false events by the total number of fields of view inspected. An FOV represents the increment of area processed by the inspection tool at one time. It also represents the increment of area by which inspection results may be reviewed. For inspection tools not based on the FOV, false event rate may be calculated by dividing the number of false events by the total area inspected (false/cm^2)

Inspections with high false event rates require longer review and classification times, increasing the time to results. Also, large false event rates may reduce the accuracy of classification as the operator tires of going through too many unimportant events. Therefore, false event rate should be considered when evaluating the capabilities of a tool.

Stability. An inspection tool's stability may be tracked by plotting the aggregate capture of a specific group of defects along with a false event rate over time as shown in figure 13. Aggregate capture is the average capture for all defects being inspected. This value is determined by inspecting an area containing a known group of defects multiple times. The capture rate of each defect is determined and averaged to obtain the aggregate capture for all defects. A sample of defects is reviewed to determine the false event rate. Each of these values is plotted versus time. Control limits for the aggregate capture rates and false event rates may be determined with a process capability study.

Because information from an inspection tool is used as a basis for decisions, awareness of the tool's stability over time is critical. For example, suppose an inspection tool is being used to monitor the defect density coming out of etch and the density seems to be improving over time as shown in figure 14a. If in reality the system's sensitivity has been decreasing as shown in figure 14b, then the engineer may wrongly interpret the results as improvement in the process. There may actually be a yield impacting problem which the tool is not detecting.

CONCLUSION

Automated wafer inspection tools have been used successfully by many semiconductor manufacturers to reduce defect levels. Three components important to the successful application of a defect reduction program are a structured methodology, organizational support, and a thorough understanding of the inspection tools being utilized.

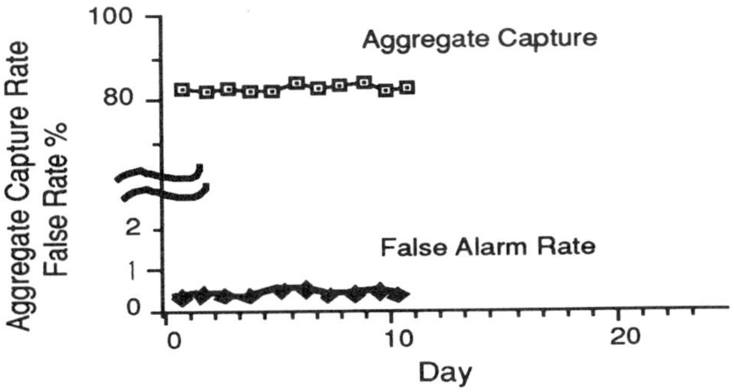

Figure 13. Stability plot of aggregate capture and false event rates over time.

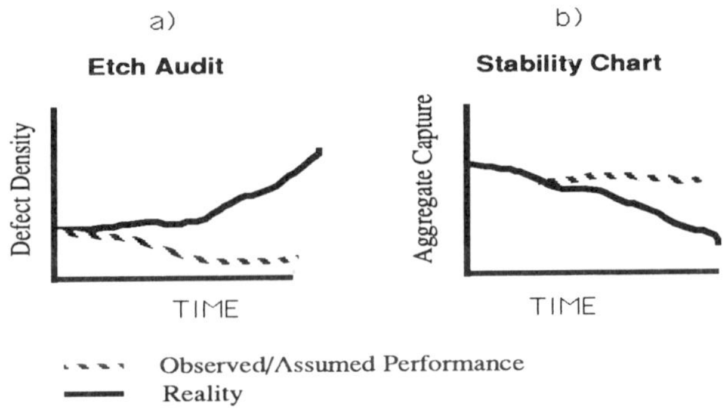

Figure 14. a) Actual and observed defect density coming out of etch over time. b) Assumed and actual capture rate performance of inspection tool over time.

REFERENCES

1. C. Radin, A defect reduction methodology for increased sort yield using automated inspection, SPIE Proc. 921, 224 (1988)
2. R. Wallace and G. Dickerson, Characterization of automated wafer inspection tools, SPIE Proc. 1087, 180 (1989)
3. K.L. Mittal, editor, "Particles on Surfaces 1: Detection, Adhesion and Removal," Plenum Press. New York, 1988
4. K.L. Mittal, editor, "Particles on Surfaces 2: Detection, Adhesion and Removal," Plenum Press. New York, 1989
5. A. Oberai, Lithography-challenges of the future, Solid State Technol. (Sept. 1987)
6. S. Strathman and S. Lotz, Automated inspection as part of a defect reduction program in an ASIC manufacturing environment, SPIE Proc. 1261, 208 (1990)
7. S. Weirauch, Background defect density reduction using automated defect inspection and analysis, SPIE Proc. 921, 241 (1988)
8. T. Ecton and K. Frazee, Methodology to reduce chronic defect mechanisms in semiconductor processing, SPIE Proc. 1261, 366 (1990)
9. T. Kobayashi and Y. Shioya, Yield improvement of submicron devices using defect source analysis on Al interconnections, SPIE Proc. 1261, 216 (1990)
10. M. Cherniawski and K. Chan, Automated defect inspection: A valuable tool for proactive defect control", SPIE Proc. 921, 234 (1988)
11. T. Ishikawa "Guide to Quality Control" UNIPUB, New York (1976)

ULTRASONIC CLEANING OF SURFACES: AN OVERVIEW

Ismail Kashkoush and Ahmed Busnaina

Center for Particulate Control in Process Equipment &
Mechanical and Aeronautical Engineering Department
Clarkson University, Potsdam, NY 13699-5725

Frederick Kern Jr. and Robert Kunesh

IBM General Technology Division
Essex Junction, Vermont 05452

Removal of submicron particles from silicon-wafers presents an enormous challenge to the semiconductor industry. Ultrasonic cleaning is an established method for removing micron-size microcontaminants or larger from hard surfaces. This paper presents an overview of the sonic cleaning research as applied to microcontamination control. The paper also presents new recent results on submicron particle removal.

In this study, the effectiveness of ultrasonic cleaning technique in removing submicron particles is investigated. The effect of the particle size, concentration, time and frequency functions, and sonication time on submicron particle removal effectiveness is studied. The effect of moving the fluid in the sonication bath is also considered. The results indicate that acoustic microstreaming produced by ultrasonics waves play the dominant role in the cleaning process. The removal efficiency is shown to be dependent on the ultrasonic time and frequency functions. Also, the liquid temperature is shown to be a major factor on the ultrasonic cleaning process. Cavitation damage to the wafer surface has been observed under certain conditions. Preliminary results indicate that the ultrasonic cleaning technique is capable of removing particle sizes of 0.3 microns or larger.

INTRODUCTION

The knowledge of the the ultrasonic cleaning mechanism is essential in understanding, improving and optimizing the cleaning process. It provides useful and much needed design guidelines for making better cleaning equipment. The formulation of the basic

Particles on Surfaces 3, Edited by K.L. Mittal
Plenum Press, New York, 1991

physical principles of ultrasonic cleaning has not been finalized yet[1–3]. It has been shown that the main mechanisms responsible for the the cleaning process in a sound field are cavitation and acoustic streaming. In fact, the role played by each of these factors is not well understood yet[3].

The objective here is to present the fundamental principles of surface cleaning by means of ultrasound. The dependence of the ultrasonic cleaning efficiency on the properties of the liquid used and the parameters of the sound field are discussed. The role played by the acoustic cavitation and acoustic streaming in the cleaning process is discussed. The analysis of an ultrasonic cleaning process is performed and the conditions under which the operational characteristics are optimized for maximum removal efficiency are determined.

Many attempts to study the bubble dynamics, as related to the surface cleaning, have been performed. Rayleigh[4] studied theoretically the collapse of a spherical cavity. Rayleigh's paper is still considered even today as the foundation of physical acoustics. Noltingk and Neppiras[5] have developed the equations that describe the motion of a gas filled and empty cavity in a liquid medium excited by a sound source. They predicted that cavitation will diminish as the frequency is raised (more than 15 MHz).

The dynamics of bubble motion has been developed[6–8] stressing on the relationship between transient and stable cavitation and their threshold. Cases of isothermal and adiabatic motion are studied. Direct verification of theories of bubble dynamics is obtained using high speed cinematography[9].

Benjamin[10,11] studied the physical mechanism of the collapse of cavitation bubbles and pressure waves arising therefrom. Once the collapse is completed, a strong shock wave is formed. This shock has a rarefaction phase behind it which might reopen a cavity.

Olaf[12] has investigated surface cleaning by the use of ultrasonic waves in the range of 15 kHz to 2.5 MHz on mechanically polished surfaces. He studied sound density and direction, time of irradiation, and liquid properties. The cavitation threshold was shown to increase with the increase of hydrostatic pressure. Experiments were performed to show the existence of acoustic streaming around the radiator. Results obtained confirmed that the cleaning is due to acoustic streaming as well as cavitation. Also, the cleaned area was shown to increase as a function of the irradiation time, see figure 1.

Rozenberg[13,14] used high speed photography to examine the mechanism of ultrasonic cleaning. Two independent mechanisms are revealed. The first is impact explosive destruction associated with the annihilation of cavitation bubbles. The second mechanism starts after the explosion is completed. Small bubbles energetically penetrate into all pores and gaps between the contaminant layer and the surface, and their vibrations peel and tear off the contamination layer. If the adhesion force between the contamination layer and surface is greater than the strength of the contamination layer, a fragment of the layer is torn loose. If the strength of the contaminant is greater than that between the layer and the surface, the contamination layer is peeled from the surface.

Plesset and Chapman[15] have modeled numerically the collapse of an initially spherical vapor cavity in the neighborhood of a solid boundary as shown in figures 2 and 3. Jet velocities up to 130 m/s have been calculated in water at atmospheric pressure. Bolle

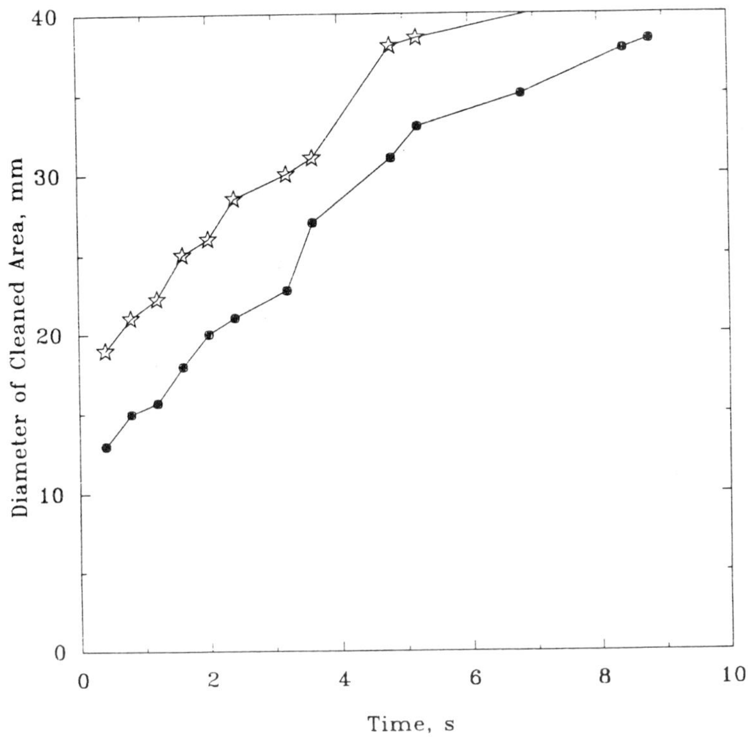

Figure 1. Diameter of Cleaned Area as a Function of the Irradiation Time (different symbols represent multiple runs, from ref. 12).

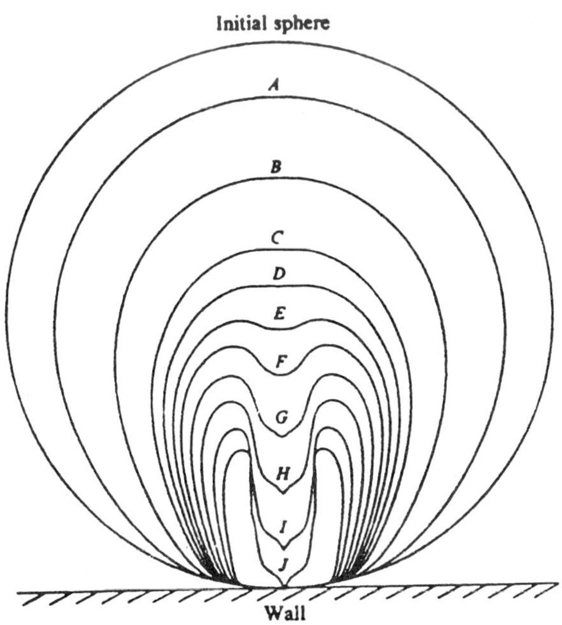

Figure 2. The Collapse of an Initially Spherical Bubble in Contact with a Solid Surface (from ref. 15).

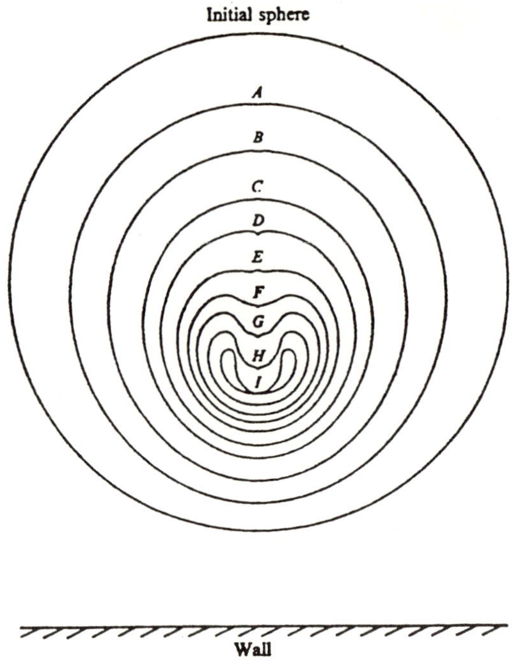

Figure 3. The Collapse of an Initially Spherical Bubble away from the Surface (from ref. 15).

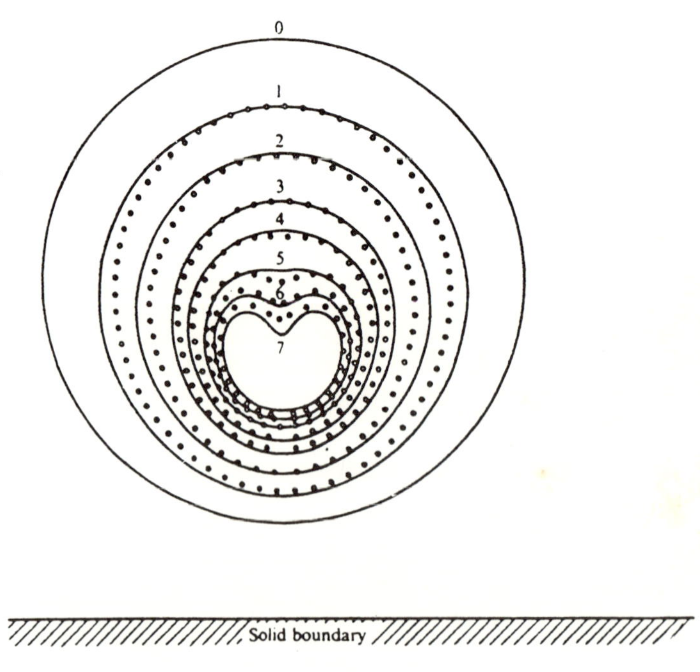

Figure 4. Comparison of Experimentally Determined Bubble Shapes (open circles) on Collapse of a Spherical Bubble near a Plane Solid Wall with Theoretical Results of Plesset and Chapman (solid curves) (from ref. 15).

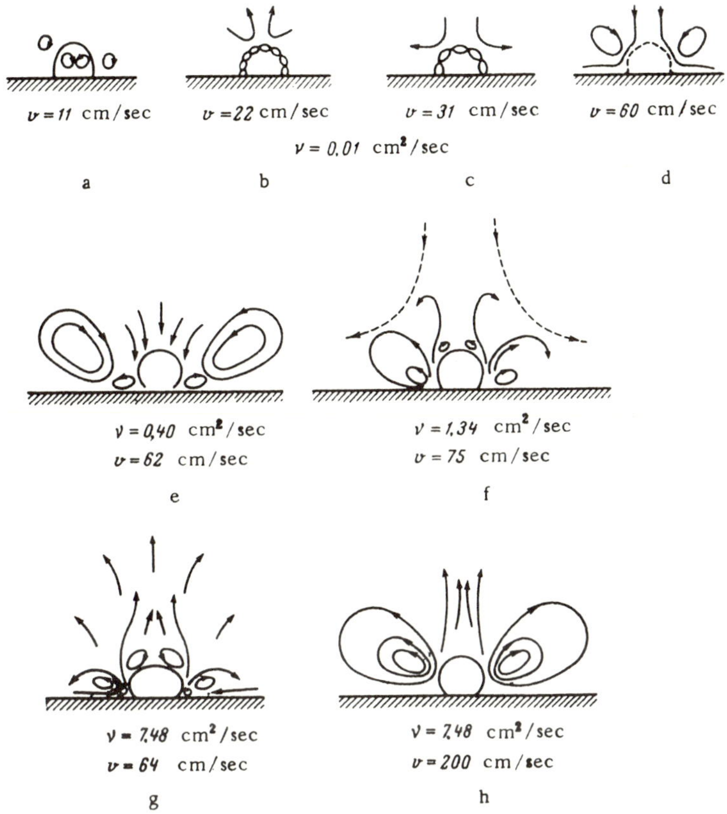

Figure 5. Microstreaming in the Vicinity of Oscillating Air Bubbles; v is the Particle Velocity at the Bubble Surface and ν is the Liquid Kinematic Viscosity (from ref. 14).

and Lauterborn[16] have quantitatively confirmed Plesset and Chapman's work. Fairly good agreement between numerical and experimental results is obtained as shown in figure 4. High velocities at the bubble surface and the associated pressure and temperature are of magnitude to explain the surface damage. Suslick[17] has reported also that a temperature of $5000°C$ and a pressure of about 500 atm can be reached inside the bubble. Also, jet velocities up to 400 km/hr have been reported.

Crawford[18] has reported that when a thin foil is immersed in an ultrasonically cavitating liquid, the stages of erosion follow a repeatable pattern. He showed that the relationship between the destroyed area and time of exposure is non-linear.

Acoustic streaming has been studied extensively by many researchers[19–25]. The first theoretical analysis of acoustic streaming phenomenon was given by Rayleigh[4]. Since then investigations have been made on related eddying motion which occurs near vibrating cylinders and spheres[3] , see figure 5. Schlichting[26] has analytically solved the Navier-Stokes equations and deduced expressions for the streaming velocity and pressures for the flow field induced by different sound fields. Zarembo[22] has studied experimentally the flow field induced by a sound field. He verified the theoretical work by Rayleigh, Eckart and Schlichting, see figure 6. With the aid of flow visualization techniques, he observed the resulting streaming and vortices in the flow field.

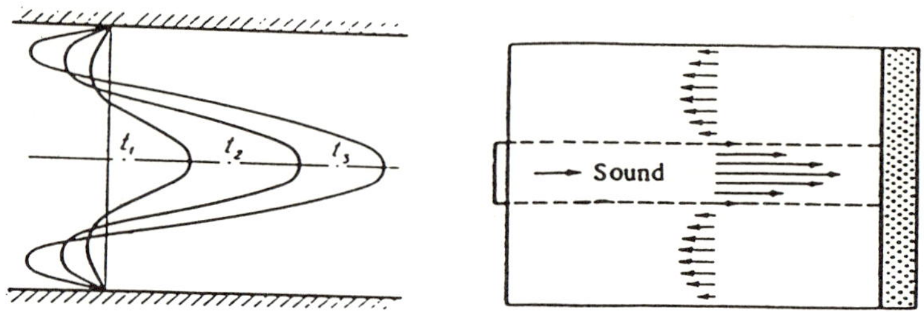

Eckart Streaming

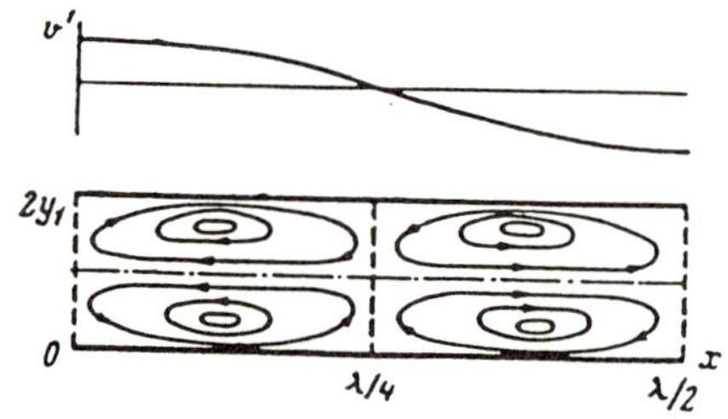

Rayleigh streaming

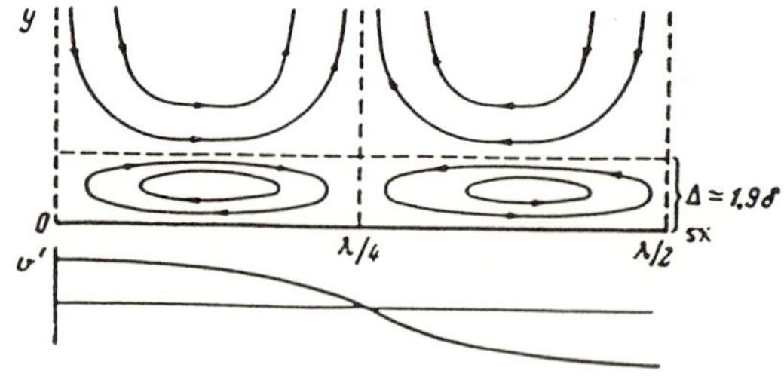

Schlichting streaming

Figure 6. Various Types of Acoustic Streaming.

The significance of these motions is that they exert viscous shear stress on the boundaries where the circulation occurs; while these stresses are not large, they should be significant in continuous removal of adhering surface contaminants, and in breaking down weak structures[3].

RECENT EXPERIMENTAL RESULTS

This section represents very recent experimental results by the authors on the removal of submicron particles from silicon wafers using ultrasonic cleaning. All experiments were performed in the Microcontamination Laboratory's Class-10 Clean Room at Clarkson University. The equipment used and experimental procedure are described earlier[1,2].

The ultrasonic unit, used in the study, is capable of controlling the center frequency, bandwidth, sweep time, train time, degas time, burst time, quiet time, and liquid temperature. Only some of the results on investigating the effect of the parameters of the sound field on the particle removal efficiency are presented here. The effects of liquid temperature as well as other liquid properties were also investigated.

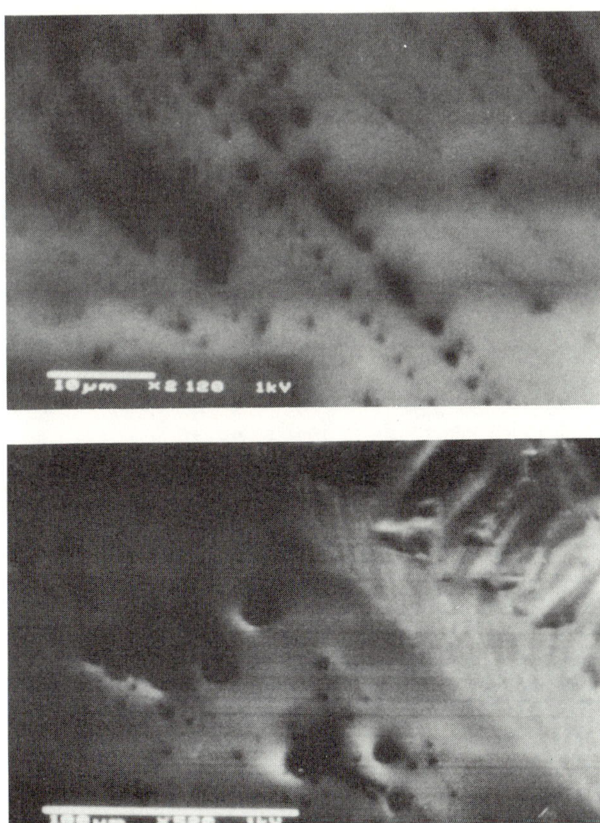

Figure 7. Surface Damage Caused by Cavitation on Si-Wafers (top and bottom figures differ in magnification).

RESULTS AND DISCUSSION

Ultrasonic cleaning can cause severe substrate damage. Figure 7 shows the *SEM* images of wafer surfaces after being sonicated for 60 *min*. The figure shows circular depressions that range between 1-10 μm in diameter. These depressions are surface pits created by cavitation erosion, since the wafer has been carefully scanned before the application of the sound waves. These observations agree with those found by other researchers[1,6,9,13,18,26,27].

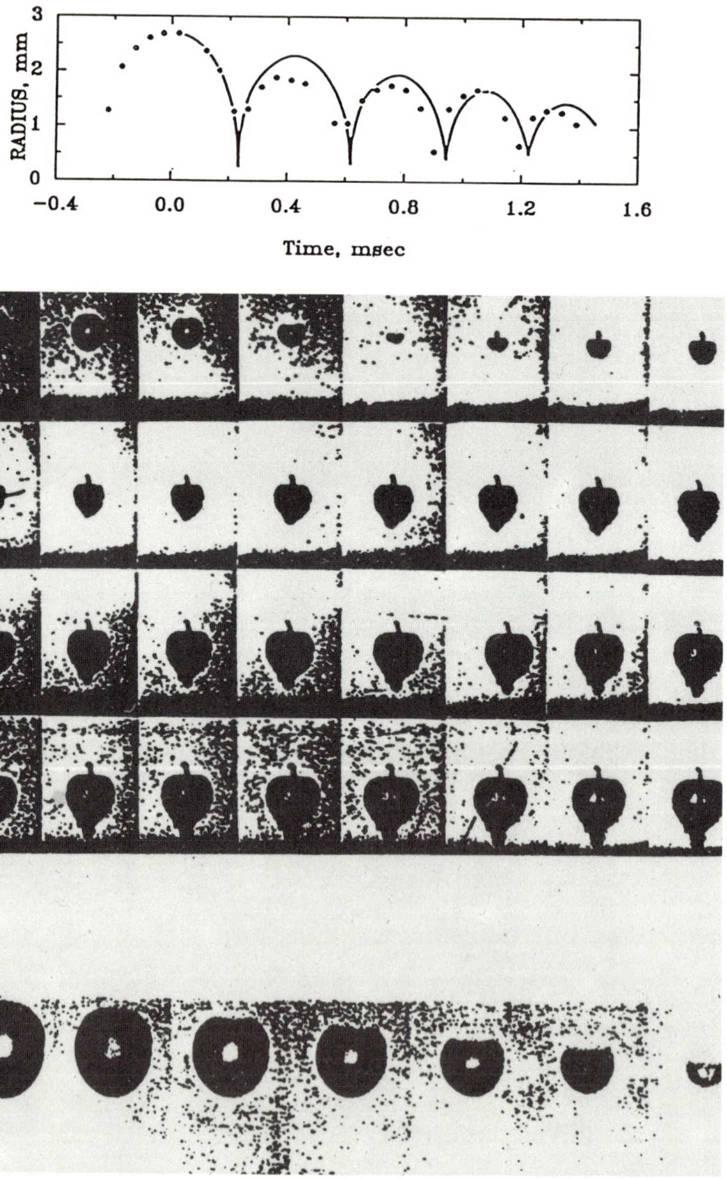

Figure 8. Bubble Collapse and Rebound near a Solid Boundary (from ref. 16).

It has been suggested that high velocity jets forming during cavity collapse will cause damages to the solid surface from their impingement on the surface, see figure 8. Liquid jets have been observed photographically and recent calculations show that they have sufficient momentum to cause observed pitting[26-28]. Flynn[9] explained that powerful hammer blows are caused by impact of high velocity liquid jets generated by asymmetrical collapse of bubbles near a solid surface, see figures 2-4.

In this study, cavitation has been eliminated by supplying the tank with fresh *DI* water (aerating the water would provide the same effect) which increases the amount of gas in the water and by adjusting frequency sweep. Once cavitation was controlled, a series of experiments were carried out to determine the effect of particle size, concentration and frequency and time functions on the cleaning efficiency. One observation that is worth mentioning here is: cleaning with zero frequency sweep time can cause severe line lifting on patterned wafers.

Figures 9-21 show the experimental results for ultrasonic cleaning after eliminating cavitation damage. The removal efficiency of submicron particles is shown versus particle diameter using high particle count in figure 9. The figure shows that the removal

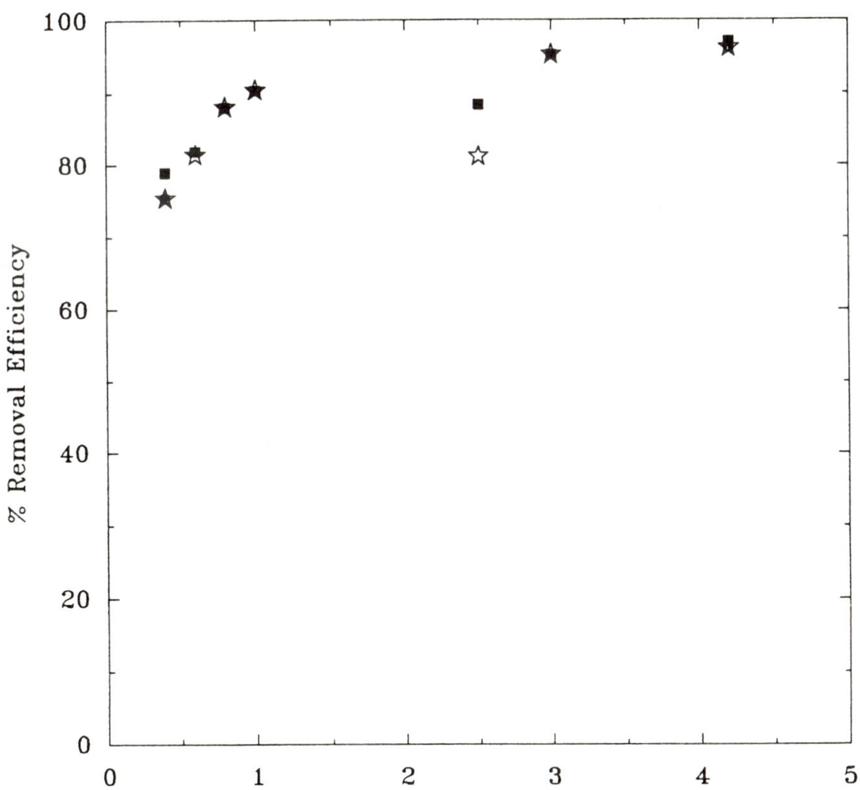

Figure 9. Removal Efficiency versus the Particle Diameter for High Initial Particle Count (different symbols represent multiple runs).

efficiency increases with the particle diameter up to 95 % at 4 μm. The interest lies, however, in the submicron range. For particles below 0.5 μm the removal efficiency averages about 75%. The ultrasonic cleaning method is shown to be capable of removing particle sizes of 0.3 μm or larger. The initial particle count for the experiments represented in figure 9 is high (on the order of 20,000-30,000). Results for low initial particle count are shown in figure 10.

Figure 10 shows the removal efficiency for different particle diameters using low initial particle count. The figure shows that the removal efficiency increases with the particle diameter up to an average of 90 % at 1 μm. For particles below 0.5 μm, the removal efficiency averages about 65%. This shows that the removal efficiency is lower when the lower particle concentration is used (specially in the submicron range). At high particle count the particles tend to agglomerate which allows for higher removal efficiency. This is due to the weak bondage between the particles compared to the adhesion force between the particles and the wafer surface. The initial particle count for the experiments shown is low (of the order of 800 to 1,500).

The cumulative removal efficiency is plotted against the sonication time as shown in figure 11. The figure shows that the removal efficiency continuously increases with

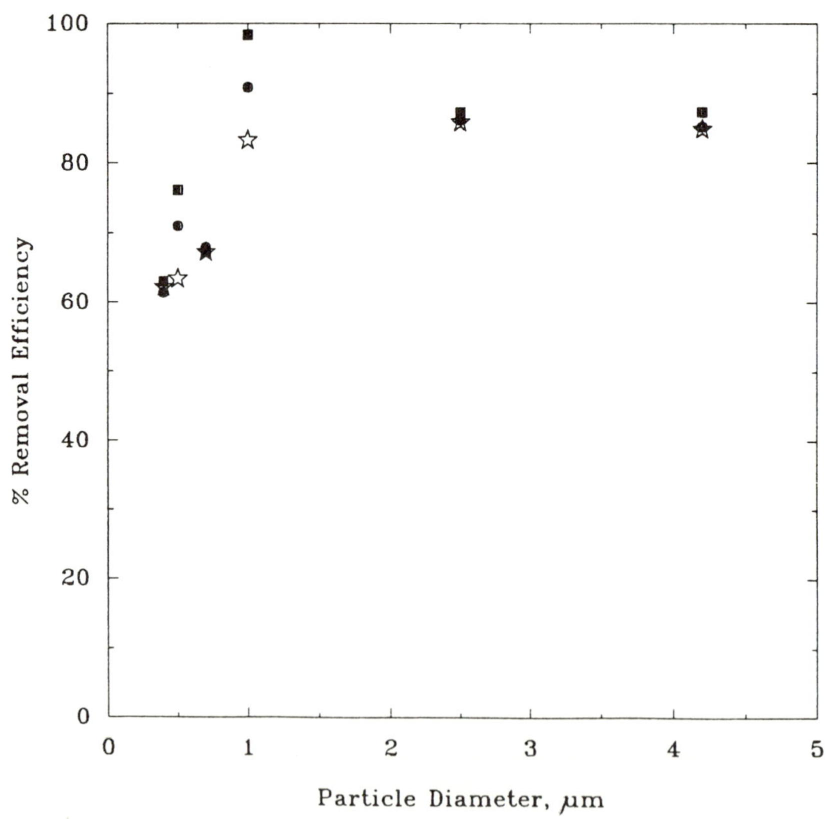

Figure 10. Removal Efficiency versus the Particle Diameter for Low Initial Particle Count (different symbols represent multiple runs).

time. This indicates that the cleaning has not reached a maximum even after one hour of cleaning. Removal efficiency results are based on the initial wafer particle count. The experiment was conducted using an initial high particle count. Olaf[12] has shown that the cleaned area is an increasing function of the sonication time. He also confirmed the role of acoustic microstreaming in the cleaning effectiveness. Results are shown again in figure 1 as the diameter of the cleaned area in millimeters versus the duration time in seconds.

Figure 12 shows the removal efficiency plotted against the train time for a particle diameter of 0.6 μm. The figure shows that the removal efficiency is a strong function of train time. The figure exhibits an improvement or increase in removal efficiency for a time interval of six minutes. This figure also shows that the initial removal efficiency can be as high as 80% simply by controlling the train time.

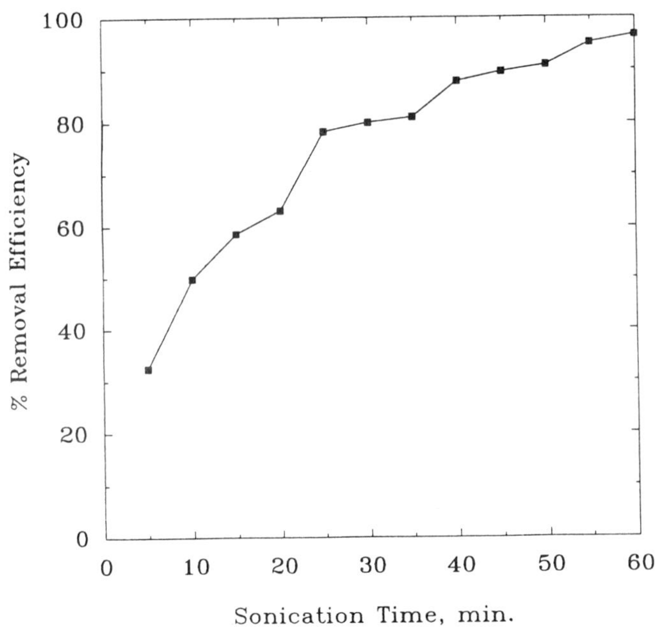

Figure 11. The Cumulative Removal Efficiency versus Sonication Time for 0.6 μm Particle Diameter.

Table I presents the effect of the initial particle count on the wafer surface on the removal efficiency for a sample of particle size of 0.4 μm. The table shows that the removal efficiency increases with the increase in number of particles deposited on the wafer surface. This may be explained that at high particle counts, there is a chance for the particles to coalesce. This allows for higher removal efficiency for the same amount of ultrasonic energy.

Table I. Removal Efficiency vs. Particle Count for 0.4 μm Particle.

Particle Dia.	*Initial Count*	*Removal Efficiency*
0.4μm	31,396	96%
0.4μm	19,218	94%
0.4μm	763	77%

Table II. Removal Efficiency for Moving and Quiescent Fluids in the Tank.

Particle Dia.	*Moving Fluid*	*Quiescent Liquid*
4.2 μm	82%	65%

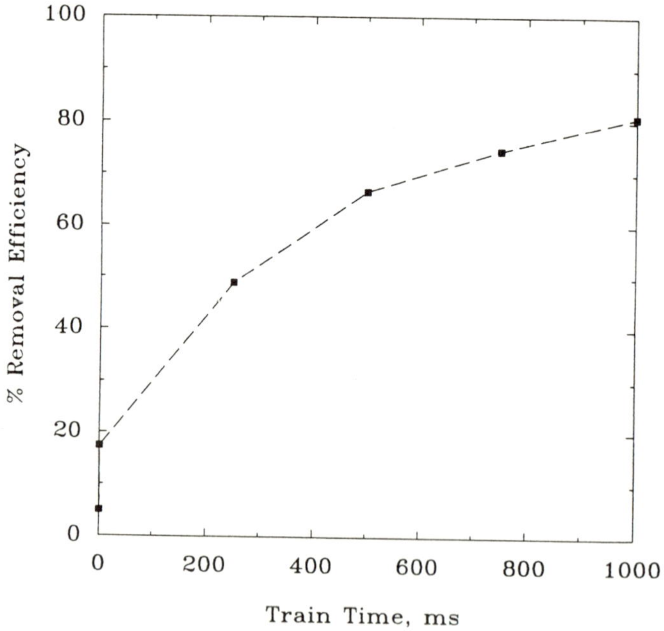

Figure 12. Removal Efficiency versus Train Time for a Particle Diameter of 0.5 μm.

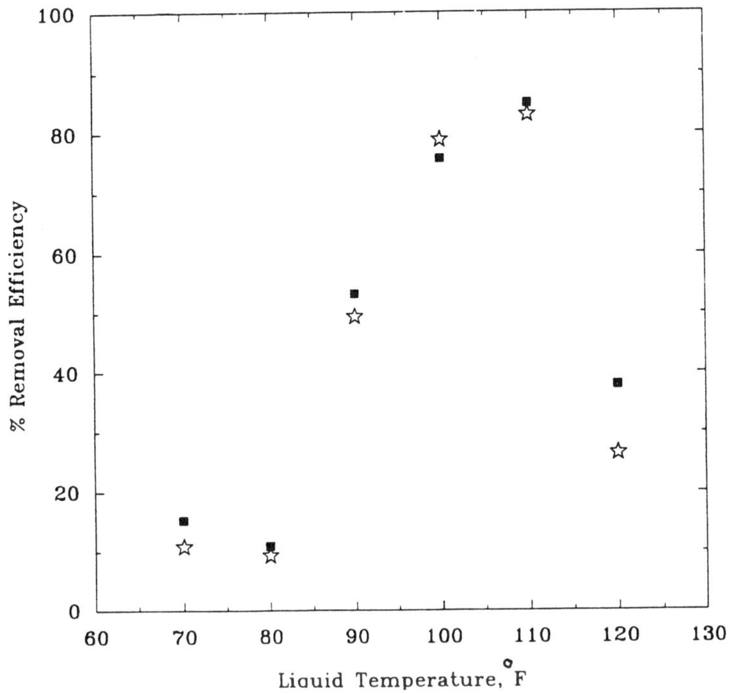

Figure 13. Removal Efficiency versus Liquid Temperature for a Particle Diameter of 0.3 μm (different symbols represent multiple runs).

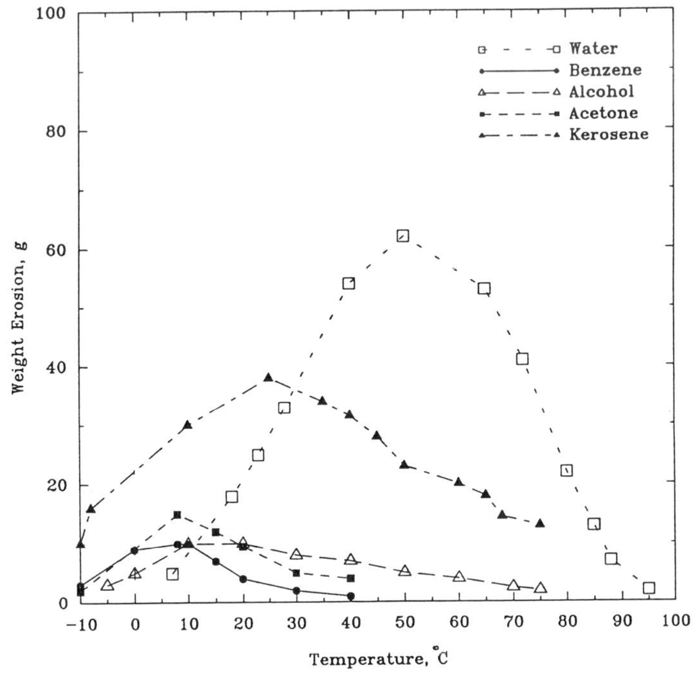

Figure 14. Cavitational Destruction as a Function of Temperature for Aluminum Foil in Different Liquids (from ref. 13).

Experiments were performed to show the effect of moving the fluid in the sonication bath. The results are presented in Table II. The removal efficiency is improved when using a moving fluid instead of quiescent liquid as can be seen from Table II. This can be explained by the fact that this prevents particles, already removed, from redepositing on the wafer surface.

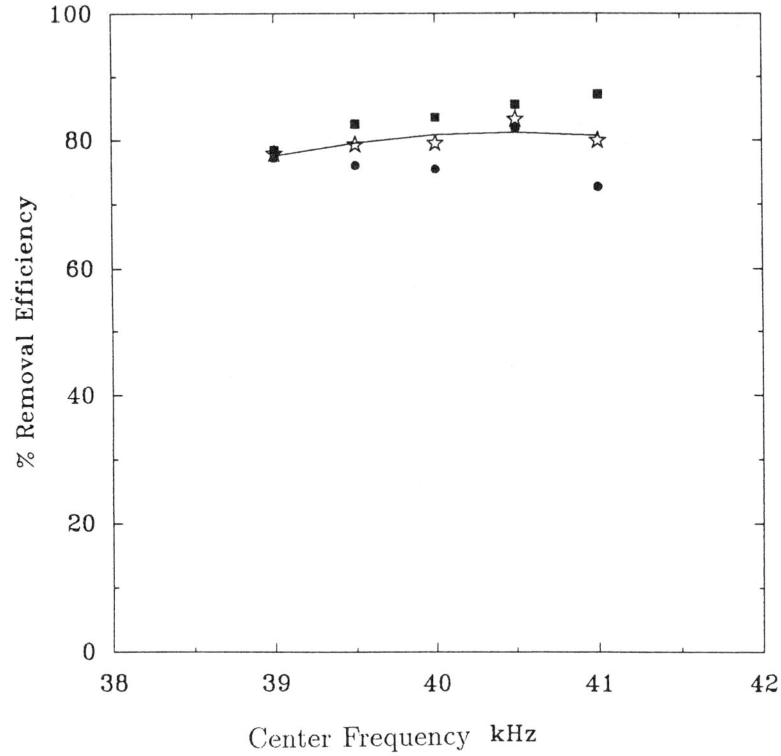

Figure 15. Removal Efficiency versus Center Frequency for 0.7 μm Particle Diameter (different symbols represent multiple runs).

The effect of liquid temperature on the cleaning efficiency is shown in figure 13. The figure shows the efficiency increases with temperature until it reaches a maximum value of 82% at $105°F$. The efficiency then drops to 30% at $120°F$. This agrees with the results obtained by Rozenberg[13] and others[29], see figure 14. Rozenberg has explained that the increase of erosion activity with increase in temperature is due to the increase in the number of cavitational nuclei. On the other hand, with increasing temperature there is a reduction in the viscosity and surface tension of the cleaning liquid and the saturation pressure rises, which leads to a decrease in the erosion activity of the liquid and a reduction in the effectiveness of the cavitational breakdown of surface contamination in the sound field.

Figures 15-17 show the effect of ultrasonic frequency functions on the removal efficiency. Figure 15 shows the effect of center frequency on the particle removal. The figure shows that the removal efficiency slightly increases with frequency. Increasing the frequency reduces the maximum bubble radius at a constant acoustic pressure amplitude. As the frequency increases, the time in which the acoustic pressure exceeds the external forces that hold the bubble in the equilibrium state decreases[3]. Agranat *et. al*[3]. have shown that an increase in the frequency leads to a linear decrease in the maximum and minimum bubble radii. Cavitation also decreases linearly with increasing frequency[3]. However, the authors[3] state that this does not apply to the entire range of acoustic frequency.

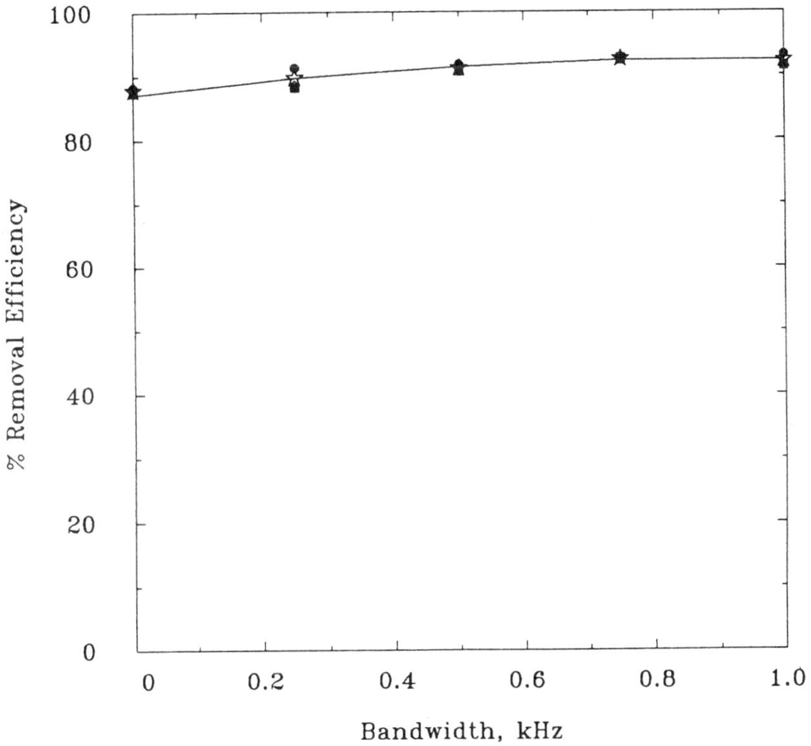

Figure 16. Removal Efficiency versus Bandwidth for 0.7 μm Particle Diameter (different symbols represent multiple runs).

Figure 16 shows the removal efficiency plotted against the bandwidth. Increasing the bandwidth will slightly raise the removal efficiency as the figure shows. This agrees with the results obtained before for the center frequency. This is because increasing the bandwidth will increase the operational range of frequency.

The removal efficiency is plotted versus the sweep time for particle diameters of 0.7 μm, see figure 17. The removal efficiency has a decreasing trend with the increase in sweep time. The sweep time control determines the time required to move through the selected bandwidth. With a fast sweep time, no single frequency can build up the energy to create the standing waves and random cavitation occur. A slow sweep moves the nodes and antinodes around the tank, like moving the object to be cleaned. This may explain the high removal efficiency obtained for short sweep time used.

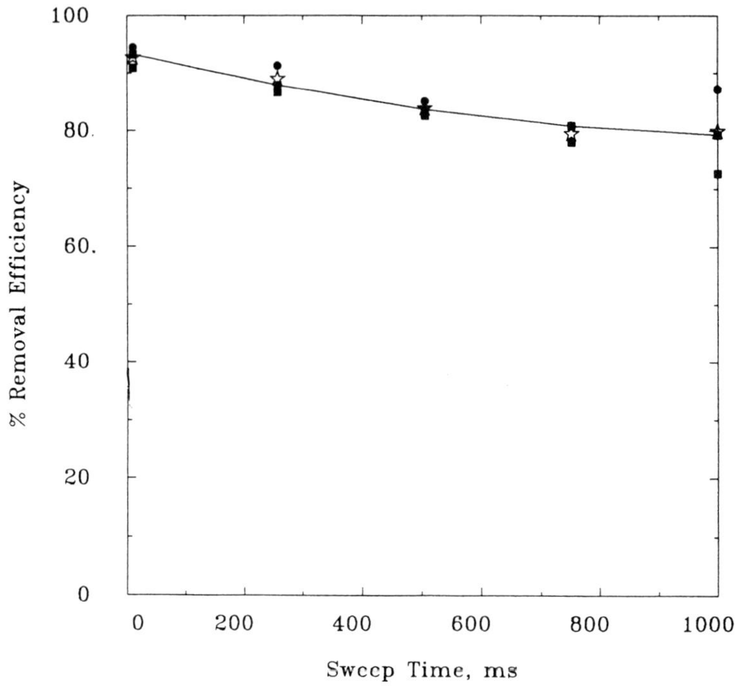

Figure 17. Removal Efficiency versus Sweep Time for 0.7 μm Particle Diameter (different symbols represent multiple runs).

The last set of results (shown in figures 18-21) shows the effect of ultrasonic time functions on the removal efficiency. Figure 18 shows the removal efficiency plotted against the train time. Train time is the parameter that establishes the length of the ultrasonic burst. Thus, increasing the train time will increase the number of bursts i.e. increases the number of bubbles and associated microcurrents responsible for cleaning. This also results in longer duration of these streaming currents induced in the sound field i.e. higher removal efficiency.

Figure 19 shows the effect of degas time. The figure shows that increasing the degas time lowers the removal efficiency. This can easily be explained by the fact that the degas time is the time during which no ultrasonic activity occurs, allowing gas to escape. In general, it can be concluded that increasing the ratio of train time to degas time results in improving the removal efficiency.

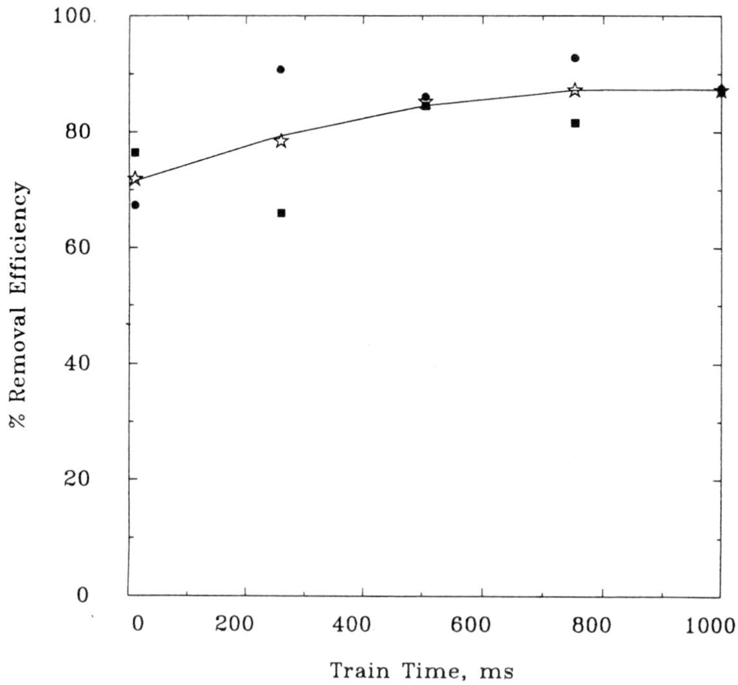

Figure 18. Removal Efficiency versus Train Time for 0.7 μm Particle Diameter (different symbols represent multiple runs).

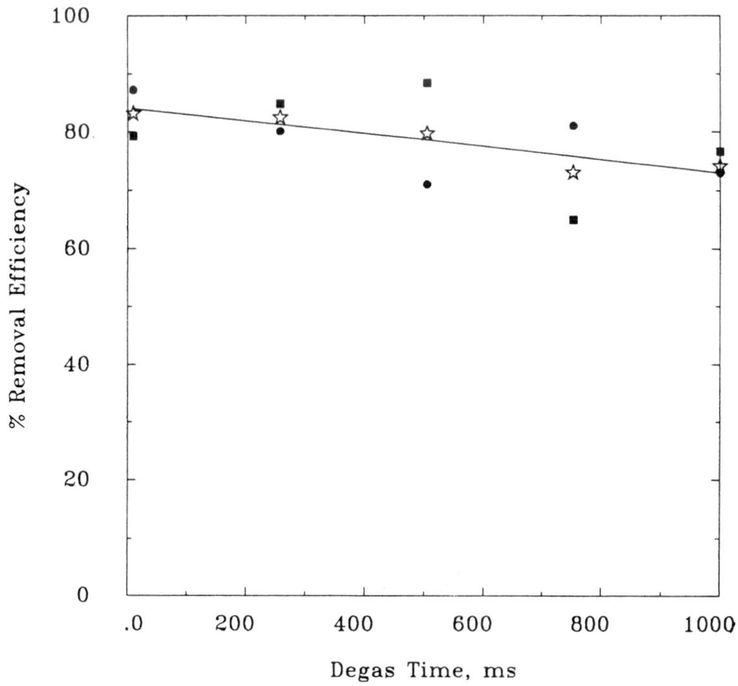

Figure 19. Removal Efficiency versus Degas Time for 0.7 μm Particle Diameter (different symbols represent multiple runs).

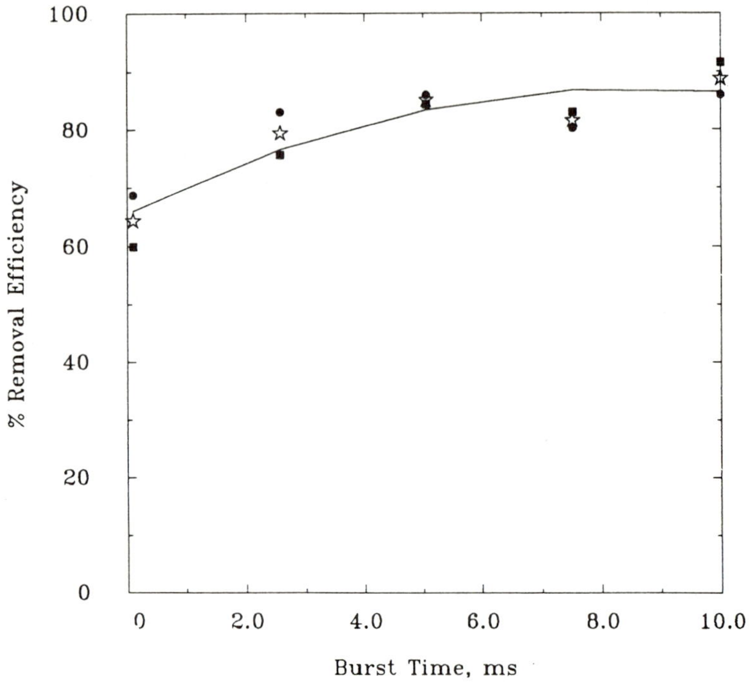

Figure 20. Removal Efficiency versus Burst Time for 0.7 μm Particle Diameter (different symbols represent multiple runs).

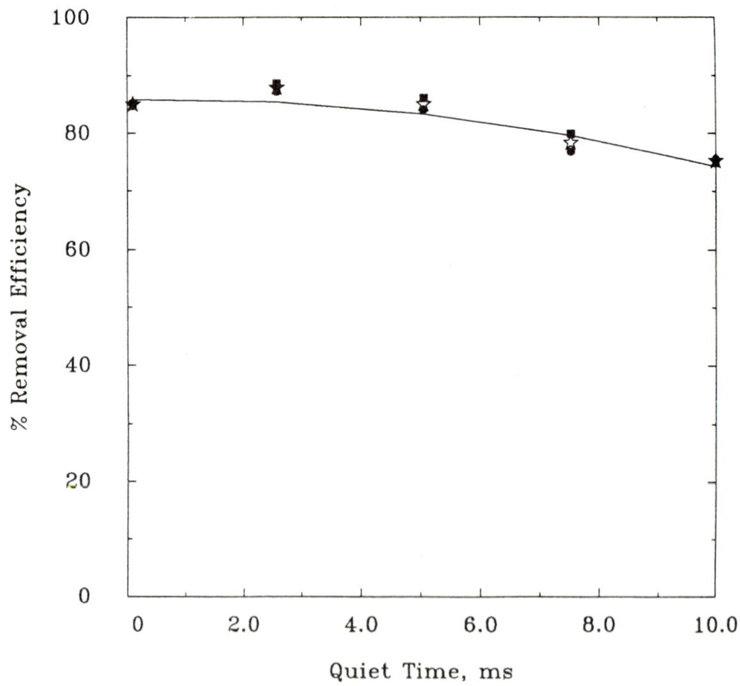

Figure 21. Removal Efficiency versus Quiet Time for 0.7 μm Particle Diameter (different symbols represent multiple runs).

Figure 20 represents the results obtained for burst time. It is shown in the figure that the removal efficiency increases with the increase in burst time. These bursts are related to the acoustic pressure and increasing the burst time will lengthen the bubble expansion phase. It will also increase the value of maximum bubble radius and the oscillation period. Forces holding the bubble in equilibrium are equalized by the acoustic pressure. Increasing the acoustic pressure increases the equalization period. As a result the bubble can experience unlimited expansion. Consequently, increasing the acoustic pressure to a certain limit has much the same effect on the bubble dynamics as a reduction in frequency[3].

Figure 21 shows that the removal efficiency decreases with increasing quiet time. This can be explained by the fact that small quiet time results in short lengths of time between the ultrasonic bursts. Once again, we can draw a general conclusion that the removal efficiency increases with the increase in the ratio of the burst time to quiet time.

CONCLUSIONS

The ultrasonic cleaning method employed in this study is an effective technique for submicron particle removal. Results indicate that high removal efficiency is obtained for large particulate sizes of about 1.0 μm or larger. For smaller sizes, relatively lower efficiency is obtained. Results show that the ultrasonic cleaning method is capable of removing particles of sizes as small as 0.3 μm in diameter with removal efficiencies in the range of 65 % to 70 %. The removal efficiency is shown to increase with the sonication time. While the removal efficiency increases with the increase in train and burst times it decreases with sweep, degas, and quiet time. The frequency used has a slight effect on the efficiency. Results indicate that initial high particle concentration produces high removal efficiency. Using moving fluids in the sonication bath is shown to improve the cleaning process significantly.

The results show that an optimum temperature for cleaning was obtained ($105^\circ F$). The removal efficiency increases as the temperature reaches $105^\circ F$ after which the efficiency drops dramatically.

The ultrasonic cleaning method should be implemented with great care to avoid cavitation surface damages. This is accomplished by avoiding cavitation erosion (bubble collapse near the surface) action altogether and utilizing acoustic microstreaming as the only cleaning mechanism.

Cavitation has been eliminated by supplying the tank with fresh DI water (aerating the water would provide the same effect) which increases the amount of gas in the water and by controlling frequency sweep and degas time.

ACKNOWLEDGMENT

This investigation was supported by the *IBM* Corporation, Essex Junction, Vermont.

REFERENCES

[1] I. Kashkoush and A. Busnaina, Ultrasonic cleaning of surfaces., Proc. IES Annual Meeting, 407-413, New Orleans, April 1990.

[2] A. Busnaina, J. Taylor, I. Kashkoush, F. Kern, and R. Kunesh, Measurement of the removal force and removal efficiency of submicron particles on silicon substrate., ICCCS $10^{th.}$ Intl. Symp. on Contamination Control, Zurich, Switzerland, Sept. 10-14, 1990.

[3] A. Agranat, V. Bashkirov, and Y. Kitaigorodskii, Ultrasonic cleaning., in "Physical Principles of Ultrasonic Technology", L.D. Rozenberg, editor, Vol 2, Part III, Plenum Press, New York, 1973.

[4] Lord Rayleigh, On pressure developed in a liquid during the collapse of a spherical cavity, Phil. Mag., 34-94(1917).

[5] B. Noltingk and E. Neppiras, Cavitation produced by ultrasonics., Proc. Phys. Soc. London, 63B, 674-685(1950).

[6] E. Neppiras, Acoustic cavitation., Physics Reports, 61, 159-253(1980).

[7] H. Flynn, "Cavitation Dynamics: A Mathematical Formulation"., Harvard University, Acoustics Res. Lab., Tech. Memo. 50, Jan. 1966.

[8] M. Plesset, Bubble dynamics., in "Cavitation in Real Liquids", R. Davies, editor, Elsevier, New York, 1964.

[9] H. Flynn, Physics of acoustic cavitation in liquids., in "Physical Acoustics", W.P. Mason, editor, Vol 1, Part 1, Chapter 1, Academic Press, New York, 1966.

[10] T. Benjamin, Pressure waves from collapsing cavities., Proc. $2^{nd.}$ Symposium on Naval Hydrodynamics, Office of Naval Research- Department of the Navy, Washington, D.C., 207-233, August 25-29, 1958.

[11] T. Benjamin and A. Ellis, The collapse of cavitation bubbles and the pressures thereby produced against solid boundaries., Phil. Trans. Royal Soc. (London), A260, pp. 221-240(1960).

[12] J. Olaf, Oberfflachenreinigung mit Ultraschall., Acoustica, 7, No. 5, 253-263(1957).

[13] L. Rozenberg, On the physics of ultrasonic cleaning., Ultrasonic News, 16-20(Winter 1960).

[14] L. Rozenberg, The Cavitation zone., in "High Intensity Ultrasonic Fields", L.D. Rozenberg, editor, Plenum Press, New York, 1971.

[15] M. Plesset and R. Chapman, Collapse of an initially spherical vapor cavity in the neighborhood of a solid boundary., J. Fluid Mech., 47, part 2, 283-290(1971).

[16] H. Bolle and W. Lauterborn, Experimental investigations of cavitation bubble collapse in the neighborhood of a solid boundary., J. Fluid Mech., 72, part 2, 391-399(1975).

[17] K. Suslick, The chemical effects of ultrasound., Scientific American, , 80-86(Feb. 1989).

[18] A. Crawford, The measurement of cavitation., Ultrasonics, 2, No. 3, 120-123(1964).

[19] W. Nyborg, Physical principles of ultrasonics., in "Ultrasound: Its Applications in Medicine and Biology", F.J. Fry, editor, Vol. 2, Part 1, Chap. 1 & 2, Elsevier Scientific Publishing Co., New York, 1978.

[20] C. Wang, Acoustic streaming of a sphere near an unsteady source., J. Acous. Soc. Amer., 71, No. 3, 580- 584(1982).

[21] C. Lee and T. Wang, Near-boundary streaming around a small sphere due to two orthogonal standing waves., J. Acous. Soc. Amer., 85(3), 1081-1088(March 1989).

[22] L. Zarembo, Acoustic streaming., in "High-Intensity Ultrasonic Fields", L. D. Rozenberg, editor, Part IV, Plenum Press, New York, 1971.

[23] J. Rooney, Other nonlinear acoustic phenomena., in "Ultrasound: Its Chemical, Physical and Biological Effects, K. Suslick, editor, VCH Publishers, New York, 1988.

[24] A. Ghabrial and E. Richardson, A study of acoustic streaming in liquids over a wide frequency range., Acoustica, 5, No. 1, 28-34(1955).

[25] O. Rudenko and S. Soluyan, "Theoretical Foundations of Nonlinear Acoustics", Ch. 8, Plenum Publishing Corp., Consultants Bureau, New York, 1977.

[26] H. Schlichting, "Boundary Layer Theory", McGraw-Hill Co., New York, 1979.

[27] F. Hammitt, Damage to solids caused by cavitation, Phil. Trans. Royal Soc. (London),A260, 245-255(1960).

[28] D. Ross, "Mechanics of Underwater Noise", Pergamon Press, New York, 1976.

[29] O. Sukhar'kov, Effect of temperature of an aqueous caustic solution on the duration of cleaning in the field of hydrodynamic radiator"., Russian Ultrasonics, 18, No. 4, 197-199(1988).

PARTICLE PROTECTION OF SEMICONDUCTOR SURFACES BY REVERSIBLE WAFER BONDING AND RELATED CONCEPTS

U. Gösele, V. Lehmann*, R. Stengl*, K. Mitani**, T. Y. Tan† and D. Feijoo

School of Engineering
Duke University
Durham, NC 27706, USA

Particle-free bonding of silicon or other semiconductor wafers to each other may routinely be accomplished in a "microcleanroom" set-up developed at Duke University. This set-up operates effectively even in a non-cleanroom environment. Without subsequent heating, room-temperature wafer bonding is reversible. Wafer bonding may, therefore, be used as a means of protecting semiconductor wafer surfaces temporarily against particle and organic vapor contamination as well as against mechanical damage during transport and storage.

The microcleanroom concept, which allows conveniently to obtain particle-free spaces between wafers, may be extended to a concept of wafer boxes which may be used as combined transport and cleaning containers between processing equipment. Consequently, one could relax the requirements for low-particle densities outside of the processing equipment.

INTRODUCTION

Putting two clean, flat, polished and hydrophilic surfaces of silicon wafers together results in the sticking together of the two surfaces due to van der Waals forces. This "wafer bonding" process appears to be a promising technique to fabricate Silicon-On-Insulator (SOI) wafers for CMOS and power device applications.[1-3] Silicon wafer bonding also allows a higher flexibility in the design of micromechanical devices.[4, 5] All these applications require that the wafer pairs after being bonded at or around room temperature have to undergo an annealing step at temperatures of 900°C or higher in order to increase the bonding strength and improve the interface quality. If the wafers are not heated, the wafer bonding process is reversible and separation of bonded wafers is easily possible by the introduction of small teflon wedges between the bonded wafers.

The present paper discusses the possible use of reversible wafer bonding for the protection of semiconductor surfaces against particles and organic contaminations.[6] We will first discuss the effect of particles on the bonding process and then describe the "microcleanroom" concept which constitutes a simple method to avoid particles in the

---

* Permanent address: Siemens Research Laboratories, Otto-Hahn-Ring 6, Munich 83, Germany

** Permanent address: Shin-Etsu Handotai, R&D Center, Gunma, Japan

† Also at: Microelectronics Center of North Carolina, P.O. Box 12889, Research Triangle Park, NC 27709

Particles on Surfaces 3, Edited by K.L. Mittal
Plenum Press, New York, 1991

bonding process without the requirement of an elaborate and expensive cleanroom. After discussing the effect of organic contamination on wafer surfaces and the use of reversible wafer bonding for protecting semiconductor surfaces, we will finally suggest that the microcleanroom concept presently used for wafer bonding may be extended to a cleaning concept in which wafers are cleaned in transport boxes and would not see the cleanroom environment between different processing steps. This would permit to tolerate much higher particle densities in cleanrooms for semiconductor processing.

INFLUENCE OF PARTICLES ON WAFER BONDING

The bonding between hydrophilic silicon surfaces appears to be due to the van der Waals interaction between monomolecular layers of water adsorbed on the native or purposely grown oxide layers on the silicon wafers.[3,7] The driving force for the bonding process is the gain in interface energy, which is in the order of $70mJ/m^2$ at or around room temperature. Unbonded areas at the interface usually called "voids" or "bubbles" may occur if the wafers are not sufficiently flat or if particles are present between the wafer surfaces during the bonding process. Elastic deformation of the wafers may overcome a certain deficiency in the wafer flatness. A simple elastomechanical calculation[8] as well as experience with the bonding of many wafers shows that the best grade commercially available silicon wafers are sufficiently flat to allow complete bonding.

Particles present between two wafers to be bonded lead to circular unbonded areas with a radius typically about 1cm for particle diameters of 1mm. These unbonded areas can conveniently be investigated by an infrared camera (S1-cathode, sensitivity 0.8 to 1.2mm) for bubbles caused by particles down to about 0.4mm diameter. Bubbles induced by even smaller particles may still easily be detected by x-ray topography. An example of an infrared image of particle-induced bubbles is shown in Fig. 1. Obviously, if too many unbonded areas are present and start to overlap appreciably no wafer bonding is possible at all, but even the presence of only a few particles with associated unbonded areas is undesirable for later device applications of wafer bonding. Particle-related bubbles can usually not be annealed out at temperatures below 1150°C. Therefore, wafer bonding should be performed under circumstances in which particles are not present during the wafer bonding process. This goal may be accomplished either in a cleanroom of class 10 or preferably of class 1 in which a low particle density is maintained within a fairly large volume or by the use of a microcleanroom in which basically only the space between the two wafers to be bonded is kept free of particles. This microcleanroom concept[9] will be described in the following section. In passing, we may mention that silicon wafer bonding appears to have a potential to measure very low effective areal particle densities in cleanrooms by determining the size and density of particle-induced interface bubbles.

THE "MICROCLEANROOM" CONCEPT

The microcleanroom concept as first suggested and realized by Stengl et al.[9,10] and later on refined by the Duke University wafer bonding group is based on the realization that wafer bonding requires only a particle-free space between the two wafers shortly before and during the bonding process. The microcleanroom set-up resembles a modified spin-dryer. As shown schematically in Fig. 2 the two wafers are stacked in a rack with the polished surfaces facing each other. Three or more spacers hold the wafers apart with a gap of about 0.5 - 2mm. After about one minute of flushing the gap with thoroughly filtered and deionized water of a velocity exceeding 4m/s, the set-up is tightly sealed with a transparent cup and rotated at about 3000 rpm. This produces a particle-free microcleanroom inside the chamber due to a centrifuge effect and also specifically between the two wafers. By illuminating the rotating wafer pair with an infrared lamp we can dry the wafers in about 5 minutes. After drying, the spacers are removed without opening the chamber so that the upper wafer drops onto the lower one. Now the chamber can be opened. With a pair of tongs with teflon tips the upper wafer, which is floating on a thin air cushion, is pressed onto the lower wafer. Using this method the bonding front spontaneously propagates outward and no air is trapped. We should mention specifically that mostly wafers directly from wafer boxes have been used without any additional precleaning procedure, after it had turned out that precleaning procedures generally do not lead to improved bonding results.

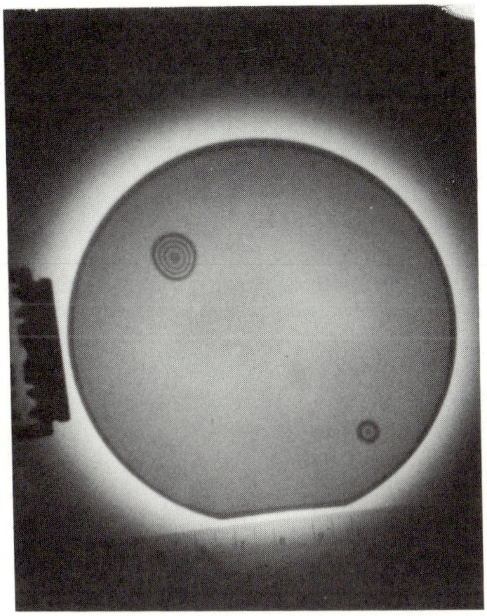

Figure 1. Example of particle-induced interface bubbles in bonded 4" silicon wafers as detected by an infrared camera.

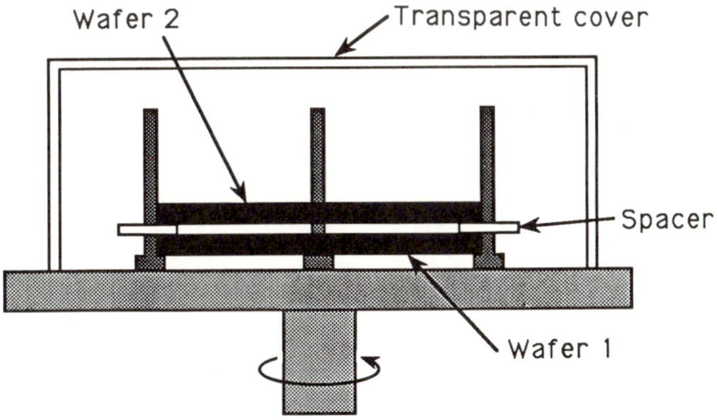

Figure 2. Schematic front view of the microcleanroom set-up. After flushing the space between the wafers with deionized water, the cover is closed and the wafers are dried by spinning in the presence of an infrared lamp.

The described microcleanroom concept was realized for the bonding of silicon wafers up to 8 inches in diameter[10] as well as for the bonding of quartz wafers, GaAs wafers, or InP wafers on silicon wafers.[12] An x-ray topography picture of an 8 inch silicon wafer pair bonded in a microcleanroom set-up situated in a normal university laboratory is shown in Fig. 3 which reveals the presence of just one particle-related bubble. All the other features are either x-ray related artifacts or due to some microscopic elastic deformation during the wafer bonding process.

The microcleanroom concept has been realized for more than one wafer pair and can be extended to the cleaning and bonding of many wafers. By the use of this concept it is possible to avoid interference from any kind of particles during wafer bonding. This particle-free bonding allowed us to investigate bubble generation due to non-particle related contamination present on wafer surfaces as described in the following section.

ORGANIC CONTAMINATION OF WAFER SURFACES

Wafer pairs which have been bonded around room temperature free of any particle-related bubbles, frequently generate interface bubbles when heated in the temperature range of 200-800°C as was first observed by Ohashi et al.[13] Ohashi et al.[13] and others[7,14] have suggested that these bubbles are related to water. Our own recent experiments rather indicate that these bubbles are due to hydrocarbon contamination adsorbed on the silicon wafer surfaces which is thermally desorbed at elevated temperature[15,16] resulting in gaseous products which are collected in the interface bubbles. The presence of high concentrations of hydrogen and carbon at the interface of bonded wafers was demonstrated by appropriate SIMS measurements.[16] The hydrocarbon contamination appears to result from the final wafer polishing or cleaning procedures and/or from the outgassing of organic vapors from the wafer boxes during transport and storage of wafers in these boxes. The development of hydrocarbon-related bubbles can be avoided by heating the wafers before bonding to 600°C in oxygen or to 800°C in nitrogen.[16] The 600°C oxygen treatment leads to a thickening of the native oxide of less than 10Å. UV-ozone cleaning procedures are under investigation.

A simple experiment demonstrated that unprotected wafer surfaces, which usually showed no bubbles after bonding and annealing, showed bubbles after storage in certain wafer containers for 15 hours at 100°C and subsequent bonding and annealing at 100°C. In a subsequent experiment two full wafers were each bonded to one half of a wafer, and then exposed to the same 15 hours/100°C treatment in the same boxes. After the treatment the half-wafers were separated from the full wafer. The two full wafers were then bonded to each other in such a way that the halves which had been protected by wafer bonding faced each other. Consequently, also the halves which had been unprotected during the 15 hours/100°C treatment faced each other. After a heat treatment of the bonded wafer pair at 500°C the halves which had been unprotected showed many bubbles due to hydrocarbon contamination from the wafer box (lower part of Fig. 4) whereas the halves which had been protected by bonded half wafers did not show any bubbles (upper part of wafer pair in Fig. 4). This experiment not only indicates that wafer boxes may be one of the sources for hydrocarbon contamination of silicon wafer surfaces but also that wafer bonding protects the wafer surfaces against such a contamination.[15] In a related experiment it could be shown that hydrocarbons present in vapor form may be deposited on silicon wafer surfaces and cause hydrocarbon-related bubbles after bonding and annealing.[17]

Organic contamination of silicon wafer surfaces has recently received an increasing interest after if became possible to measure contamination levels far below one monomolecular layer.[18] It appears that silicon wafer bonding with subsequent annealing to 200 - 800°C and inspection by infrared or x-ray topography is an alternative and extremely sensitive method of investigating hydrocarbon contaminations on silicon wafer surfaces.

CONTAMINATION PROTECTION OF SEMICONDUCTOR SURFACES BY REVERSIBLE WAFER BONDING

Semiconductor wafer bonding can be performed easily around room temperature without the interference of contaminating particles. Since the bonding process is reversible the

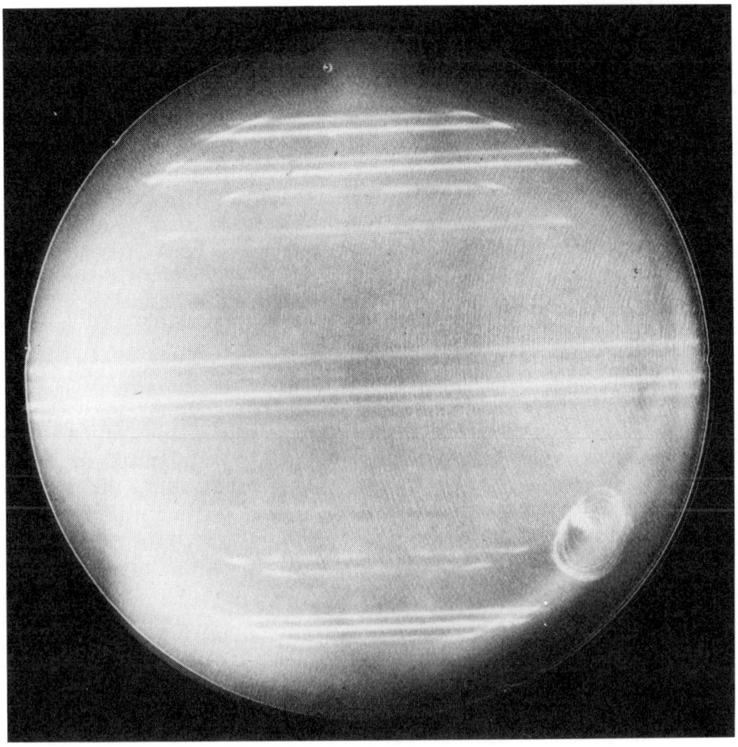

Figure 3. X-ray topography of a room temperature bonded eight-inch wafer pair using the microcleanroom set-up. The small circle (top) is probably due to a small particle, the shell-like structure (lower right) and the circular structure are due to elastic deformations induced by pressing the wafers using a pair of tongs and from thickness variations due to grinding. The horizontal lines are artifacts.[11]

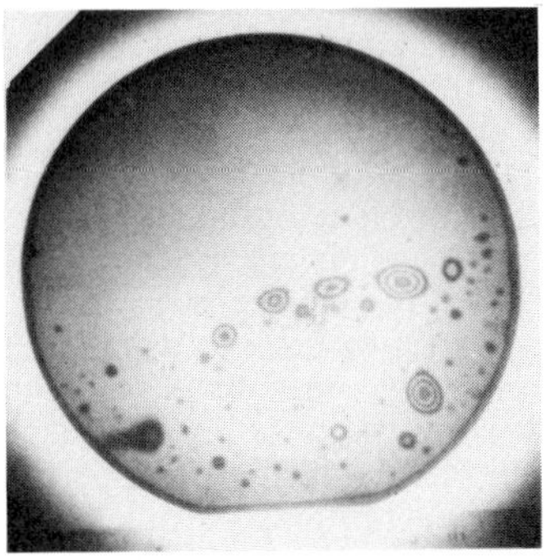

Figure 4. Infrared images of wafers which were exposed to a wafer box for 15 hours at 100°C, subsequently bonded to each other at room temperature and annealed at 600°C for 1 hour to show the effect of organic contamination. One half of each of two full wafers was protected by bonding a half wafer to them. These half-wafers were removed after the 15 hour/100°C treatment. The unprotected halves of the full wafers were facing each other in the subsequent bonding and show hydrocarbon-related bubbles (lower wafer part) whereas the parts protected by reversible wafer bonding showed no contamination related bubbles (upper wafer part).

particle-free state of the two wafer surfaces can be maintained until the wafers are separated for final use in a semiconductor processing facility by the introduction of tiny wedges at the rim between the bonded wafers.[6,19] Therefore, wafer manufacturers could bond wafers after the final cleaning which could possibly include an UV-ozone cleaning step to ensure surfaces free of hydrocarbon contamination[20] and thus preserve the ultra-clean surfaces of two wafers in terms of a bonded wafer pair. These surfaces are then protected against particle and organic contamination from the surroundings or more specifically from the wafer boxes during transport and storage. Since the wafers are then transported and stored in bonded pairs, only about half the space would be needed in conventional wafer boxes. The wafer surfaces are not only protected against contamination (see discussion of Fig. 4 in previous section) but also less sensitive to mechanical shock during transport. The latter point might be of special interest for brittle III-V compound wafers for which the transport might be facilitated by bonding these wafers to much cheaper but mechanically more favorable silicon transport wafers.

Separation of bonded wafers by the introduction of two or three tiny teflon wedges at the rim between wafers is an easy procedure and does not leave any marks on the wafer surfaces since only the bevelled part of the wafers comes into contact with the teflon wedges. The delivery of bonded wafer pairs would allow the users of wafers to inspect the wafers for the presence of particles at the bonded interface by an infrared camera. The presence of hydrocarbons at the surface could also be determined by annealing at about 500°C and then checking for bubbles with an infrared camera. Since a 500°C heat treatment appreciably increases the bonding strength, the latter procedure is destructive in the sense that the wafers cannot be separated again after this procedure which, therefore, might be useful only for random testing purposes. The acceptance of bonded semiconductor wafers by integrated circuit manufacturers will mainly depend on the demonstration that such wafers may increase the yield of device processing. Such data are not yet available.

EXTENSION OF THE MICROCLEANROOM CONCEPT

In this section we will suggest an extension of the microcleanroom concept not concerning wafer bonding. It is a well-established trend that decreasing device dimensions lead to an increasing demand on the control of particle densities in the processing facilities with a corresponding dramatic increase in the cost of cleanroom facilities. It has repeatedly been discussed[21] that restricting the requirement of extremely low particle densities to within the processing equipment might be much less expensive than extending this requirement to the whole cleanroom fabrication facility. A weak point within this concept is the wafer transport from one processing step to the next which only in a few cases can be performed within UHV equipment clusters with appropriate vacuum transport facilities. In most cases wafers have to be transported between the different processing equipments through the environment and are thus exposed to particles in the surrounding air. For a particle-free transport with built-in cleaning capability an extension of the microcleanroom concept might be helpful. Contrary to the case of cleaning and drying in conventional spin-dryer set-ups the enclosure of the microcleanroom set-up moves simultaneously with the wafers to be dried. This has the advantage that the whole set-up acts as a kind of centrifuge and avoids turbulences which might drag particles back to wafer surfaces. We therefore suggest that transparent wafer transport boxes are constructed in such a way that they may be mounted easily and temporarily on a rotating motor, allowing the flushing of the spaces between the wafers and subsequent drying under the influence of an infrared lamp. In order to check for how long the space between wafer remains particle-free after spinning and drying in the microcleanroom we kept wafer pairs separated for different periods of times after spinning and drying before taking the spacers between the wafers out and allowing bonding to occur. The transparent cover remained closed during these waiting periods. Contrary to our initial expectations, no significant enhancement of the particle density as evidenced by particle-related bubbles at the bonding interface could be observed even after a waiting period of 64 hours (Fig. 5). The single particle seen after the 64 hour treatment is within the statistical distribution of what can happen even after 10 seconds. The results of Fig. 5 demonstrate that our microcleanroom set-up is capable of creating an essentially particle-free space between two wafers which can be kept particle-free for days. The fact that the wafers are closely spaced to each other (having a gap of only 0.5 to 1mm) appears to be at least as important for keeping the space between the wafers particle-free as keeping the cover closed.

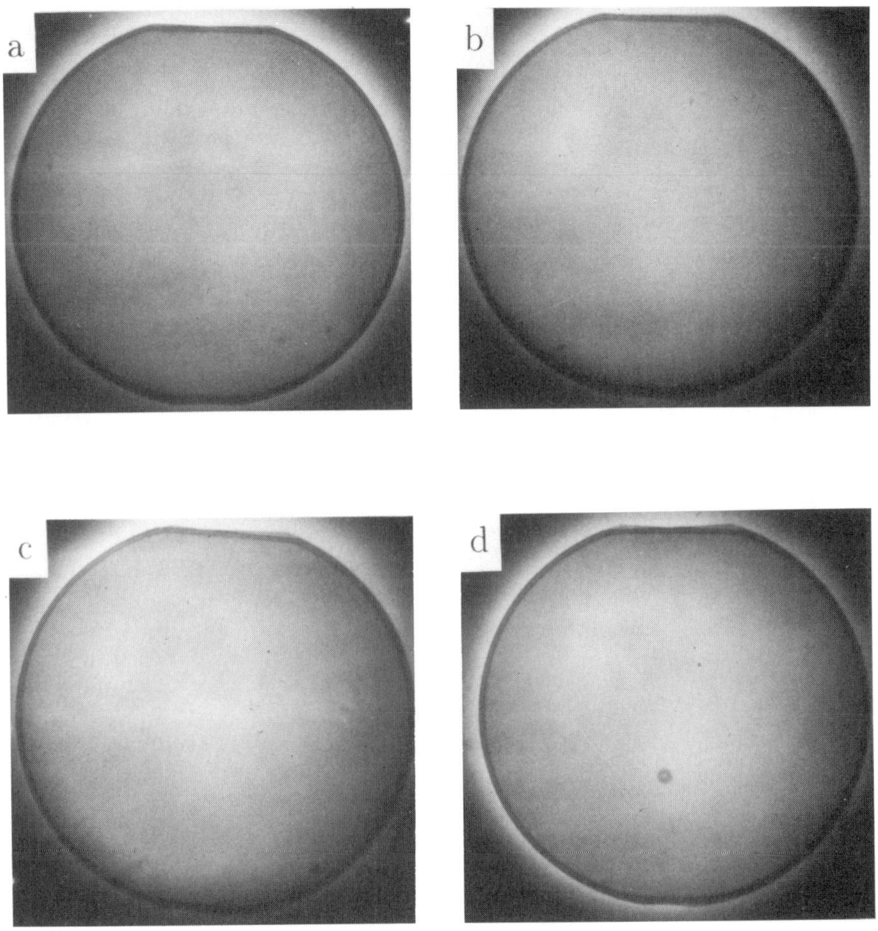

Figure 5. Infrared images of silicon wafer after taking out the spacers and performing the bonding procedure 10 seconds (a) 10 minutes (b), 1 hour (c) and 64 hours (d) after finishing the spinning and drying step in the microcleanroom set-up. No statistically significant increase in particle densities can be seen even after 64 hours.

This could be demonstrated in an experiment in which the cover was removed after spinning and drying and the wafers were exposed to the normal laboratory air for one hour before removing the spacers and bonding the wafers. No particle related bubbles could be detected. We speculate that the close spacing between the wafers produces stagnant layers which prevent air flow between the wafers which would drag particles from the outside environment into the space between the wafers. Therefore, we conclude that the spaces between wafers in the suggested wafer transport and cleaning boxes may be kept particle free for many hours or days provided that the spacing between the wafers is kept to values of 1 mm or below. These wafer transport boxes can, therefore, be used in the cleaning and drying cycle conveniently some time before the next processing step and then opened within the cleanroom environment of the processing equipment. After the processing step the wafers are stored back into the transport boxes within the processing equipment. The transport box will see the outside air only in a closed state except possibly to a certain extent in the deionized water cleaning step. Our experiments with wafer bonding have shown that even in a normal laboratory (no cleanroom at all) the space between two wafers may reproducibly be kept free of any particles down to the submicron range by the use of the microcleanroom set-up described earlier. Therefore, it appears worthwhile to check whether the extended microcleanroom concept in terms of wafer transport boxes, which may allow cleaning and spin-drying in an almost closed state, could possibly lead to much less stringent cleanroom requirements in processing facilities.

SUMMARY

Particle-free bonding of semiconductor wafers can be performed in a normal laboratory environment by using a simple "microcleanroom" set-up. Wafer bonding performed at room temperature is reversible and can be used to protect semiconductor surfaces against particles and against organic vapor contamination from wafer boxes.

The microcleanroom set-up allows to conveniently create particle-free spaces between wafers which can be kept particle-free for many days. Therefore, we suggest to look into the possibility of constructing wafer transport and cleaning boxes in which in a cleaning and drying step a transparent wafer box spins together with the wafers for creating particle-free spaces in wafer boxes instead of in the whole processing facility. The processing equipment itself would still need to be kept superclean and with an extremely low particle density, but these wafers transport and cleaning boxes would permit to tolerate much higher particle densities in the area between the processing equipment.

ACKNOWLEDGEMENTS

We appreciate the financial, material and technical support by the Alexander von Humboldt Foundation, the McCloy Foundation, Shin-Etsu Handotai, Texas Instruments, Sumitomo Electric Industries, Siemens Corporation, and Ford Motor Company. We thank A. Muller and H. Krautter from AT&T for collaboration in some of the organic contamination experiments and Dr. H. Z. Massoud for valuable discussions.

REFERENCES

1. J. B. Lasky, Appl. Phys. Lett. 48, 78 (1985).
2. W. P. Maszara, G. Goetz, A. Caviglia and J. B. McKitterick, J. Appl. Phys. 64, 4943 (1988).
3. J. Haisma, G. A. C. M. Spierings, U. K. P. Biermann and J. A. Pals, Jpn. J. Appl. Phys. 28, 1426 (1989).
4. K. Petersen, P. Barth, J. Poydock, J. Brown, J. Mallon, Jr., and J. Bryzek, IEEE Solid-State Sensors Workshop, Technical Digest, Hilton Head Island, South Carolina, USA, p. 144 (1988).
5. P. W. Barth, Sensors and Actuators A21-A23, 919 (1990).
6. V. Lehmann, U. Gösele and K. Mitani, Solid State Technol. 91 (April 1990).
7. R. Stengl, T. Tan and U. Gösele, Jpn. J. Appl. Phys. 28, 1735 (1989).

8. R. Stengl, V. Lehmann, K. Mitani and U. Gösele, Proc. IEEE SOS/SOI Techn. Conference, Stateline, Nevada, Sept. 1989, p. 123.
9. R. Stengl, K.-Y. Ahn and U. Gösele, Jpn. J. Appl. Phys. 27, L2364 (1988).
10. R. Stengl and U. Gösele, U. S. Patent No. 4,883,215.
11. V. Lehmann, K. Mitani, I. Ong and U. Gösele, Advanced Materials 2, 372 (1990).
12. V. Lehmann, K. Mitani, R. Stengl, T. Mii and U. Gösele, Jpn. J. Appl. Phys. 28, L2141 (1989).
13. H. Ohashi, K. Furukawa, M. Atsuta, A. Nukagawa and K. Imamura, Proc. Int. Electron Device Meeting, 678 (1987).
14. T. Abe and M. Nakamura, in "Silicon-On-Insulator Technology and Devices", D. N. Schmidt, Ed., p.61, The Electrochem. Soc., (1990).
15. V. Lehmann, K. Mitani, I. W. V. Ong, R. Stengl and U. Gösele, in "Silicon-On-Insulator Technology and Devices", D. N. Schmidt, Ed.,p. 213, The Electrochem. Soc., (1990).
16. K. Mitani, V. Lehmann, D. Feijoo, U. Gösele, and H. Z. Massoud , Jpn. J. Appl. Phys., submitted.
17. V. Lehmann, K. Mitani, D. Feijoo, U. Gösele, A. Muller and H. W. Krautter, unpublished.
18. K. Budde, "Analytical Techniques for Semiconductor Materials and Process Characterization," B. O. Kolbesen, D. V. McCaughan and W. Vandervorst, Eds., The Electrochem. Soc. PV 90-11, p. 215 (1990).
19. V.Lehmann and U. Gösele, U. S. Patent No. 4,962,879.
20. S. R. Kasi and M. Liehr, Appl. Phys. Lett. 57, 2095 (1990).
21. R. A. Hughes, G. B. Moslehi, D. M. Campbell, K. Radigan, W. Lukasek, and E. D. Castel, Microcontamination 35 (April 1990), and references therein.

ULTRA-CLEAN AIR IONIZERS FOR SUPPRESSION OF PARTICULATE SURFACE CONTAMINATION

Philip C. D. Hobbs and J. Samuel Batchelder

IBM Research Division
Thomas J. Watson Research Center
P.O. Box 218, Yorktown Heights, NY 10598

Vaughn P. Gross and Kenneth D. Murray

IBM General Technology Division
1000 River St., Essex Junction, VT 05452

We describe two air ionization techniques for use in semiconductor processing, which do not themselves cause contamination. The first is a method for eliminating the characteristic emissions of metal and ammonium nitrate particles from corona air ionizers; it consists of a very simple modification to a standard bipolar DC corona ionizer, permitting the discharge to occur in a clean, dry environment in which ammonium nitrate cannot form and corona point corrosion is minimized. The particulate emission is essentially eliminated, and the point life greatly prolonged, allowing these units to be used in Class 1 clean rooms and reducing maintenance costs.

The second method is the use of a small, inexpensive pulsed Nd:YAG laser to ionize the air near a wafer by dielectric breakdown. This method is primarily intended for point-of-use applications inside process tools where turbulence makes it essentially impossible to transport balanced ionized air, and restricted space precludes locating a corona ionizer close enough to the wafer to be effective. We present data which demonstrate the effectiveness of both methods.

INTRODUCTION

Air ionizers are becoming more and more important in semiconductor manufacturing as device feature sizes shrink. Their use greatly reduces both the incidence of electrostatic discharge (ESD) damage and the settling rate of submicron particles onto wafers[1,2] The two most common operating principles of air ionizers are corona discharge and impact ionization due to charged particles from radioactive decay. Radioactive ionizers are relatively rare in semiconductor lines nowadays, because of safety concerns, fear of soft errors from trace radioactive contaminants in the product wafers, and their short range, but nevertheless have been the only technology available for use in restricted spaces.

Particles on Surfaces 3, Edited by K.L. Mittal
Plenum Press, New York, 1991

Corona discharge ionizers are very effective at neutralizing surface charges, at least in areas of unimpeded laminar air flow, and are quite safe; unfortunately, they usually generate large numbers of small (<0.1 μm) particles, primarily metal eroded from the corona points themselves[3,4].

Recently, Murray and Gross[5] and Murray, Ainsworth and Gross[6] have shown that proper point design and material selection, in particular the use of plain tungsten corona points instead of thoriated tungsten ones, and careful control of the tip shape, can reduce the eroded metal particles to insignificant levels (although the points still erode gradually), and have suggested that after this improvement, the major remaining source of contamination is ammonium nitrate (NH_4NO_3) precipitated on to the negative points from the ambient air. They were unable to collect enough of the airborne particles to demonstrate that they were ammonium nitrate, but energy-dispersive X-ray analysis (EDX) ruled out tungsten as a major contributor. FTIR and micro-Raman analysis of the white precipitate showed it to be mostly NH_4NO_3; this made it seem plausible that the particles were as well. Previous work by Blitshteyn and Shelton[7] found silicon, sulphur, oxygen, and phosphorus, which seemed to indicate that the white precipitate was mostly glass fibers from the clean room HEPA filters; however, their analysis was based on EDX using windowed detectors, which are insensitive to elements lighter than sodium (Z=11), and so missed the hydrogen, nitrogen, and most of the oxygen (Z=1, 7, and 8, respectively).

Besides generating contamination, this precipitation necessitates replacing the points every month, which contributes significantly to the cost of maintaining corona ionizers. It is therefore very worthwhile to seek a method of eliminating these ammonium nitrate particles.

The particles are not always present in significant quantities; during several months of monitoring in IBM's Essex Junction, Vermont facility, we have observed that they may be absent for months, and then suddenly millions are emitted in a short time. Their highly intermittent character makes the emission episodes difficult to analyze, particularly since we do not have fine control over several relevant variables, such as humidity and temperature. If the particles are indeed ammonium nitrate, then their formation requires a source of hydrogen. The most plausible one is atmospheric water vapour. We have shown in a recent paper[8] that excluding water vapour from the vicinity of the corona points (by flowing dry gas over the points) completely eliminates the ammonium nitrate buildup and the bursts of particles.

CLEAN CORONA IONIZER

The experimental apparatus for the clean corona ionizer is shown in Fig. 1. We modified one half of a commercial bipolar DC air ionizer (Semtronics, Inc. Model 2001) by covering the right-hand half of the 2 meter long extrusion with PVC tape. A hole of $\approx$1 cm diameter was cut in front of each corona point. Each hole had a 1 cm long sleeve (made from 0.5 inch OD Tygon tubing) pushed into it to prevent moist air from being entrained into the neighbourhood of the point by turbulence. The space under the PVC cover was continuously flushed with clean, dry N_2 gas via a perforated Teflon tube with a high-efficiency in-line filter. The corona points were pure tungsten wire. In order to provide a control, we left the left-hand piece of the ionizer uncovered. Sleeve dimensions and placement are important; if the sleeve is too long or too small in diameter, the discharge is suppressed, and if the corners of the sleeves are too close to the discharge, they erode, producing large particle fluxes. In the setup described, there was no sleeve erosion, the point currents of the covered points were within specification, and the air ionization (measured by a

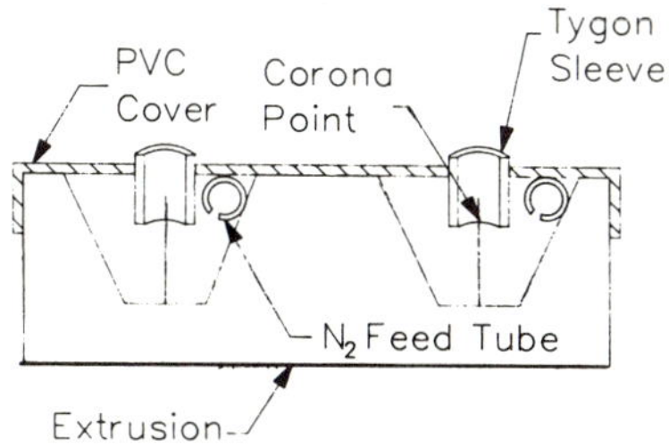

Fig. 1. Cross-section of the modified end of the corona ionizer, showing the PVC cover, corona point sleeve, and perforated N_2 supply tube.

charged plate monitor) was within 10% of that on the control side; the automatic balancing circuitry of the ionizer was apparently able to cope with the perturbation due to the sleeves.

In our original experiment, the bar hung 60 cm from the clean room wall, in unobstructed 50 cm/s vertical air flow about 20 cm below the ceiling HEPA filters. The particle concentrations were monitored with three TSI 3760 condensation nucleus counters (CNCs), fed with identical Tygon tubes of 1.5 m length, whose ends were placed in free air, respectively, approximately 15 cm below a negative corona point with flowing nitrogen; the same place relative to a negative point in air; and at the same height, but 15 cm behind the ionizer bar (for a background count).

Most of the time, the system was quiescent, with only gradually increasing particle emission from the air-immersed corona points, and a roughly constant count from the nitrogen-immersed points, due to particles released from the new in-line filter (this gradually subsided to the background level over a few weeks, and was constant irrespective of whether or not the ionizer was operating). The characteristic white NH_4NO_3 deposits collected on the air-immersed points as usual, but there was no visible contamination of the N_2-immersed point.

One afternoon, we experienced an unusually intense rain storm, which caused a humidity spike of a few hours' duration. Within an hour or so of the humidity peak, the particle counts from the air-immersed point rose by four orders of magnitude, to a high of over 50 000 per cubic foot. No change was observed in either the N_2-immersed or background counts. Fig. 2 compares the data from the three counters; the fourth curve shows a run of typical data from the N_2-immersed point from earlier the same day (normal temperature and humidity), which demonstrates that the particle count from the N_2-immersed point was not affected by the high humidity.

It is clear from these results that eliminating the water vapour in the vicinity of the corona points suppresses the particle generation mechanism very effectively. Our long term follow-up experiments[9] have demonstrated that dry compressed air works at least as well as nitrogen, that the particle counts from the dry-air-immersed corona points remain at the background level (<1 particle/cu.ft.) once the in-line filter has flushed itself, and that the life of the corona points is greatly extended. Sputtering is the physical ejection of material from a surface due to high-energy bombardment; since an energetic ion is not significantly impeded by a few monolayers of water, one would not expect the sputtering rate to be strongly dependent on relative humidity, and so it appears that the loss of point material represents corrosion rather than sputtering as was previously believed.

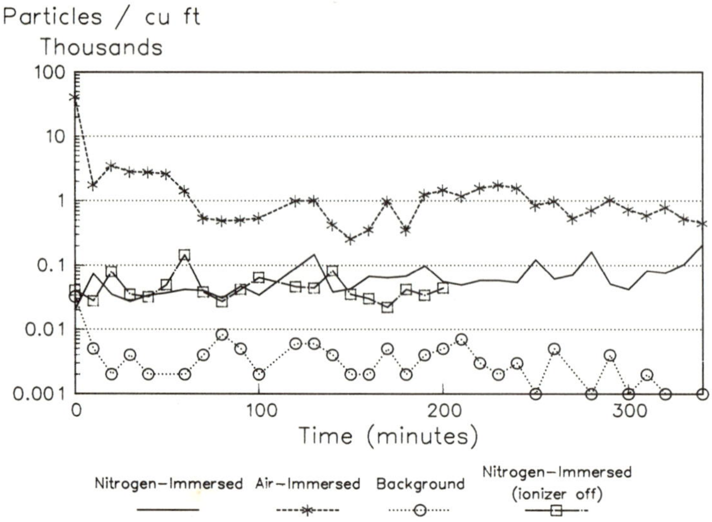

Fig. 2. Measured particle counts/cu. ft. 15 cm below corona points during high humidity episode, compared with background counts and a typical run of counts from the N_2-immersed corona point taken under quiescent conditions.

LASER AIR IONIZER

The modified corona ionizer is a highly effective way to generate large volumes of ultraclean ionized air in unimpeded laminar air flow. The major remaining problem is what to do about static charges inside semiconductor manufacturing tools. In most vacuum process tools, the wafer is held in a grounded stainless steel chuck, and so does not charge appreciably. In a fluorocarbon wafer boat or in a stepper, on the other hand, most of the surfaces are nonconducting, and triboelectricity is important. An oxidized silicon wafer sliding on a Teflon surface can develop 10 kV in a reasonably dry environment, and wafers transported on air tracks or rubber belts charge too. Masks have been known to acquire static charges during handling, which may result in ESD damage to the chromium pattern or in more mask contamination by small particles.

It is very difficult to apply corona air ionization inside process tools like steppers, for several reasons. The air flow inside a stepper is slow and highly turbulent, because of the many obstructions it encounters; the resulting mixing and ion recombination makes it essentially impossible to transport balanced ionized air any distance whatsoever. There is very little space available close to the wafer, which makes it hard to install corona points close enough to the wafer to help. There are many moving parts, operating in a tightly confined volume, so finding space for any added objects is difficult.

There are a few ways out of this dilemma. One is to use a pulsed DC ionizer with a slow repetition rate, so that the positive and negative ions are separated in time and do not mix; unfortunately, in a complicated environment like a stepper, it is difficult to achieve charge balance on all surfaces, due to the different mobilities of positive and negative ions, and to know just where the ions will be transported in the slow, random air currents. Even when the average voltage is zero, pulsed-DC systems cause a triangle-wave oscillation of the surface voltages, which may interfere with the proper operation of the stepper. Another way is to use radioactive sources inside the stepper itself. This alternative is not very attractive because of the danger of soft errors due to contamination of the wafer with minute quantities of radioactive material. Ionization is rarely used inside process tools, primarily because of these problems.

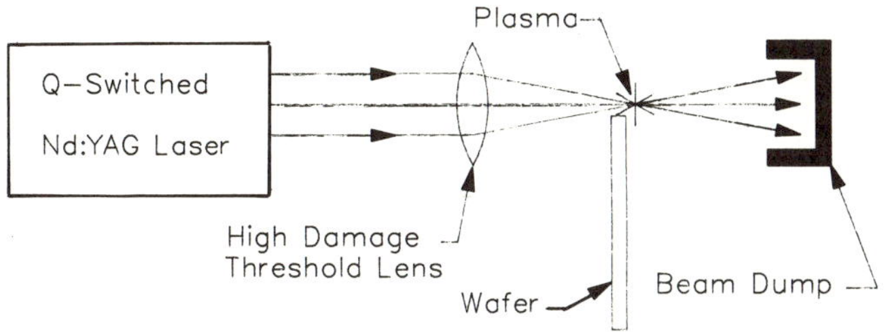

Fig. 3. Schematic diagram of the laser air ionizer.

We have developed a method for overcoming most of these difficulties, namely the laser plasma air ionizer shown in Fig. 3. In its most basic form, it consists of a passively Q-switched Nd:YAG laser, a single-element lens, and a beam dump. The laser is one of a number of very small (9 X 5 X 3 cm), inexpensive ($6-10K in unit quantity), commercially available units, originally designed for opthalmological applications, which have pulse energies of 10-20 mJ, pulse widths of 5-15 ns, and repetition rates of 0.5-5 Hz. The laser pulse has a peak power of 2-5 MW, which when focussed to a 10 μm diameter spot, gives a flux of the order of 10^{13}W/cm^2, which is sufficient to cause dielectric breakdown of the ambient air[10-12]. The dielectric breakdown results in very efficient absorption of the laser energy, and produces an air plasma which cools and expands rapidly with a small flash and a noise like fingers snapping, achieving a maximum apparent diameter of a millimeter or so. The conductivity of the air lasts for only 2 ms or so, because the ions recombine rapidly when the temperature drops. In the presence of an applied electric field, for example from a nearby charged wafer, ions drift out of the plasma. During the few-millisecond persistence of the plasma, some of the ions will drift far enough away that they will not encounter oppositely charged ions, and hence will not recombine. Fig. 4 illustrates this process. The wafer, shown negatively charged, attracts positive ions from the plasma, and repels negative ones. The positive ions drift toward the wafer, eventually hitting and partially neutralizing it.

The neutralization process continues on subsequent laser pulses. The total charge extracted from the plasma on each pulse is approximately proportional to the strength of the applied field, so that the charge on the wafer decreases approximately geometrically with the number of pulses (the decay would be exponential in time if the plasma were continuous rather than episodic). Figure 5 shows the decay of the voltage on a wafer attached to a charged-plate monitor, with a capacitance of about 20 pF. The laser was a 10 mJ, 0.4 Hz rep rate unit made by Laser Photonics. The plasma was about 3 mm from the edge of the wafer, and about the same distance below the back side. The wafer was initially charged up to 200 V. The geometric decay has about the same time constant for voltages up to the +- 5 kV limit of the charged plate monitor, and exhibits no discernible polarity dependence. Although the low rep rate of this particular laser makes the decay relatively slow, lasers with more pulses per burst or a higher rep rate will accelerate the decay considerably. We are currently conducting tests with a 1 Hz burst rep rate, 5 pulse/burst (total 5 pulses per second), 20 mJ laser made by Kigre Inc.

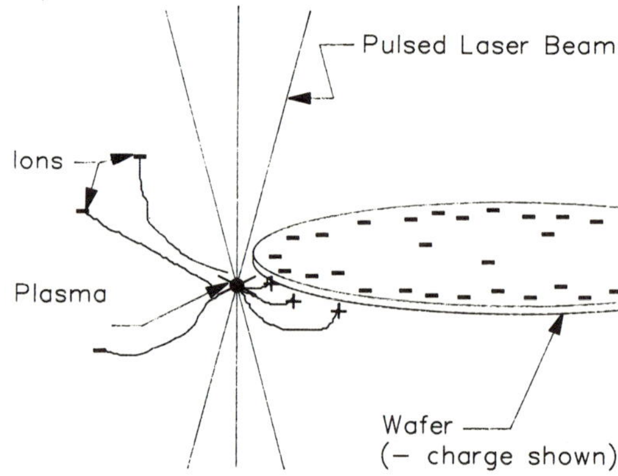

Fig. 4. Detail of the laser-produced air plasma and the wafer, showing the ions drifting under the applied E field of the charged wafer.

The laser ionizer exhibits no measurable particle production. To test it, we positioned the sampling tube of a CNC a few centimeters directly downstream of the laser plasma, oriented at 90 degrees to the beam axis so that the beam did not hit the tube. Another CNC was placed a few feet away to give a background count. We let the laser pulse continuously at 0.4 Hz for half an hour, then turned it off for fifteen minutes, while measuring the particle counts. Figure 6 is a plot of the resulting data. The data points are 20 second totals, plotted as counts per minute; thus a single particle appears as a short triangular spike of 3 particles/minute. There was no appreciable change in the particle count when the laser was turned off, leading us to conclude that the laser produced plasma was clean down to the level of one particle (of 0.03 μm or larger) every few minutes. Dividing the background count rate by the small CNC sampling flow (5 ml/s) gives an unrealistically pessimistic upper limit of around 400 particles per cubic foot. The actual sensitivity of the measurement is considerably higher, because the sampling tube was positioned close by and directly downstream of the plasma, so that the CNC should have intercepted a large fraction of any particles produced. Even assuming a 10% capture probability for each particle, the laser plasma is shown to be producing much less than one particle per pulse.

Besides cleanliness and effectiveness, the laser air ionizer has a number of other attractions. It is relatively inexpensive, even with each tool having its own laser, being roughly competitive with a high performance hood ionizer such as the unit mentioned above. Costs should decrease with manufacturing volume, and could be reduced further by using a single larger laser to serve an entire tool cluster. The laser specifications are quite loose: great pulse energy repeatability and low beam divergence are not required. The laser ionizer's most salient advantages, however, are that it works at a distance, does not require laminar air flow, and is inherently charge balanced; thus it can be used in a crowded tool interior where mechanical parts must share the same space as the laser beam. The laser light can be piped around with mirrors if desired, so that one laser can serve several ionization sites, and can even be located outside the tool for easy maintenance. Due to the competing effects of peak power density and the more rapid cooling of a smaller plasma, the optimal focal spot size, and hence the numerical aperture (NA), for air ionization depends on the pulse energy, duration, and wavelength; the laser used here worked best at about 0.05 NA. Such a low NA allows the lens to be placed at a distance of up to 50 cm from the ionization site, further easing the mechanical constraints. The beam dump also can be located remotely if required.

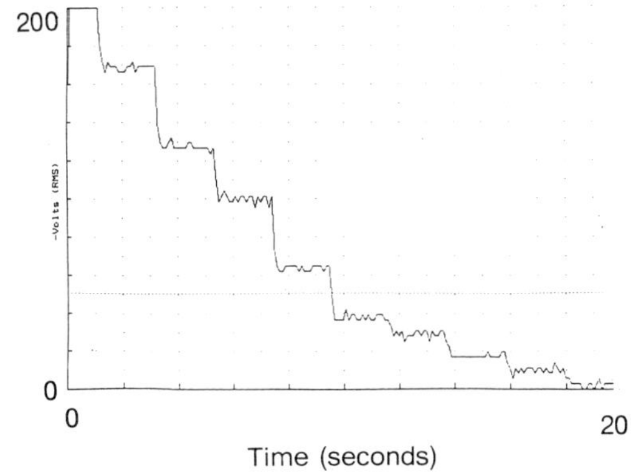

Fig. 5. Plot of the voltage on a wafer attached to a charged-plate monitor (20 pF) vs. time, showing the effectiveness of the laser plasma air ionizer. Note the roughly geometric staircase decay of the wafer voltage with the number of pulses, which indicates that the plasma acts like a resistor to ground while it is present.

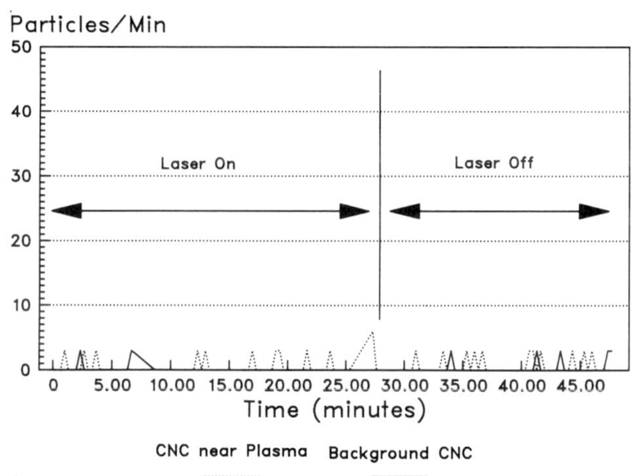

Fig. 6. Particle counts per minute (20 second totals) from a CNC placed just downstream of the laser plasma. Triangular spikes of 3 particles per minute represent single counts in a 20 second counting interval. No change is evident at the level of one particle every few minutes (>= 0.03 μm, 5 ml/s sampling flow) when the laser is switched off, which indicates that the laser plasma ionizer is clean down to this level.

There are a few drawbacks to this system as well. Laser safety is a concern; the system must be integrated into the tool with appropriate light shields and interlocks to prevent operator injury even under fault conditions, and to conform to occupational safety regulations. The laser beam must not hit the wafer or nearby parts; the wafer could be damaged and particles released from surfaces under high energy density bombardment. When undeveloped photoresist is present, the plasma should probably be positioned below the wafer in order to avoid resist exposure by the plasma flash, although since the plasma is so small and short-lived, such exposure would not be very great.

CONCLUSION

After eliminating the emission of particles of eroded point metal by proper point design, the residual particle generation from a commercial bipolar DC air ionizer is highly dependent on humidity in the immediate vicinity of the corona discharge. A simple modification of the ionizer design to exclude water vapour essentially eliminates this source of contamination, so that for the first time, one can regard a corona discharge ionization technology as really clean. Corona ionizers using this principle are currently under development at several vendors.

The problem of providing ionization in tight spaces with weak, turbulent air flow (such as the interior of process tools), which has hitherto gone almost completely unaddressed, can be solved in an effective, clean, and economical way through the use of laser-produced air plasmas. Although there are some engineering difficulties involved in applying this technology to pre-existing tools, it is the only non-radioactive technology available for the purpose, and its integration into the fab line is being pursued actively.

REFERENCES

1. K. Dillenbeck, Selection of air ionization within the cleanroom, in Proc. Ann. Tech. Mtg. Inst. Envir. Sci., pp.387-392, (1986).
2. R. P. Donovan, A. C. Clayton, and D. S. Ensor, The dependence of particle deposition velocity on surface potential, Proc. Ann. Tech. Mtg. Inst. Envir. Sci., pp. 473-478, (1987)
3. B. Y. H. Liu, D. Y. Pui, W. O. Kinstley, and W. G. Fisher, Aerosol charging and neutralization and electrostatic discharge in clean rooms, J. Envir. Sci., March/April, pp. 42-46 (1987)
4. M. Suzuki, H. Matsuhashi, and Y. Izumoto, Effectiveness of air ionization systems in clean rooms, in Proc. 34th Ann. Tech. Mtg. Inst. Envir. Sci., pp.405-412 (1988)
5. K. D. Murray, and V. P. Gross, Ozone and small particle production by steady state dc hood ionization: An evaluation, in 1989 EOS/ESD Symposium Proceedings, pp. 18-22 (1989)
6. K. D. Murray, G. F. Ainsworth, and V. P. Gross, Hood ionization in semiconductor wafer processing: an evaluation, in 1988 EOS/ESD Symposium Proceedings, pp. 195-200 (1988)
7. M. Blitshteyn, and S. Shelton, Contamination and erosion of cleanroom air ionizer emitters, Microcontamination, 3, 8 (August 1985)
8. P. C. D. Hobbs, V. P. Gross, and K. D. Murray, Suppression of particle emission in a modified clean room corona air ionizer, J. Aerosol Sci., 21, 3, pp.463-465 (1990)
9. K. D. Murray, V. P. Gross, and P. C. D. Hobbs, Clean corona ionizers, to be published in 1990 EOS/ESD Symposium Proceedings
10. C. G. Morgan, Laser-induced breakdown of gases, Rep. Prog. Phys., 38, pp.621-665 (1975)
11. R. J. Dewhurst, Comparative data on molecular gas breakdown thresholds in high laser-radiation fields, J. Phys. D: Appl. Phys., 11, pp. L191-L195 (1978)
12. D. E. Lencioni, Laser-induced air breakdown for 1.06 micron radiation, Appl. Phys. Lett., 25, 1, pp.15-17 (1974)

THE COLD JET PROCESS - AN ENVIRONMENTALLY SOUND ALTERNATIVE FOR PARTICLES REMOVAL FROM ADVANCED SUBSTRATES

Laurence C. Archibald
Pera International
Melton Mowbray
Leics LE13 OPB
United Kingdom

Dan Lloyd
Cold Jet Inc.
55W Techne/Centre Drive
Suite F
Cincinnati, Ohio 45150, USA

A process is described for cleaning surfaces by blasting with solid pellets of carbon dioxide. The cleaning mechanism is explained in terms of thermal shock and the effectiveness of contamination removal is assessed by a technique involving the measurement of optically stimulated electron emissions. Case studies illustrating the successful removal of particulates from various substrates are presented.

INTRODUCTION

The removal of contaminants, both particulate and film/coating type, can be effected by a multitude of techniques[1] involving physical or chemical actions. The success of any technique in removing a given contaminant depends upon a complex relationship between the various materials involved, but it is fundamentally dependent upon the bonding between contaminant and substrate.

For many applications, the difficulty in removing contaminants results in cleaning procedures which are so aggressive that the substrate surface is damaged, e.g., shot blasting. In addition, many chemical processes are now seen as being environmentally unacceptable and require expensive control procedures.

The "Cold Jet" process was developed to overcome the above problems and has found a number of applications[2] and some are presented here. The industrial grade carbon dioxide used in the process is originally extracted from the atmosphere and therefore there is no net increase in atmospheric CO_2 levels when using the equipment.

THE COLD JET PROCESS

The "Cold Jet" cryogenic blasting process involves the use of solid pellets of carbon dioxide as the blasting media for removing contaminants and coatings from substrates. The propellant stream may be air or inert gas, other than CO_2, such as nitrogen. Material removal occurs as a result of the thermal shock action of pellets in addition to a minor abrasive contribution.

Particles on Surfaces 3, Edited by K.L. Mittal
Plenum Press, New York, 1991

Figure 1. The Cold Jet Unit.

Figure 1 shows the "Cold Jet" unit and Figure 2 shows schematic diagrams of the equipment. Liquid CO_2 is injected into the hydraulic cylinder of the pelletizer assembly. The pellets produced are collected in a hopper and introduced into a dry compressed air stream by a rotating feed assembly. Pellets are then expelled through a gun and nozzle arrangement. Blast pressure, pellet feed rate, pellet density and pellet production rate are all variable.

The unit is designed to operate at blast pressures between $0.52 MN/m^2$ and $2 MN/m^2$, and pellet density may vary between $1.3 g.cm^{-3}$ and $1.6 g.cm^{-3}$

PELLET FORMATION PROCESS

The method of pellet formation is schematically represented in Figure 3 which shows the die plate in the hydraulic extrusion cylinder of the pelletizer assembly.

Liquid CO_2 is injected into the cylinder on the backstroke of the extrusion cylinder piston, and CO_2 "snow" is formed (i.e.low density solid CO_2). Upon compression by the piston, a "cake" of solid CO_2 is formed at the die plate. Under further compression (at around $14 MN/m^2$) solid CO_2 is extruded through the die plate.

At the edge of the extrusion an additional concurrent process occurs. At position A CO_2 vapour forms due to the temperature difference between the die plate and the solid CO_2. As extrusion proceeds, CO_2 liquid forms at position B due to increasing pressure. Upon exiting the

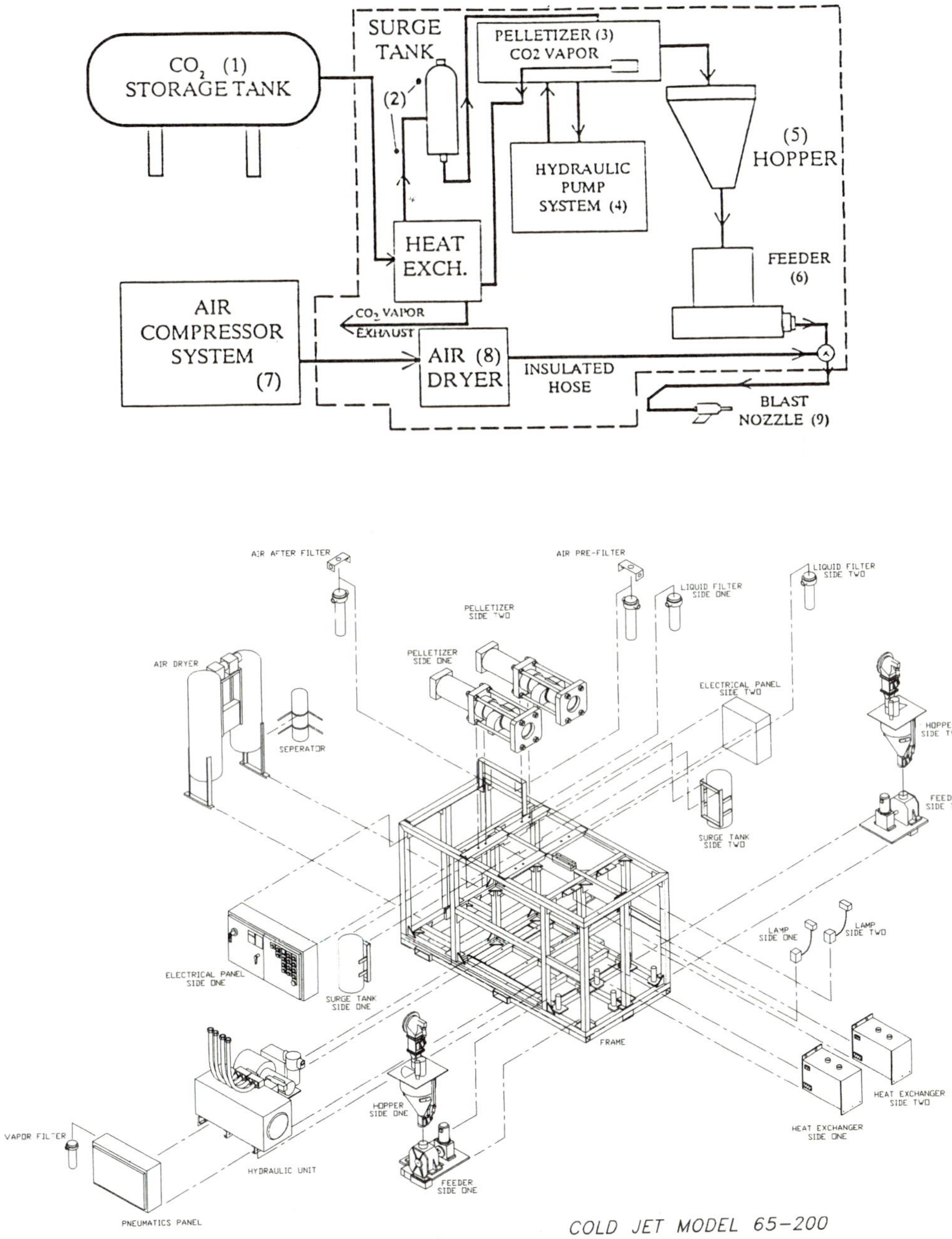

Figure 2. Schematic diagrams showing the constituent parts of the Cold Jet unit.

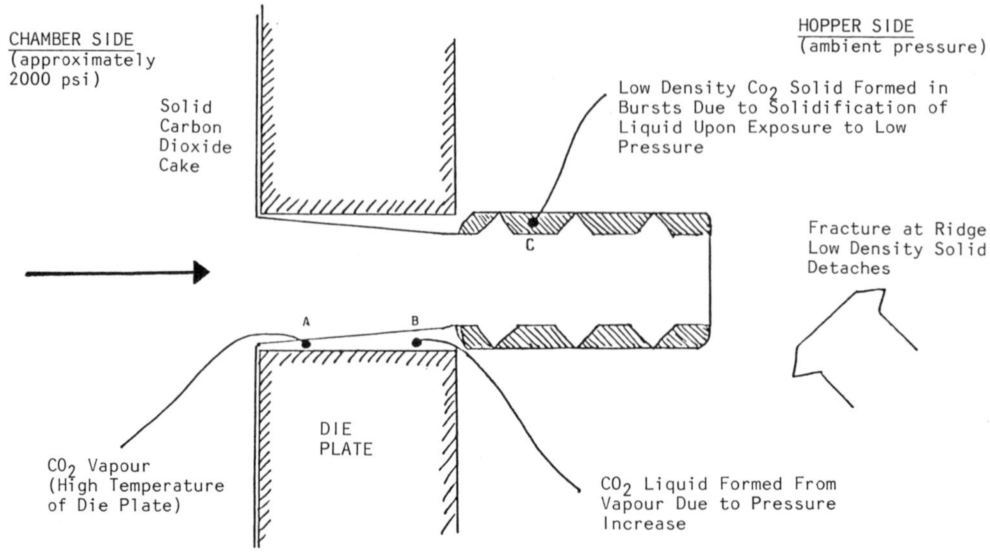

Figure 3. Solid carbon dioxide pellet formation process.

die plate, pressure on the liquid CO_2 drops to atmospheric and low density solid CO_2 is formed which coats the outside of the high density solid CO_2 extrusion.

This is a cyclic process, and regular small areas of low density CO_2 form to coat the extrusion. As the extrusion grows it fractures under its own weight, and a "pellet" drops into the hopper assembly. As this occurs, the low density solid detaches, and leaves a pellet of dimensions typically 5 mm x 3 mm with a characteristic ridged appearance (Figure 4). It is estimated that the final pellets comprise approximately 50% by weight of the liquid gas feed to the equipment, the remainder being exhausted to air.

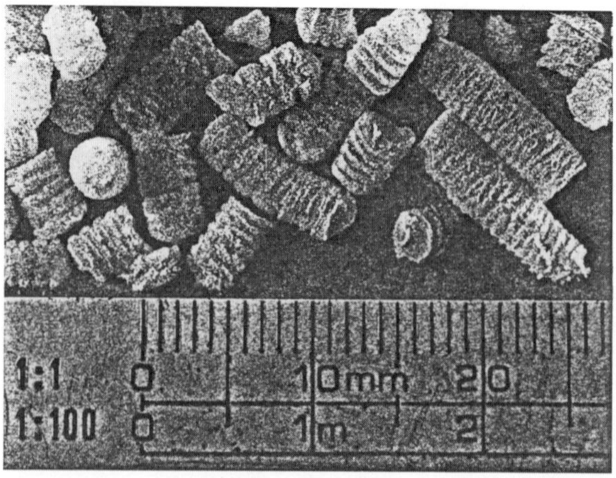

Figure 4. Typical pellets of solid carbon dioxide as used in the Cold Jet process.

CLEANING EFFECTIVENESS

The effectiveness with which cryogenic blasting will clean substrates has been investigated by using a Patscan OP1020 surface contamination monitor, supplied by Photo Acoustic Technology Inc, Westlake Village, CA, USA. This relatively new technique was selected as it has a proven ability to qualitatively assess levels of cleanliness on a range of substrate materials [3,4].

The principle of the process is that if a surface is illuminated by UV light, electrons will be stimulated to leave the surface. This is known as Optically Stimulated Electron Emmission (OSEE). These electrons are collected by a positively biased sensor and converted to a measurable voltage. Different materials emit different amounts of electrons under the influence of UV light and these differences permit the detection of surface contaminants. For example, the presence of oils, greases, fingerprints etc. will greatly reduce the degree of electron emission from a metal surface. The technique is sufficiently sensitive to detect layers of such contamination down to nanometers in thickness.

The equipment consists of a sensor containing the UV light source and a collector, an X-Y table to traverse the specimen beneath the sensor, and a computer to control the equipment and display the results either as a colour map or topograph. Local colour changes on the computer printout, or peaks and troughs on the topograph displays indicate the presence or absence of contaminants.

Results on a variety of different substrates and contaminants such as automotive oil, silicone release agents and fingerprints have demonstrated the relative performance of cryogenic blasting compared to other cleaning processes involving solvent degreasing. In all cases, cryogenic blasting reduced the detected levels of residual contamination below that which could be quantifiably measured by using control samples to generate a calibration curve.

It has therefore been concluded that cryogenic blasting is capable of extremely thorough removal of contaminants such as oils and greases and is equally effective on difficult to remove materials such as silicone release agents.

CLEANING MECHANISM - THERMAL SHOCK EFFECT

The surface temperature during cryogenic blasting was monitored by means of thermocouples mounted on a test piece, as shown on Figure 5.

The test-piece, made from 10 mm thick steel plate, had five thermocouples positioned in it; one level with the surface, the others at depths of 0.5 mm, 1 mm, 1.5 mm and 2 mm below the surface.

In an initial control experiment, the cryogenic blasting jet was moved at a constant speed across the uncoated surface of the test piece during which the temperature changes at the five thermocouple positions were monitored.

The test piece was then given a single, 100 micron thick, coat of alkyd paint. When this paint film was fully dry, the cryogenic blasting jet was traversed across the surface at the same rate as in the initial experiment. The temperature changes were again monitored as the paint was removed from the surface. In both experiments, a blast pressure $1MN/m^2$ and a feed rate of "60" (equivalent to 455 kg/hour) was used.

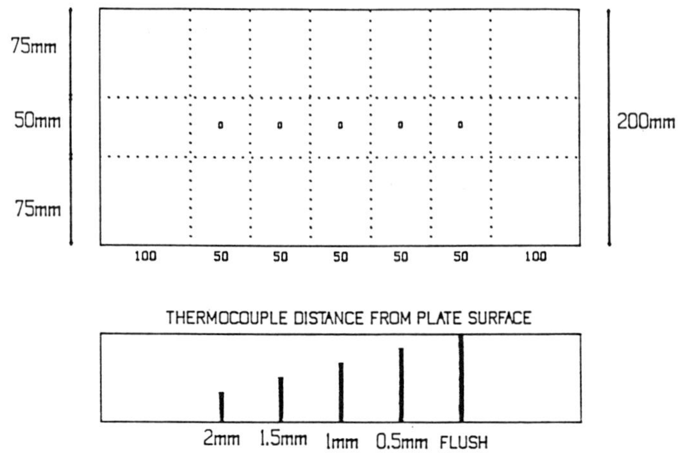

Figure 5. Location of thermocouples within steel test plate for thermal shock tests.

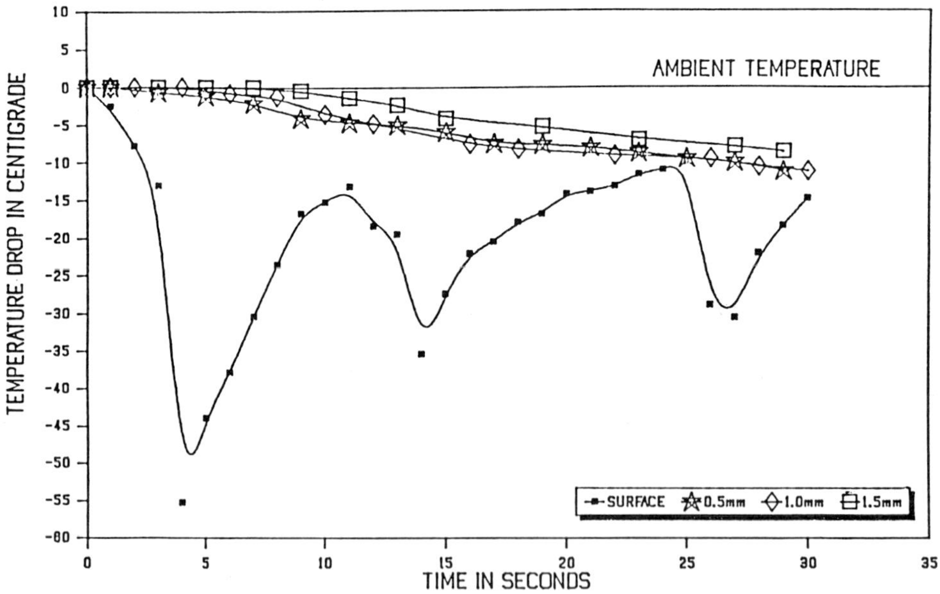

Figure 6. Temperature change versus time of blasting for bare plate with thermocouples at various depths.

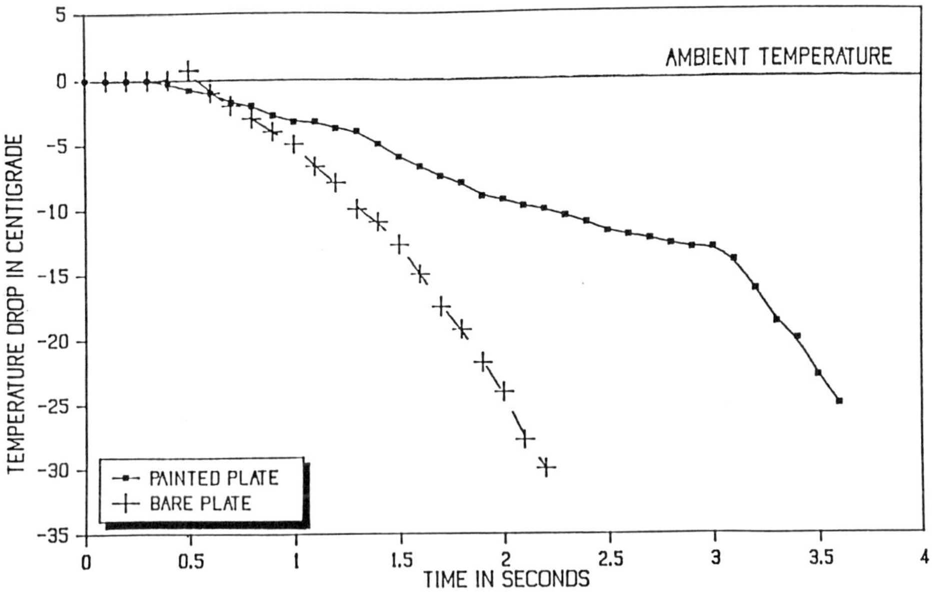

Figure 7. Temperature change versus time of blasting for surface mounted thermocouples on bare and painted plates.

Data obtained from these experiments are shown in Figures 6 and 7, respectively.

In the initial control experiment, the surface mounted thermocouple showed a large drop in temperature on each pass of the blast jet. In the second experiment, the same thermocouple with a coat of paint showed a much lower rate of drop in temperature until all the paint was removed. At this stage, the uncovered thermocouple reacted as in the previous experiment. Other thermocouples, in both experiments, showed general lowering of temperatures corresponding to the overall cooling of the test piece as the blasting process continued.

The results demonstrate that rapid temperature drops of the order of 60°C can be experienced at the surfaces of substrates, whilst in the body of the work pieces, at depths as low as 0.5 mm below the surface, such drops in temperature are not experienced, Thus, the "thermal shock" effect is confined to a thin surface layer.

THE MECHANISM OF CONTAMINATION REMOVAL

The thermal shock exerted by the impact of solid carbon dioxide pellets at -80°C on the workpiece has been shown to be confined to a thin surface layer less than 0.5 mm in depth. This has the effect of setting up differential thermal contraction stresses between the substrate and the contamination layer or coating. If the magnitude of this stress differential is greater than the interfacial bond strength between the substrate and the surface layer(s), then, the bond will be broken and cleaning effected.

This mechanism applies whether the contamination is a solid coating (e.g. paint), a "liquid" coating (e.g. grease) or particulate-based (e.g. burnt-on carbon). The hardness of the coating is of less importance than its adhesion to the substrate. It should be noted that the crucial factor is the presence of a clearly defined interface and that where this is absent, as in the case of advanced stages of corrosion, surface removal of integral films will not occur.

It should also be borne in mind that liquid carbon dioxide is an excellent solvent for organic materials. Currently, unpublished work involving high speed photography has indicated that, at impingement pressures, the solid pellets can be converted to liquid which could assist in the cleaning process. This would also explain the proven ability of the process to clean small holes of the order of 50-100 micron diameter.

Loose material is removed by a combination of thermal shock, air flow (up to 8.5 m^3/min.) and mechanical impingement/abrasion. This latter effect has been found to be of minor importance and to result in no surface damage to substrates of hardness in excess of 135 Vickers [5]. Softer materials, such as polymers, are affected although to a lesser degree than occurs by other, abrasive blasting processes. Even this effect is eliminated when cryogenic blasting is used to remove coatings from soft substrates. This is illustrated in Figures 8 and 9.

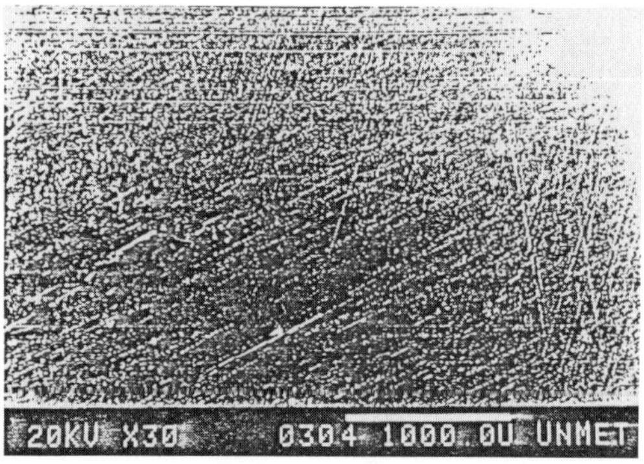

Figure 8(a). PEEK/carbon fibre material as received.

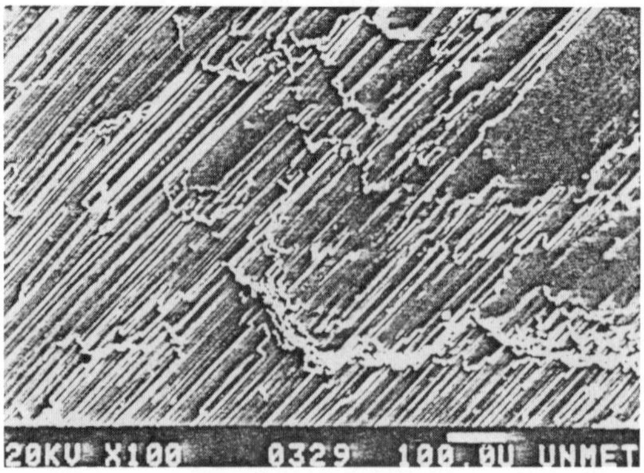

Figure 8(b). PEEK/carbon fibre material after cryogenic blasting for 10 seconds.

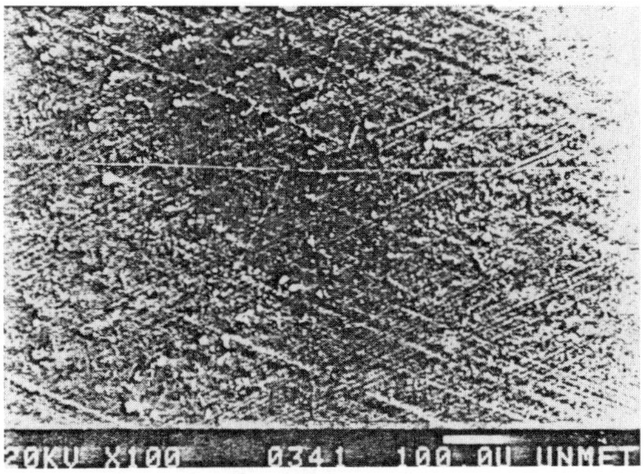

Figure 9(a). PEEK/carbon fibre material after cryogenic blasting to remove paint coating (x100).

Figure 9(b). As Figure 9(a) - Magnification x 10000.

Figure 8 shows the effect of 10 seconds continuous blasting on the surface of a PEEK/carbon fibre composite material as examined by a scanning electron microscope (SEM). This results in removal of the resin-rich surface although little damage or separation of the reinforcement is evident. However, when cryogenic blasting is used to remove a paint coating from the same material, no substrate surface damage can be detected by the SEM (Figure 9). This damage effect does not occur on tougher polymeric materials such as epoxy/carbon fibre, as shown in Figure 10.

EXAMPLES OF SURFACE CLEANING BY CRYOGENIC BLASTING

One of the beneficial aspects of cryogenic blasting is that, because the CO_2 particles are non-conductive, the process can be used on live electrical equipment. During trials, it has been found that even delicate components such as printed circuit boards (PCB) can be successfully cleaned.

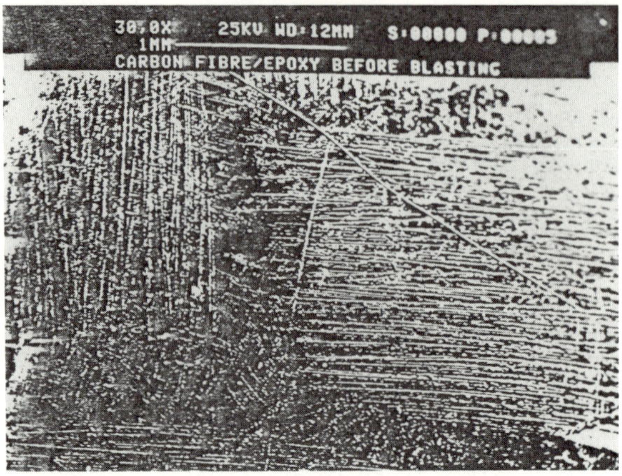

Figure 10(a). Epoxy/carbon fibre material before cryogenic blasting (x30).

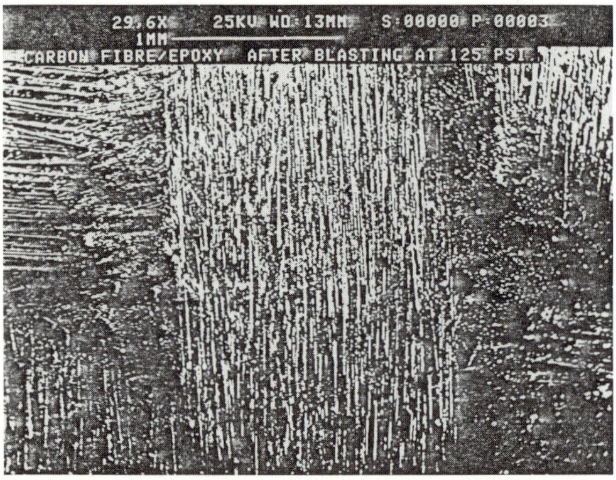

Figure 10(b). Epoxy/carbon fibre material after cryogenic blasting (x30).

Figure 11 shows optical micrographs of a PCB track which had been badly contaminated by hard, metallic particles which had resisted attempts to remove them by brushing or blowing with air. The component was treated for a few seconds at 0.67 MN/m^2 blast pressure in order to achieve a clean surface.

Another example of PCB cleaning is illustrated in Figure 12. Initially, the board was dirty and what appeared to be light corrosion of the tracks was present. All of this contamination was removed by cryogenic blasting without damaging the structure and even delicate connection wires were left intact, as shown on the bottom right-hand corner of the optical micrograph.

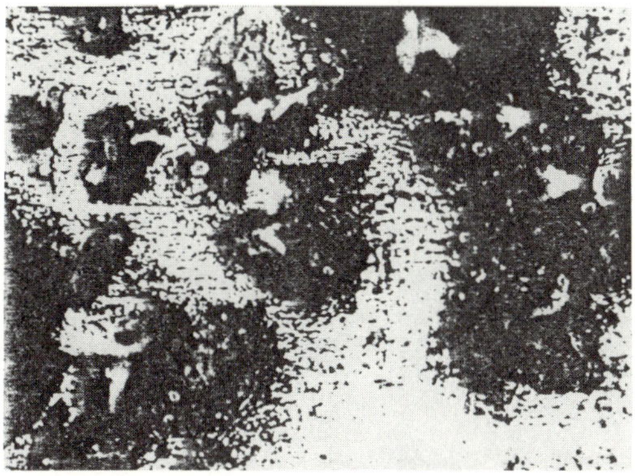

Figure 11(a). Metal particles on copper PCB track (x420).

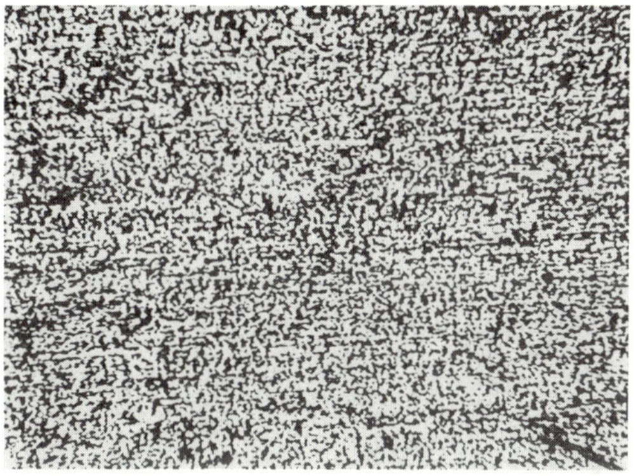

Figure 11(b). Clean copper PCB track after cryogenic blasting (x420).

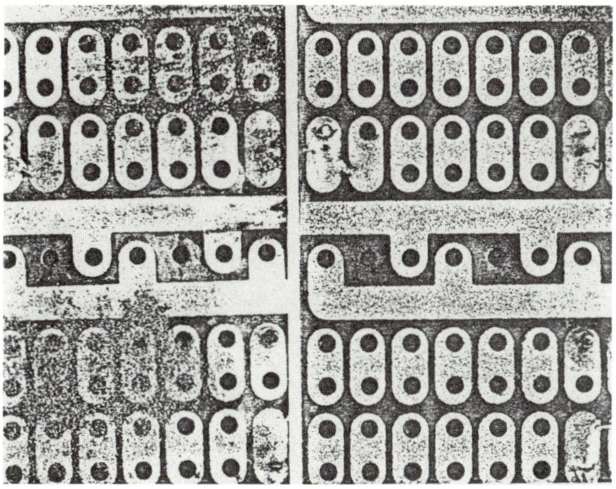

Figure 12. Section of PCB before (left) and after (right) cryogenic blasting (x3).

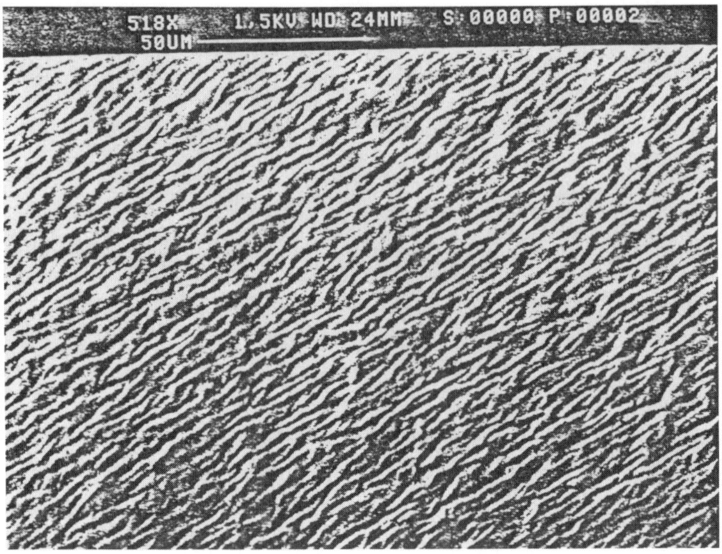

Figure 13(a). As-received PVC tube (x518).

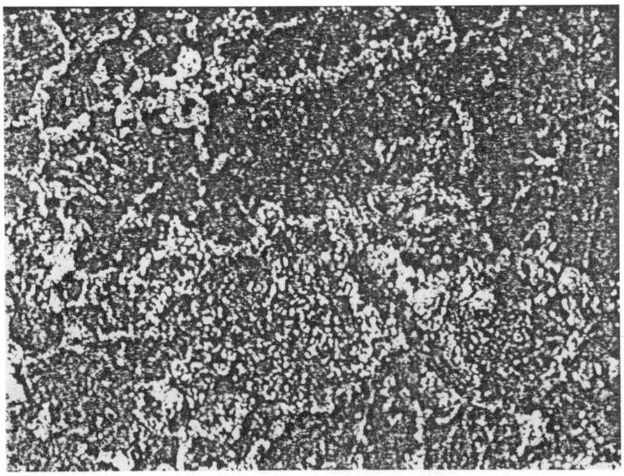

Figure 13(b). PVC tube contaminated with calcium-rich particles (x75).

Figure 13 shows SEM micrographs of a PVC tube which was used to pump a fluid containing calcium-rich particles. The fluid flow induced a charge pick-up on the particles which resulted in their adhering to all surfaces of the equipment, including the tubing. Cryogenic blasting removed the deposits shown in Figure 13(b) and caused no erosion of the PVC surface (Figures 13(a) and 13(c)).

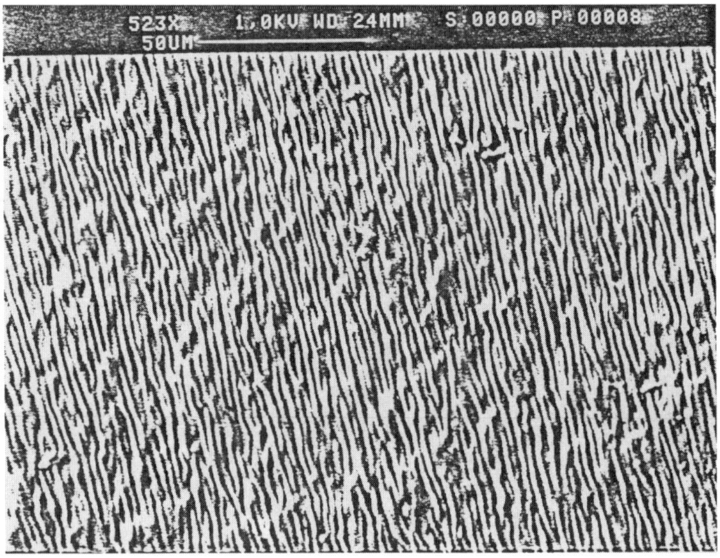

Figure 13(c). As Figure 13(b) after cryogenic blasting (x523).

CONCLUSION

Cryogenic blasting using solid pellets of carbon dioxide as the blasting media has been found to be an effective way of removing a range of contaminants including particulate-type deposits from a variety of substrates whilst incurring zero (or minimal) damage to the underlying surfaces. The process does not cause environmental problems either by polluting the air or waterways and no clean-up of used media is required due to sublimation of the pellets directly to gas.

The combination of the three factors of cleaning ability, lack of damage, and environmental-friendliness gives "Cold Jet" cryogenic blasting unique industrial application areas when compared to competitive processes such as solvent cleaning and plastic media or other blasting processes.

REFERENCES

1. Metals Handbook, Volume 5. "Surface Cleaning, Finishing, and Coating". American Society for Metals, 1982.

2. L. C. Archibald, Cryogenic blasting as a tool cleaning process, Composite Tooling Conference, Amsterdam, 1989.

3. T. Smith, NASA Report NA58 33694, 1980.

4. M. K. Chawla, A non-destructive technique for surface contamination/quality monitoring, Proc. Inst. Metal Finishing Conference, April 11-14, 1989.

5. Pera International Research Report 73283, 1990.

IDENTIFICATION AND REMOVAL OF STORAGE INDUCED PARTICLE CONTAMINATION ON SILICON WAFER SURFACES

Brent Carlson, Lillian Gulbrandsen and Donald C. Grant

FSI International
322 Lake Hazeltine Drive
Chaska, MN 55318

Control of contamination on the surface of wafers during the manufacture of state-of-the-art microcircuits is imperative if high yields are to be obtained. Hence, wafers are often cleaned using a modified RCA clean to remove organic, metallic and particle contamination prior to microcircuit manufacture. Recently, wafers obtained from a particular vendor were found to have heavy particle contamination after a modified RCA clean. The wafers were also found to have a hydrophobic surface indicating the presence of heavy organic contamination. Analysis of the wafer surface revealed that it was covered with film of a polydimethylsiloxane-like compound. The source of the compound was found to be a gasket in the storage boxes used to transport the wafers. This paper describes the methodology used to identify the source of the contaminant and the cleaning methods developed to eliminate the particle contamination which it caused.

INTRODUCTION

From crystal growth to a finished semiconductor device, the fabrication and surface preparation of silicon proceeds through a host of tightly controlled process steps before the end result – a working semiconductor microcircuit. Silicon wafers are traditionally fabricated by outside vendors and shipped to semiconductor microcircuit manufacturers. Before use, these wafers may be stored for indefinite periods of time, possibly incurring unwanted surface contamination which may affect circuit fabrication.

In order to ensure that the wafers have minimal contamination prior to processing, the wafers are commonly cleaned using a modified RCA type clean which creates a passive surface with minimal metallic and ionic contamination[1,2]. A typical modified RCA clean consists of four steps:

Particles on Surfaces 3, Edited by K.L. Mittal
Plenum Press, New York, 1991

Step	Abbrev.	Volume Ratio	Time, sec.	Purpose
Sulfuric Acid/ Hydrogen Peroxide	SPM	4:1	45	Removal of heavy organic residues
Dilute Hydrofluoric Acid/DI Water	DHF	1:100	30	Removal of mobile ions and thin layers of silicon dioxide
Ammonium Hydroxide/ Hydrogen Peroxide/ DI Water	APM	1:1:5	90	Removal of lighter organics, particles and surfaces passivation via regrowth of a thin protective oxide
Hydrochloric Acid/ Hydrogen Peroxide/ DI Water	HPM	1:1:5	90	Removal of metallic impurities

Recently, a lot of wafers which was stored for approximately one year was received and processed using a modified RCA clean. Results of this clean revealed that large numbers of particles were added to the wafer surface after processing as shown in Table I. In this paper we will address the progress made to date in identifying and removing this storage induced particle contamination.

PROBLEM IDENTIFICATION

The hydrophobic nature of the wafers as received indicated that the poor cleaning performance was a result of a thick layer of some type of organic contamination. Therefore, Static Secondary Ion Mass Spectrometry (Static SIMS) was used to gather information on the surface properties of an unprocessed wafer. This is a recently developed technique used mainly for detection of impurities on polymer surfaces[3,4]. It uses a low dose of primary ions to bombard the surface. During bombardment secondary ion fragments representative of the surface chemistry are released and analyzed in a mass spectrometer. Static SIMS is an extremely surface sensitive technique which yields data from the top monolayer of the surface, yet sensitive enough to detect impurities at low concentrations.

Table I. Comparison of Particle Levels Before and After Cleaning on Typical and "Problem" Wafers.

Particle Size (μm)	Particles per 200mm Wafer					
	Typical Wafer			"Problem" Wafer		
	Pre	Post	Delta	Pre	Post	Delta
>0.15	885	393	-492	1645	4873	+3228
>0.17	423	125	-298	1084	3831	+2747
>0.20	228	63	-165	661	2735	+2074
>0.23	103	33	-70	358	1817	+1459
>0.25	70	23	-47	253	1521	+1268
>0.30	33	15	-18	80	933	+853
>0.32	29	14	-15	53	813	+760
>0.35	17	8	-9	24	610	+586
>0.38	11	5	-6	6	386	+380
>0.40	9	4	-5	3	314	+311

Results of the Static SIMS analysis indicated the presence of a siloxane-like compound on the wafer surface as shown in Figure 1. Since there are no siloxane compounds in the cleaning equipment or cleaning solutions used, the storage or shipping boxes became suspect. Strips of silicone rubber on the inside cover of the storage boxes used for transporting the wafers were analyzed by Attenuated Total Reflection (ATR).

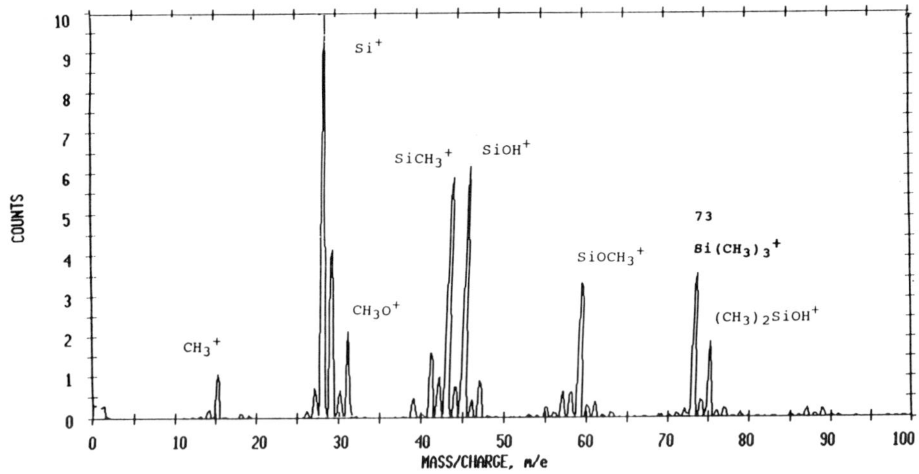

Figure 1. Positive Static SIMS Spectrum of "Problem" Wafer.

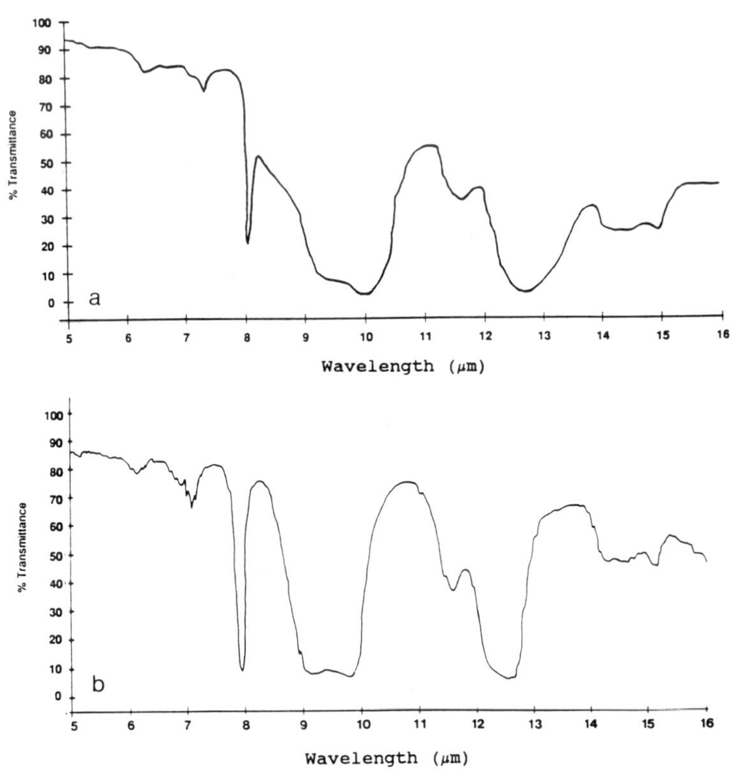

Figure 2. a) ATR Spectrum of Storage Box Gasket, b) ATR Reference Spectrum of Polydimethylsiloxane.

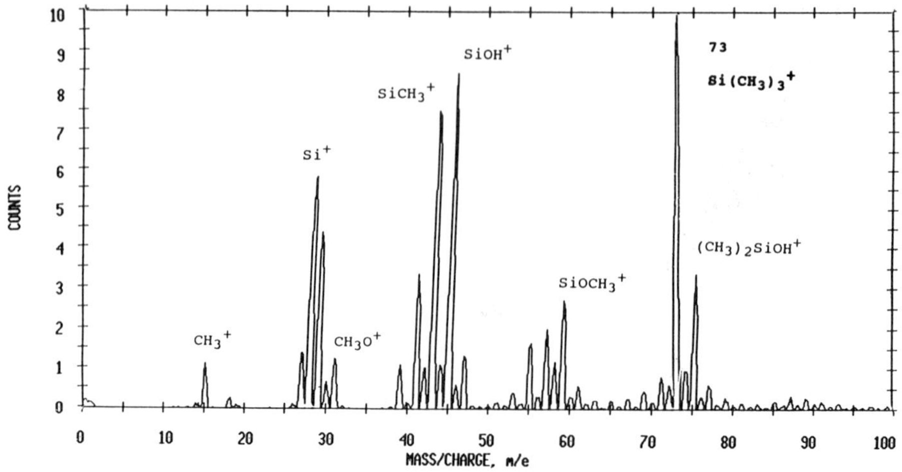

Figure 3. Positive Static SIMS spectrum of Wafer from Heated Storage Box with Siloxane Gasket.

ATR is a type of internal reflection spectroscopy (IRS)[5]. The sample being analyzed is pressed against a special crystal, termed an internal reflection element(IRE). When the angle of incidence of the infrared radiation at the sample/IRE interface is greater than or equal to the critical angle, total internal reflection occurs. The radiation exiting the IRE, after multiple internal reflections, is attenuated by the absorption characteristics of the sample. When this spectrum is ratioed against a background spectrum taken through the IRE without a sample present, a spectrum similar to a transmission spectrum is obtained. Attenuated Total Reflectance analysis revealed an exact match of the absorbance spectrum of polydimethylsiloxane[6] as shown in Figure 2. These results indicate that the probable cause of the contamination on the wafers is the result of outgassing from the silicone rubber material.

PROBLEM REPLICATION

An experiment was performed to confirm the outgassing suspicion in which clean wafers were intentionally exposed to the silicone rubber strips. Sample wafers known to be contamination free were first cleaned using a modified RCA process which yields wafers with a hydrophilic surface. These wafers were then placed in a storage box with the silicone rubber strips overnight in an oven set at 50°C in order to accelerate the outgassing process. This rendered the wafers hydrophobic. Static SIMS of these wafers again showed siloxane like compounds on the wafer surface as shown in Figure 3. Figure 3 indicates that the level of contamination was greater than that seen with the original lot of contaminated wafers indicating that contamination increases with heat and time.

EVALUATION OF WAFER CLEANING METHODS

RCA based cleans were developed to remove trace quantities of organic, metallic and particle contamination. These wafers were obviously heavily contaminated; therefore, we first evaluated successive modified RCA cleans to see if the contamination could be removed. Figure 4 shows that the initial modified RCA process added many particles. However, subsequent cleans brought particle levels back down to their initial levels indicating that this contamination can be successfully removed by using modified RCA clean chemistries. These observations led us to believe that the related PDMS organics were deposited on the wafers as a film, which was partially removed by the modified RCA process resulting in particle contamination.

Since multiple RCA cleans were successful in reducing particle formation, an investigation was done to determine which steps in the modified RCA clean were responsible for removing the contamination. From examination of the polydimethylsiloxane chemical structure it was thought that the organic portion of the formula was being removed by the sulfuric-peroxide portion of the process, leaving behind silicon dioxide. The silicon dioxide was subsequently being removed by the dilute hydrofluoric acid. Hence, cleaning methods were examined which included extended sulfuric-peroxide and dilute hydrofluoric acid steps.

Table II shows four cleaning methods evaluated and the effects of these cleans on particle reduction with heavily contaminated wafers. Particles were counted before (Pre) and after (Post) processing and the difference in the number of particles (Delta) was calculated for different particle sizes. When the SPM or dilute HF dispense times were increased alone there was only a slight reduction in particle contamination. Increasing both the SPM and dilute HF steps together, as well as the other two process steps, had a significant effect in reducing particle contamination. In fact, these cleaning methods significantly reduced the level of particles initially on the wafer.

No attempt was made to optimize the length of the sulfuric-peroxide and dilute hydrofluoric acid steps used. It is likely that shorter times could be used and still obtain sufficient cleaning.

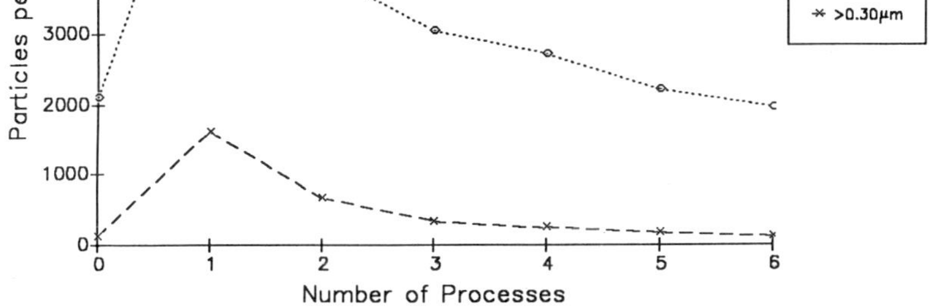

Figure 4. The Effect of Successive Modified RCA Cleans on Wafer Particle Level.

Table II. Effect of Chemical Dispense Times on Particle Reduction

Process Time (sec)				Particles per 200mm Wafer					
				>0.15µm			>0.30µm		
SPM	HF	APM	HPM	Pre	Post	Delta	Pre	Post	Delta
135	30	90	90	2572	4943	+2371	104	108	+4
45	300	90	90	1620	2721	+1101	72	493	+421
300	250	180	180	846	522	-324	71	31	-40
300	250	180	180	1232	1060	-172	72	57	-15

CONCLUSION

The outgassing of organic materials during the storage and shipment of silicon wafers is a concern to semiconductor manufacturers. This outgassing has the potential for contaminants to congregate and collect on wafer surfaces creating a host of process related problems. This study was undertaken to investigate one example of storage induced contamination and its removal from silicon wafer surfaces. Polydimethylsiloxane gaskets used in storage boxes were found to produce heavy organic contamination on wafers due to outgassing. This contamination was not removed by standard modified RCA cleans.

Polydimethylsiloxane gaskets should be eliminated from storage and wafer shipping boxes. Precautions should be taken when choosing storage boxes so this type of contamination can be avoided. When this contamination does occur it can be removed using RCA based chemistries with extended sulfuric-peroxide and dilute hydrofluoric acid steps.

REFERENCES

1. D.S. Becker, W.R. Schmidt, and C.A. Peterson, Effects of various chemistries on silicon-wafer cleaning, in "Microelectronics Processing: Inorganic Materials Characterization," ACS Symposium Series No. 295, pp. 336-376, American Chemical Society, 1984

2. D. Burkman, Optimizing the cleaning procedure for silicon wafers prior to high temperature operations, Semiconductor Intl., 103-116, (July 1981)

3. Charles Evans & Associates, "Static Secondary Ion Mass Spectrometry (Static SIMS)," Technical Bulletin# SSIMS-AN-001-01, 1990

4. J. M. Walls, Editor, "Methods of Surface Analysis," Cambridge University Press, Cambridge, 1989

5. N.J. Harrick, "Internal Reflectance Spectroscopy", John Wiley & Sons, New York, 1967

6. C. Pouchart, Aldrich Library of I.R. Spectra, Ed. 3, Z10,750-6, Aldrich Chemical Company, 1981

PARTICLE REMOVAL FROM OXIDE, NITRIDE, AND BARE SILICON SURFACES USING DIRECT-DISPLACEMENT ISOPROPYL ALCOHOL (IPA) DRYING

Chris McConnell

CFM Technologies, Inc.
501 Gordon Drive
Lionville, PA 19353

INTRODUCTION

The effectiveness of particle removal depends upon a number of factors, including the source of the particles, the adhesion mechanisms, and the removal techniques. A number of researchers[1-4] have explored the impact of varying the source of the particles (i.e. alumina, silica, and polystyrene). The present study examines another contributing factor--how the surface chemistry of the substrate itself impacts particle removal performance.

Using direct displacement isopropyl alcohol (IPA) drying, particle removal data were gathered on bare silicon wafers as well as on wafers having thin films (approximately 250 Å thick) of silicon dioxide and silicon nitride. These findings suggest a strong relationship between removal performance and the chemical nature of the substrate.

Although silicon wafers are commonly dried using spin rinser dryers, these dryers rarely remove particles. Conventional IPA vapor dryers, which immerse wafers into a vapor zone and then gradually withdraw the wafers, tend to be "particle neutral" at best. In contrast, direct displacement IPA drying typically removes a high percentage of particles in all measurable size categories (submicron and larger)[5].

Prior work[6,7] has shown that direct displacement IPA drying removes a high percentage of particles from wafer surfaces. Several factors may contribute to the particle removal mechanism:

1. Filtered IPA vapor introduced into the top of the vessel results in an exceptionally clean gas phase.

2. The IPA condensate continuously covers any floating particles, much like a CVD process.

3. The wafers reach particle-equilibrium with a freshly-deposited, low-surface-tension liquid.

4. Entrapped particles are transported out of the vessel.

Particles on Surfaces 3, Edited by K.L. Mittal
Plenum Press, New York, 1991

In this study we have attempted to more closely examine the details of the tertiary interface between IPA, water, and wafer. Although a complete understanding is far from being realized, a more comprehensive picture is now available.

EXPERIMENTAL

Figure 1 shows the apparatus[8,9] used in this study. The system totally encloses process wafers, thereby eliminating external sources of contamination. Hot and cold water are blended together to achieve the desired temperature and flow rate. This water flow fills the chamber completely. Overflow accumulates in the drain line where in-line temperature, resistivity, and conductivity monitors measure the quality of the water or aqueous reagent flowing through the vessel. Chemicals may be injected through a special manifold, although this capability was not used in the present study.

Drying is accomplished by shutting off the water supply and the drain, isolating the wafer chamber, yet leaving it full of pure water. A dedicated drain then opens in the bottom of the chamber and shortly thereafter warm, submicron-filtered isopropyl alcohol vapor enters the top of the chamber to fill the void created by the descending water level.

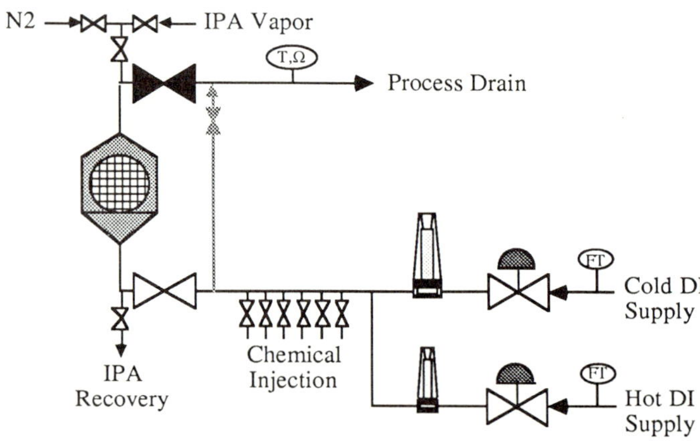

Figure 1. Schematic diagram of the experimental apparatus.

No nitrogen or air is allowed to enter the chamber at any time during the dry. A metering pump draws water out of the bottom of the vessel at a precise rate. A pressure controller insures that, above the water, system pressure remains at 1.5 ($\pm$ 0.1) psig.

During the IPA dry, a distinct liquid layer of IPA forms on top of the descending water level. This layer results from condensation on the cooler (57 $\pm$ 1 C) water. Figure 2 shows a front view and a side view of the direct displacement IPA drying process. Although IPA and water are totally miscible, interfacial tension between the water and the liquid IPA acts to maintain the distinct liquid layer floating above the water.

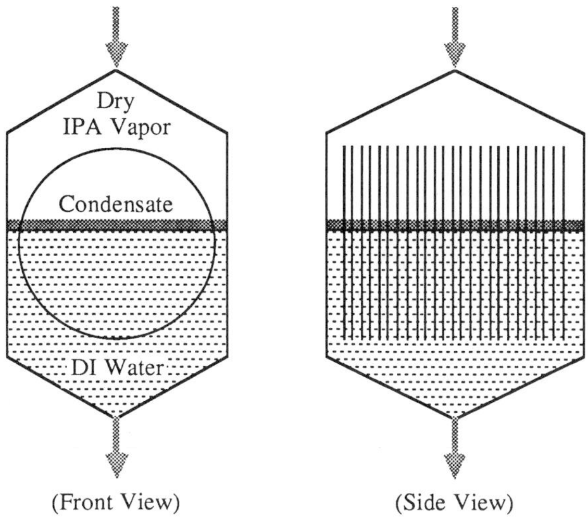

Figure 2. Cross-sectional diagrams of direct-displacement isopropyl alcohol (IPA) drying.

The wafers used in this study were 125 mm silicon wafers. Some of the wafers had films approximately 250 Å thick of thermally grown silicon dioxide and silicon nitride. Particles, observed initially on the wafers as opened from shipping containers, and after various rinse-dry treatments, were counted on a laser surface scanner (Estek WIS 150) calibrated weekly at 0.3 μm.

RESULTS AND DISCUSSION

Figures 3 through 6 show the particle counts initially and after several sequential rinse-dry runs. The performance on bare silicon wafers showed high removal on the first rinse-dry process (Figure 3). Subsequent rinse-dries showed little further removal. Prior work[6] has suggested that a pseudo-equilibrium is reached with respect to particles on the wafer surface. Under these conditions, the pseudo-equilibrium count would be approximately 10 particles per wafer (≥ 0.3 μm).

An alternative hypothesis is that these last ten particles are simply "glued" to the surface, and direct displacement IPA drying cannot remove them. Initial data on wafer maps, where particles were tracked by x-y coordinate, suggested that these particles do indeed relocate. Approximately 50% of the particles moved from run to run even though the total count did not change. Although this tends to support the pseudo-equilibrium argument, we suspect that the "glued" hypothesis is also applicable to at least some extent.

Some of the new bare silicon wafers had extraordinarily high initial particle counts (5,000 to 10,000 per wafer). Again, the first rinse-dry process exhibited very high removal efficiency, resulting in a final particle count of roughly 10 particles per wafer (Figure 4). Subsequent runs did not change the final particle count significantly.

Rinse-dry performance on wafers with a 250 Å thermally grown silicon dioxide film showed very different results (Figure 5). There was essentially no particle removal, and in some cases, there was slight particle addition. No convergence to equilibrium occurred.

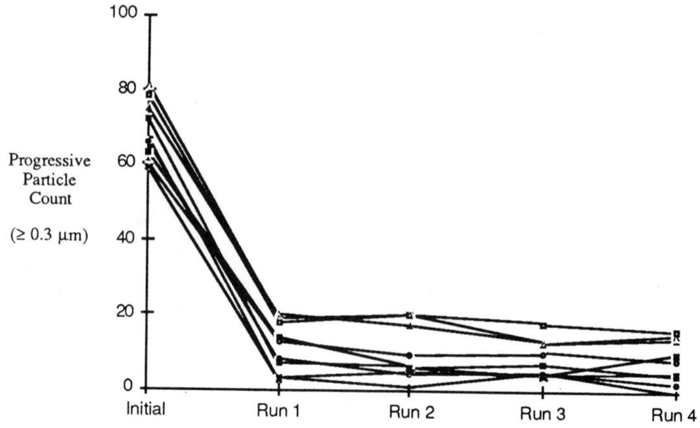

Figure 3. Rinse-dry particle performance using bare silicon wafers. Each symbol corresponds to a different wafer within the same batch.

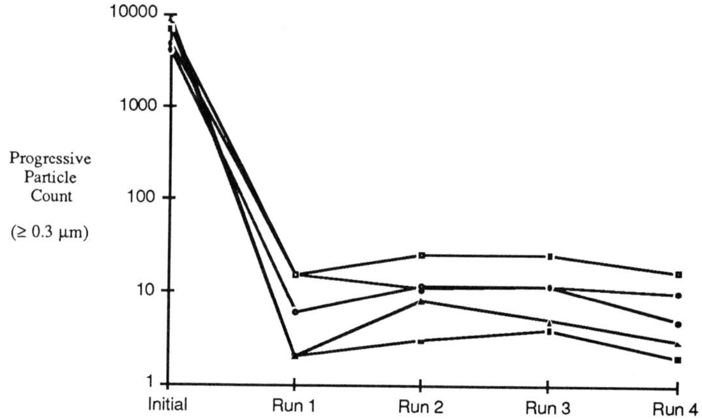

Figure 4. Rinse-dry particle removal performance using hazy bare silicon wafers. Each symbol corresponds to a different wafer within the same batch. (Note logarithmic scale.)

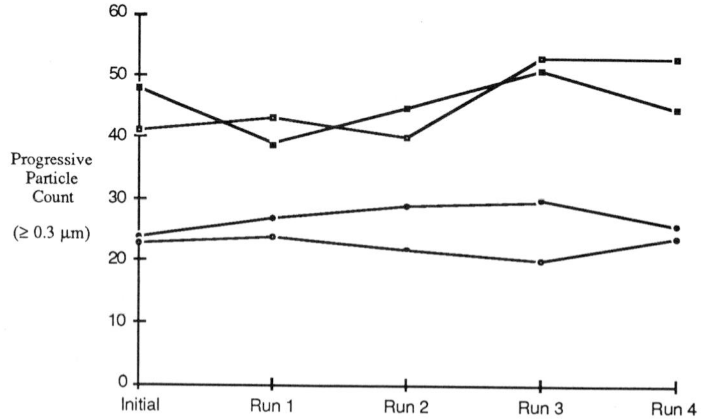

Figure 5. Rinse-dry particle removal performance using wafers having a 250Å thick silicon dioxide film. Each symbol corresponds to a different wafer within the same batch.

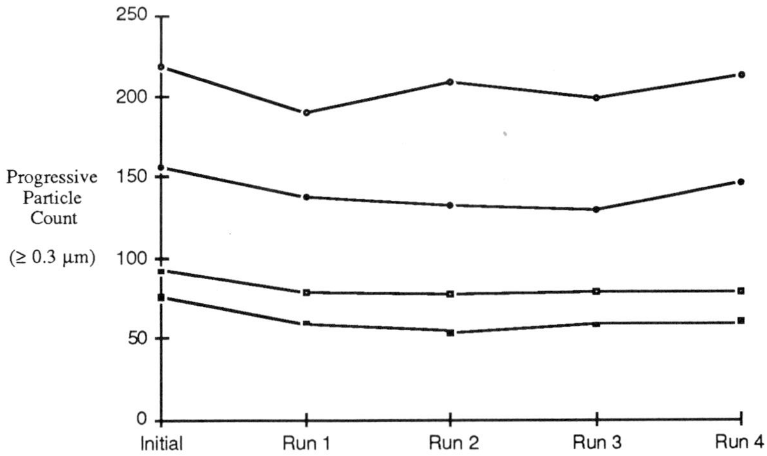

Figure 6. Rinse-dry particle removal performance using wafers having a 250Å thick silicon nitride film. Each symbol corresponds to a different wafer within the same batch.

One suggestion is that these particles are actually film defects and, therefore, cannot be removed by the direct displacement rinse-dry process. The surface is also extremely hydrophilic and, as will be discussed, this may contribute to the less effective particle removal performance.

Figure 6 shows the performance on wafers having 250 Å of silicon nitride film. Again the performance is relatively particle neutral: there was no conversion to a pseudo-equilibrium particle count. In this case there was some removal; however, the high absolute particle counts obscure the delta values.

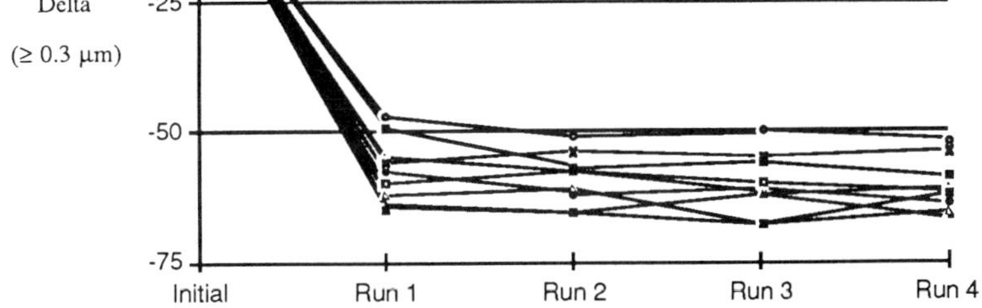

Figure 7. Rinse-dry particle removal performance using bare silicon wafers (replotted from Figure 3 using cumulative particle removal rather than absolute particle counts).

To more clearly see the relative particle removal performance on each of the three different types of substrates, the data were replotted as shown in Figures 7 through 10. Instead of showing absolute pre-counts and post-counts, these figures show the particle delta values, or removals, occurring during each run. By definition, the initial particle delta is zero. The advantage of plotting the data in this way is that all of the results are put on an equivalent scale despite the large differences in absolute counts for the different types of substrates.

Figure 7 shows the cumulative particle removal from bare silicon wafers using direct displacement IPA drying. The first run removed approximately 60 particles, while subsequent runs had little additional impact. Equilibrium convergence is readily apparent.

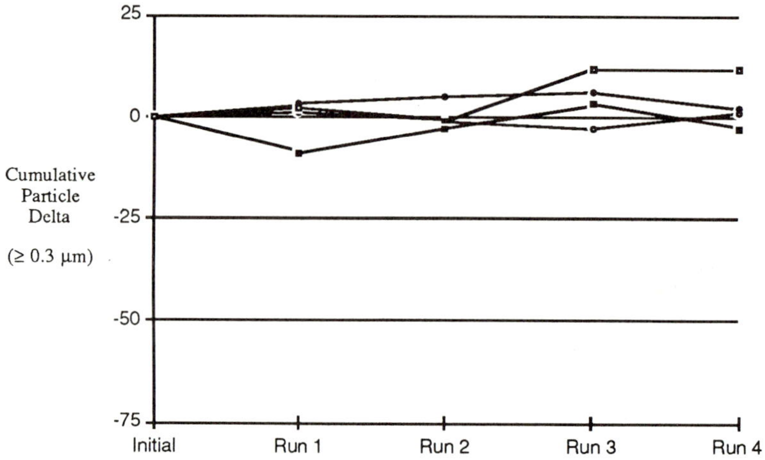

Figure 8. Rinse-dry particle removal performance using silicon dioxide wafers (replotted from Figure 5 using cumulative particle removal rather than absolute particle counts).

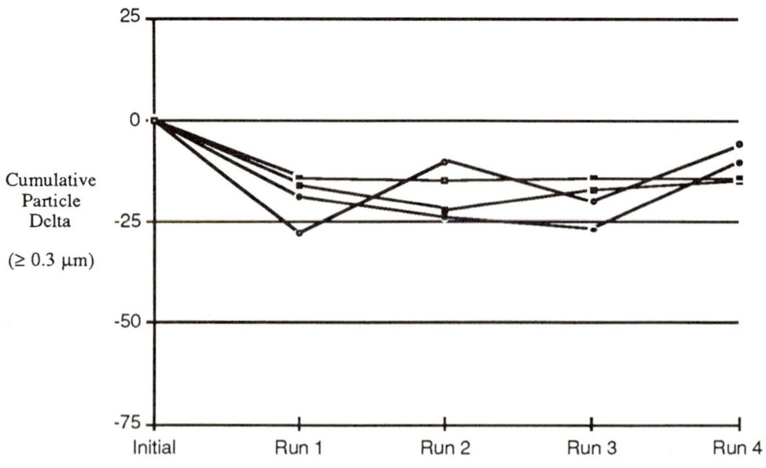

Figure 9. Rinse-dry particle removal performance using silicon nitride wafers (replotted from Figure 6 using cumulative particle removal rather than absolute particle counts).

In Figure 8, the particle neutrality is clearly shown with respect to particle removal on the oxide wafers. For three of four wafers, the final particle count was essentially unchanged after four runs. One wafer showed minor addition on Run 3.

Figure 9 reveals the particle removal from the wafers having a silicon nitride film. Each wafer had approximately 20 particles removed on the first run. Most of the subsequent runs had relatively little impact; however, there was some addition to a single wafer on Run 2 and to a pair of wafers on Run 4.

The particle removal data from the three different surfaces are all plotted on the same graph in Figure 10. Here it becomes evident that the particle removal performance depends upon the chemical nature of the substrate. While bare silicon surfaces seem to offer the greatest potential for particle removal, silicon nitride and silicon dioxide have lower tendencies for particle removal. In fact silicon dioxide seems to be essentially neutral with respect to particle addition and removal.

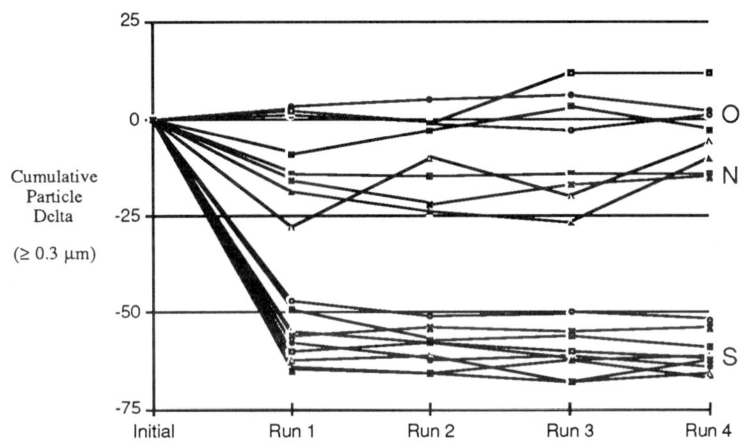

Figure 10. Comparison of rinse-dry particle performance on all three different wafer surfaces (plotted from Figures 7, 8 and 9).

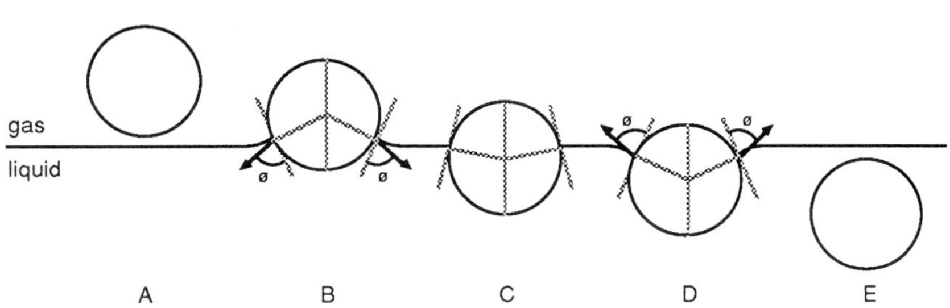

Figure 11. Interfacial tension forces exerted on a particle.
A--Completely hydrophobic particle (no interfacial forces).
B--Typical particle displaced upward (net downward force).
C--Typical particle suspended at equilibrium (zero net force).
D--Typical particle displaced downward (net upward force).
E--Completely hydrophilic particle (no interfacial forces).

Prior researchers[3,4] have pointed out that interfacial phenomena, as wafers are inserted or withdrawn from liquids, play an important role in particle *addition*. Similarly, these interfacial phenomena may impact particle *removal*. Leenaars, through elegant mathematics and detailed experimentation, proved that interfacial forces can remove particles[10]. Forces, caused by the tendency for a particle to remain at the interface between a gas and a liquid, are perpendicular to the plane of the gas-liquid interface. If the liquid level is horizontal, then the net interfacial tension force vector will be vertical (Figure 11).

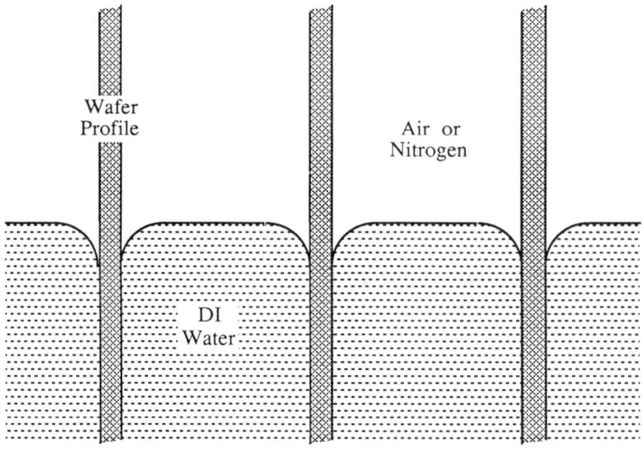

Figure 12. Cross-sectional drawing of the tertiary gas-liquid-solid interface between air or nitrogen, water, and a completely hydrophobic wafer.

Close to an extended surface, the plane of the gas-liquid interface curls upward for hydrophilic surfaces and downward for hydrophobic surfaces. A cross sectional, close-up view of the idealized case for hydrophobic silicon wafers is shown in Figure 12. Instead of the gas-liquid interface being completely straight across, this interface intermittently extends downward, along the surfaces of each wafer.

On withdrawal through this interface, particle removal forces are generated through interfacial tension. As the liquid surface separates from the wafer and particles are caught in the meniscus, there is a tendency to strip them away from the wafer surface. The interfacial tension vector points sideways, generating a powerful force opposite to the adhesion force between the particle and the surface. According to Leenaars, the efficiency of particle removal will depend upon the magnitude of the adhesion force versus the magnitude of the horizontal vector of the surface tension force.

Several researchers[3,4,6] have described the difficulties in processing wafers during an HF treatment. The hydrophobic surface is described as "activated" or "magnetic" with respect to particle addition. Figure 12 suggests another possibility. As hydrophobic wafers are immersed into an aqueous solution, the air-liquid interface may be stretched downward along the side of the wafers. Particles on this interface may be dragged down with the wafer and pressed against the wafer surface via the same interfacial tension as described above to remove these particles. In fact, particle addition and removal may be reversible: upon immersion particles are added to the wafer and upon withdrawal particles are removed.

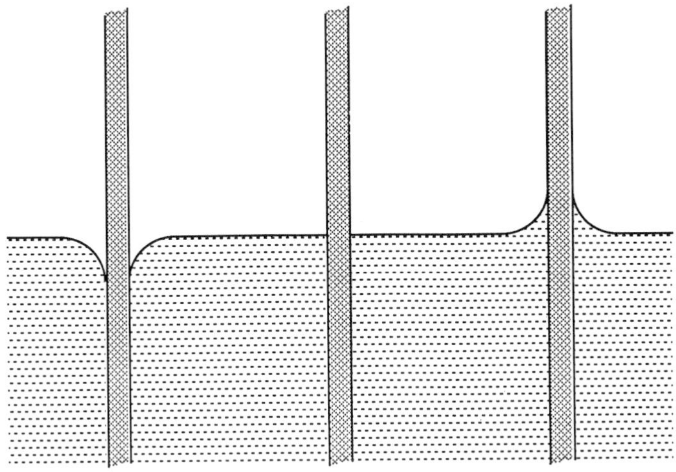

Figure 13. Comparison between the gas-liquid-solid interfaces associated with a completely hydrophobic wafer, a neutral wafer, and a completely hydrophilic wafer.

Figure 13 contrasts the interfacial profile of a hydrophilic wafer with that of a hydrophobic wafer. The hydrophilic surface better represents the wafers used in this study having silicon dioxide and silicon nitride films. In the hydrophilic case, water wicks up the wafer surface creating a concave, upwardly-curved meniscus. As in the hydrophobic case, the liquid surface is extended, but this time instead of extending down along the wafer surface, it extends up.

Through a series of careful experiments, Riley and Carbonell[11,12] have shown that withdrawing a hydrophilic wafer through a gas-liquid interface causes high levels of particle addition. They essentially proved that a particle-laden film remained on the wafer as it was withdrawn. Subsequent evaporation and deposition caused the high particle counts.

Leenaars[10] demonstrated the opposite effect. By slowly *immersing* a hydrophilic wafer into water, he achieved astonishingly high levels of particle *removal*. It is important to note that Leenaars' immersion procedure required nine hours. This ensured that the concave curvature of the gas-liquid interface was maintained. Using this slow immersion technique, he obtained particle removal efficiencies as high as 90%; however, with faster immersion times the particle removal efficiency fell dramatically. Presumably this was due to the hydrophilic wafer failing to reach equilibrium with the gas-liquid interface, thereby collapsing the upward curvature of the meniscus. The curvature of a gas-liquid-solid interface is dependent upon the direction of substrate movement and also the speed of immersion or withdrawal.

Figure 14 shows the cross-sectional view of hydrophobic wafers with the addition of an IPA layer floating on top of the water. It should be remembered that this IPA liquid layer is exceptionally pure since it is freshly condensed from a gas filtered through a submicron element. Isopropyl alcohol has very low viscosity and low molecular weight. Bolster[13] showed that these properties enhance surface chemical displacement. He found that certain liquids will effectively compete for contact with a solid surface and displace another already-present liquid by penetrating between that liquid and the solid surface. This tendency is referred to as spreading pressure. Lower surface tension allows the displacing fluid to spread, or drive into the interface between the substrate and the prior fluid.

Figure 15 shows both the hydrophilic and hydrophobic cases with an IPA layer included. Since none of the wafers examined in this study were truly hydrophobic, we believe that the center and right hand wafers represent a closer approximation to our experiments than the left hand view. By qualitatively comparing a bead of water on each of the three surfaces, it was apparent that the oxide surface was the most hydrophilic. This would be represented by the rightmost drawing of Figure 15. The wafers having the silicon dioxide film were also the most neutral with respect to particle removal.

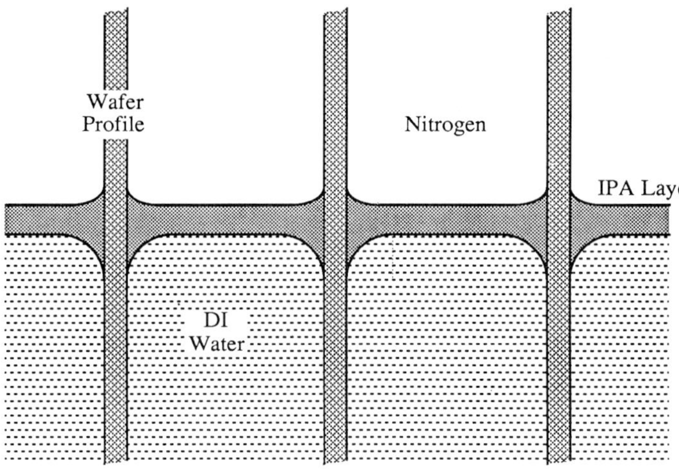

Figure 14. Cross-sectional drawing of the interface modified by a layer of liquid IPA between nitrogen, water, and a completely hydrophobic wafer.

We hypothesize that because of the strongly hydrophilic nature of these wafers, the water may have effectively competed for surface contact on the oxide film versus the IPA layer. In fact, if the IPA layer was too thin, there might not be any contact whatsoever of IPA on the wafer surface. In this case there would be no mechanism for particle removal and perhaps even cause for particle addition.

In another experiment, the length of time to perform the IPA drying was extended from three minutes to eight minutes. This effects a substantially thicker IPA layer. Figure 16 shows the results of this experiment. In this case the particle removal from relatively dirty oxide wafers was dramatic. Also, the ultimate count converged to an equilibrium particle count of approximately 30 particles per wafer. Using a thicker IPA layer, therefore, appears to engage the particle removal mechanism that otherwise might not occur.

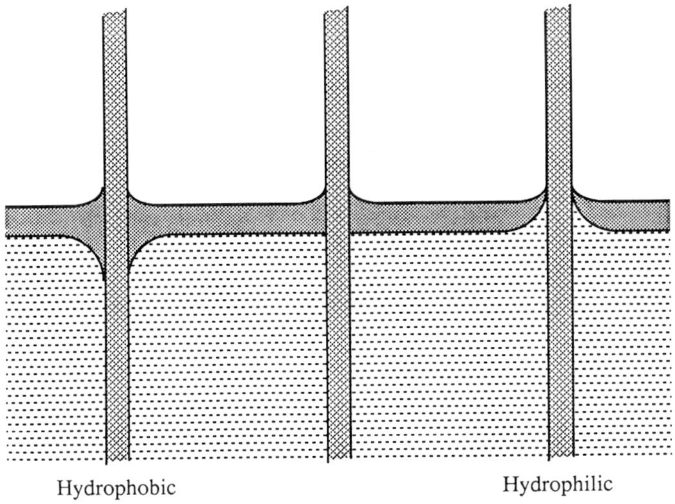

Figure 15. Comparison between the interfaces modified by an IPA liquid layer associated with a completely hydrophobic wafer, a neutral wafer, and a completely hydrophilic wafer.

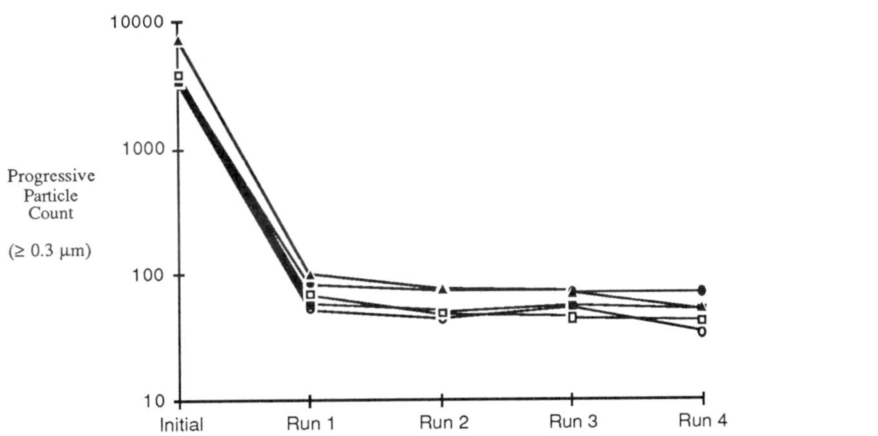

Figure 16. Rinse-dry particle removal performance on wafers having a 250Å thick silicon dioxide film using an extended 8 minute IPA dry to increase the thickness of the IPA layer.

Because wafers having a silicon nitride film are less hydrophilic than those having a silicon dioxide film, we expect a less pronounced curvature in the water-wafer interfacial meniscus. Our hypothesis is that this property enhances particle removal performance. Finally the bare silicon surfaces, which would be the least hydrophilic, should offer the greatest opportunity for capillary particle removal. We imagine that the interfacial cross section might look most like the center drawing of Figure 15. Although we show a 90° contact angle, it may be that the low surface tension and high spreading pressure of IPA will cause the meniscus to curve downward like the left hand drawing of Figure 15. This would certainly explain the outstanding particle removal from bare silicon surfaces. Leenaars[10] showed that higher contact angles result in larger horizontal surface tension force vectors, resulting in higher removal efficiencies. Since IPA penetrates and increases the contact angle at the interface, it should enhance the particle removal efficiency of the direct displace drying.

In contrast to other particle removal techniques, the interfacial mechanism described in this paper is equally powerful for both large particles and small particles. While adhesion forces are proportional to particle radius, r, removal forces diminish with r^2 for drag force and with r^3 for centrifugal force[2]. Submicron particles adhere tenaciously; however, interfacial tension forces are directly proportional to particle radius and therefore will maintain the same magnitude of strength relative to adhesion forces regardless of particle size[10].

CONCLUSIONS AND SUGGESTIONS FOR FUTURE RESEARCH

Previous studies[6,7] indicate that in direct displacement drying, the nature of the wafer surface influences particle removal performance. We suggest that the relative hydrophobicity or hydrophilicity of the surface results in different shapes in the cross section of the tertiary gas-liquid-solid interface. Hydrophobic wafers create a meniscus which curves down, resulting in high particle removal efficiency due to a strong horizontal vector of interfacial tension. This same tendency may cause high particle addition during the immersion step of the hydrophobic wafer. Hydrophilic wafers, on the other hand, create a meniscus that curves upward, resulting in the potential for particle addition upon withdrawal. Addition of an IPA layer may modify this interface so as to cause removal; however, the IPA layer must be thick enough to contact the wafer surface and modify the interfacial profile.

REFERENCES

1) A.A. Busnaina, M.A.R. Sharif, G. Gale and F.W. Kern, Jr., A fluid dynamic study of a microcontaminant particles removal process, in "Particles in Gases and Liquids 1: Detection, Characterization and Control," K.L. Mittal, Editor, pp. 235-257, Plenum Press, New York, 1989.

2) J. Taylor, A.A. Busnaina, F.W. Kern, Jr. and R.F. Kunesh, Measurement of removal force of submicron particles on a silicon wafer, Proceedings of the Institute of Environmental Sciences, 422-426 (1990).

3) L.D. Michaels, V.B. Menon, and R.P. Donovan, Particle deposition at the solid-liquid interface, Proceedings of the Institute of Environmental Sciences Annual Technical Meeting, 438-442 (1988).

4) R.G. Carbonell, A.C. Clayton, R.P. Donovan, V.B. Menon and D.J. Riley, Investigating particle deposition mechanisms on wafers exposed to aqueous baths, Ultrapure Liquids, 25-29 (August 1990).

5) J.W. Koppenbrink, C.F. McConnell and A.E. Walter, Particle reduction on silicon wafers as a result of isopropyl alcohol vapor displacement drying after wet processing, in "Particles on Surfaces 2: Detection, Adhesion and Removal," K.L. Mittal, Editor, pp. 235-243, Plenum Press, New York, 1989.

6) A.E. Walter and C.F. McConnell, Direct displacement wet processing: how it affects wafer surface phenomena, Microcontamination, 35-38, 60-61 (January 1990).

7) C.F. McConnell, Investigating the impact of wafer history on the removal of particles from silicon surfaces, Microcontamination, 45-50 (August 1990).

8) C.F. McConnell, Vessel and system for treating wafers with fluids, U.S. Pat 4,757,650 (1986).

9) C.F. McConnell and A.E. Walter, A process and apparatus for treating wafers with process fluids, U.S. Pat 4,778,532 (1988).

10) A. Leenaars, A new approach to the removal of sub-micron particles from solid (silicon) substrates, in "Particles on Surfaces 1: Detection, Adhesion and Removal," K.L. Mittal, Editor, pp. 361-372, Plenum Press, New York, 1988.

11) R.G. Carbonell and D.J. Riley, Direct measurement of particle concentration at the air-liquid interface, Proceedings of the Institute of Environmental Sciences Annual Technical Meeting, 426-427 (1989).

12) D.J. Riley and R.G. Carbonell, The deposition of liquid-based contaminants onto silicon surfaces, Proceedings of the Institute of Environmental Sciences, 224-228 (1990).

13) N. Bolster, Removal of fluid contaminants by surface chemical displacement, in "Surface Contamination: Genesis, Detection and Control," K.L. Mittal, Editor, Vol. 1, pp. 359-368, Plenum Press, New York, 1979.

ELIMINATION OF FRETTING WEAR PARTICLES FROM THE SURFACE OF A POWER CABLE ON A DISK DRIVE ACTUATOR: A CASE STUDY

Walter Prater, Monica Vargas and John McDowell

IBM Corporation
Storage Systems Products Division
5600 Cottle Road
San Jose, CA 95193

This paper discusses in detail the SEM/EDS failure analysis as well as the thorough evaluation of the mechanical design and its contributing factors to fretting wear. The experimental results of the solution to the fretting problem are presented.

Early in the development of an actuator, fretting at the strain relief and power cable interface was discovered. Fretting wear was observed on five out of eight actuators on life-test and on six out of eight on HDA life-test. Scanning-electron microscopy revealed power cable wearing and shedding of fine particulate debris. This fretting caused no functional problems to the disk drive.

The wear-scar pattern and its location indicated that mechanical misalignment was a contributing factor to fretting. Six dimensional tolerance-analyses were performed to assess the misalignment of the power cable to the strain relief. These analyses uncovered three design shortcomings that would cause high contact-stress concentrations or looseness of the components which could promote fretting wear. This was remedied by making a dimensional change to the strain relief. The wear-scar-pattern analysis also revealed four contributing factors to fretting wear that were attributed to the injection mold design of the strain relief.

Eight experimental actuators were tested for fretting by changing the strain relief material from Ultem-1000 to carbon-filled polycarbonate. The material change so reduced the fretting of the power cable that this source of particulate contamination within the disk drive was virtually eliminated.

INTRODUCTION

Particulate contamination in magnetic disk drives is a major concern for the design engineer. Magnetic disk drives cannot use conventional lubricants because of the strict cleanliness required for the head-disk interface. With the exception of the head, disk, and ball bearings, a sliding or rolling material pair must be selected where the interfacing materials have the required wear resistance or self-lubricating properties. Often the base materials have specific physical properties needed to satisfy functional requirements, but they do not meet tribological requirements. When this happens wear can occur and release particles into the airstream of the HDA (Head Disk Assembly).

Particles suspended in the airstream during the operation of a magnetic disk drive can have adverse effects on the performance and reliability of the drive. The interface between the read/write head and the disk itself is the heart of any disk drive. Plastic debris does not typically get imbedded in the disk when encountering the head-disk interface at full rotational speed, but instead can melt or vaporize from the kinetic energy and become deposited on the head and disk. This can later lead to a friction build up causing the heads to adhere to the disk at start up resulting in distortion to the head-flexure assembly. This is known as stiction. Another concern is that large agglomerates of fine particles can be readily broken up by the impact of the head and then be redistributed on the disk surface. Impact of the flying head against these imbedded particles can drastically affect the aerodynamics, thereby destabilizing the head and forcing it into contact with the disk. Even extremely small particles are dangerous, since they are able to pass between the head and the disk and can then become caught up in the airstream eddies behind the head. Once a single small particle becomes attached to the head, then the attractive forces increase to capture more. They accumulate there in large thin agglomerates, called whiskers, until they fall onto the disk, potentially causing a head crash. Because the head/disk interface is the main source of disk drive frailty, it follows then that the most productive way to enhance the reliability is to keep the interface clean.

Early in the development of an actuator design, eight actuators were removed from an access-life tester and analyzed for wear. Careful visual examination of the actuators revealed the early stages of fretting wear of the power cable and strain relief. A white powdery residue was readily visible on the surface of the strain relief and on the surfaces of the power cables. The power cable is composed of copper conductors laminated between two sheets of Kapton®, a polyimide. The strain relief is made of Ultem-1000®, a polyetherimide, which serves two functions. First, it acts to relieve the strain in the cable near the solder joint on the coil where it is bent during normal actuator accessing, and second, it serves as a guide to the cable allowing it to roll unrestricted along a flat surface without being abraded by the metallic actuator parts during actuator accessing, see Figure 1.

Fretting wear is a term that describes severe material removal caused by oscillatory movements between the mating surfaces of a pair of materials.[1] Fretting has been studied most commonly on metals,[2] and ceramics[3] but it can be a major problem for plastics as well. The movements that initiate the damage may be minute, but the cumulative damage to the surfaces after many millions of cycles can be extensive. A fretting system requires a clamping load and an oscillating tangential force. The key parameter, within a fretting system, is the peak-to-peak amplitude of the relative movement between the two surfaces. Rapid degradation of high quality surfaces can be produced by very small amounts of slippage.

RESULTS

The Ultem-1000 strain relief promoted fretting in 62.5% of the parts tested, while the carbon filled polycarbonate nearly eliminated the fretting, see Table I. Use of the carbon filled polycarbonate strain relief reduced actuator particle counts significantly.

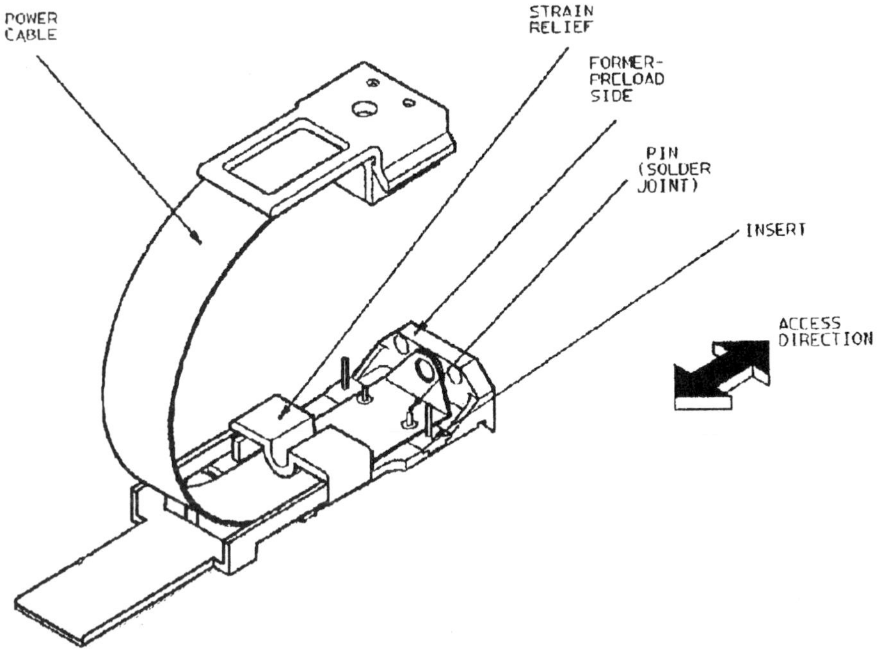

Figure 1. Assembly showing power cable, strain relief and former.

Table I. Fretting Incidence and Particle Count for Polycarbonate and Ultem-1000 Strain Reliefs.

Strain Relief Material	Actuator Sample Size	Millions of Seeks	Fretting Incidence, %	Actuator Particle Count Average	
				≥ 0.3 μm	≥ 0.19 μm
Polycarbonate	8	27.5	12.5	225	1200
Ultem-1000	8	8	62.5	625	1781

Six dimensional tolerance analyses were performed to evaluate the mechanical alignment of the power cable to the strain relief. There are three assembly tolerances identified as contributors to the fretting problem, see items 2, 3 and 5 of Table II.

DISCUSSION

Eight actuators removed from an access-life-tester were examined for wear. The access-life-tester exercised these actuators through their full stroke similar to an actuator in an HDA seeking from the innermost track to the outermost track and back again. The Visual examination revealed an aggressive fretting wear problem at the interface between the power cable (made of Kapton) and the strain relief (made of Ultem-1000). Prior field experiences have shown airborne plastic particles from fretting wear to cause reliability problems. Even though these particles were not shown to be airborne and caused no functional problems, they are considered to constitute an unacceptable risk.

Table II. Summary of Strain Relief to Power Cable Alignment.

Tolerance Analysis	Number of Tolerances	$\bar{x} \pm$ Statistical Tolerance Range mm	Comments and Conclusions
1. Strain relief tooth to former side clearance	3	$+0.370 \pm 0.093$	Should be a loose fit. Always is a loose fit.
2. Strain relief tang to former side clearance	5	$+0.420 \pm 0.145$	Should be a tight fit. A problem Always is a loose fit.
3. Snap fit engagement	4	-0.105 ± 0.124	Should be a tight fit. Sometimes is a loose fit. A problem
4. Retention of power cable	3	-0.143 ± 0.069	Should be a tight fit. Always is a tight fit.
5. Edge clearance strain relief & power cable	22	-2.705 ± 1.304	Should be a worst case clearance. Always is overlapping. A problem
6. Cable twist against strain relief	12	0 ± 0.644	Nominally should be zero. Nominally is zero.

The wear scars on the parts can be categorized into: strain relief edge-induced wear, and power cable edge-induced wear, denoted respectively by 'A' & 'B' in Figure 2. Table III shows that five of eight actuators quickly experienced severe fretting in only 8-million full-stroke length seeks; that is approximately one percent of the actuator-life-design goal.

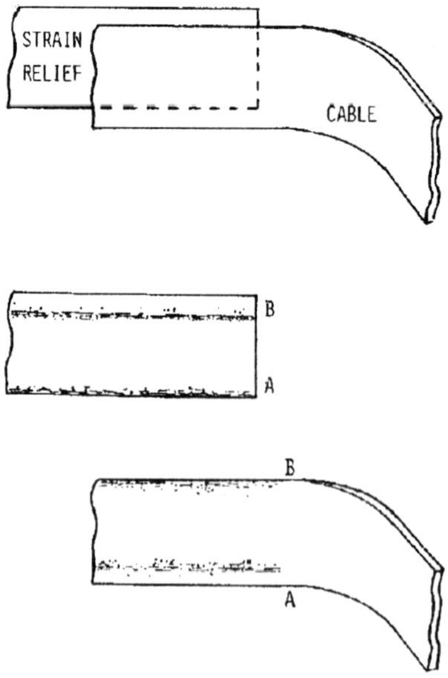

Figure 2. Fretting scar nomenclature of strain relief and power cable: Type-A is strain relief edge -induced wear, and type-B is power cable edge-induced wear.

Table III. Fretting Incidence and Particle Count of Actuators with the Ultem-1000 Strain Relief.

Actuator Serial Numbers	Millions of Seeks	Strain Relief Material	Visible Fretting
828	8	Ultem-1000	Fret
829	8	Ultem-1000	No
826	8	Ultem-1000	Fret
827	8	Ultem-1000	No
818	8	Ultem-1000	Fret
819	8	Ultem-1000	Fret
814	8	Ultem-1000	No
815	8	Ultem-1000	Fret

Wear scar dimensions, from cases of severe fretting, measure several millimeters in length as evident in Figures 3 and 4. The power cable fretting wear scar of a 'type-A' category, from actuator 815, is shown on an SEM micrograph in Figure 3. This wear scar was produced when the power cable overlapped the edge of the Ultem-1000 strain relief and was fretted at the location of high contact stress. The SEM micrograph of the power cable shows the wear scar to be long and narrow and surrounded by fine particles of Kapton. Figure 3, Photo 1 shows loose fretting particles that formed in the wear scar which were dispersed by both mechanical motion and the ensuing air turbulence and were then redeposited a short distance away. These particles settled on an area of the power cable that was never in contact with the strain relief. Figure 3, Photo 2 reveals the pitted microtopography of the fretting scar resulting from the scouring action of the harder strain relief against the softer power cable. Notice how few particles are in the wear scar proper. Large agglomerates of compacted fretting particles found outside of the wear scar are examined in Photo 3 of Figure 3. Their morphology is plate-like with a typical size range of 1-10 μm. EDS analysis identifies a trace of copper observed on both the clusters of particles and the unworn area free of particles, see Photos 4 and 5 of Figure 3. Small concentrations of copper ions in the Kapton from prior immersion in the copper plating baths provide a unique EDS signature. This confirms that the particles are Kapton and not Ultem-1000. Notice that both the border of the wear scar and the large clusters of particles show distinct scalloping with a periodic spacing of 0.11 mm.

The strain relief fretting wear debris of a 'Type-A' category, from actuator 815, is shown in Figure 4. Numerous vertical striations evident on the surface were, in fact, impressions made by the injection mold, that were created by the cutting tool during the machining of the mold. Striation periodicity of 0.11 mm matches the scalloping periodicity that was on the mating power cable wear scar seen previously in Figure 3. Evidently, the Ultem-1000 strain relief wears material off the Kapton power cable for three reasons: (1) The high points on the ridges, seen as striations, cause contact stress concentrations when in contact with the power cable. (2) Mechanical misalignments between the two parts cause the power cable to contact the strain relief edge, which also causes high contact stresses. (3) The Ultem-1000 is harder and stiffer than Kapton, consequently the weaker material will wear, see Table IV. Both Ultem-1000 and polycarbonate have common properties that make them good choices for this application. For example, they are low in outgasing, have no abrasive filler, contain no blooming fire retardants or plasticizers, are thermally stable, and are nonreactive.

SEM examination of clusters of particles on the Ultem-1000 strain relief shows that they are made of Kapton coming from the power cable. They have the same platelet shape, a similar 1-10 μm size range, see Photo 1 of Figure 4, and they have the identical copper peak seen in the EDS spectrum, see Figure 4, Photo 2. What appeared to be a wear scar, in Photo 3 of Figure 4, was proven by SEM/EDS analysis to be compacted particles of Kapton creating an irregular surface which gives the illusion of material removal. Ultem-1000 has no copper peak, see Figure 4, Photo 4, providing a means to distinguish particles of the two plastics.

Further understanding of the fretting problem increased when 12 actuators from six life-test HDAs were disassembled and visually examined for power cable fretting. All 12 actuators were built with Ultem-1000 strain reliefs. All nine power cables showed severe fretting wear after extensive customer seek-pattern testing amounting to 179 to 326 million seeks, see Table V.

Actuator 648 was extensively analyzed to better understand the fretting mechanism. This examination uncovered a second and a previously unknown problem with the design of the injection mold used to manufacture the strain relief. Specifically, the problem was with the two ejector pins which were located so they could upset the still soft Ultem-1000 on a critical surface. In fact, the deformation of the strain relief was so large that it prevented the tang teeth from engaging the former, resulting in the absence of a snap fit. This allowed the loose strain relief to slide against the power cable during every seek, thereby inducing fretting wear. Figure 5 is an SEM micrograph of the strain relief clearly showing the ejector pin impression and the high spot raised in the upper left corner. Fretting-wear debris located there, shown in Photo 1B of Figure 5, has the typical structure of compacted particles. Details of these particles, seen in Figure 5, Photo 1A, have the familiar thin, platelet morphology.

Table IV. Strain Relief Material Properties.

Material	Electrostatic Potential, Volts	Measured Shore Hardness, D scale
Polycarbonate, Carbon Filled	2000	82
Ultem-1000	2100	87

Table V. Actuators from HDA Life Test Showing Fretting Incidence.

Millions of Seeks	Actuator Serial Number	Strain Relief Material	Power Cable Fretting
179	648 647	Ultem-1000 Ultem-1000	Yes Yes
326	650 649	Ultem-1000 Ultem-1000	Yes Yes
366	631 630	Ultem-1000 Ultem-1000	Yes Yes
320	607 606	Ultem-1000 Ultem-1000	Yes Yes
355	672 671	Ultem-1000 Ultem-1000	Yes No
310	593 592	Ultem-1000 Ultem-1000	No No

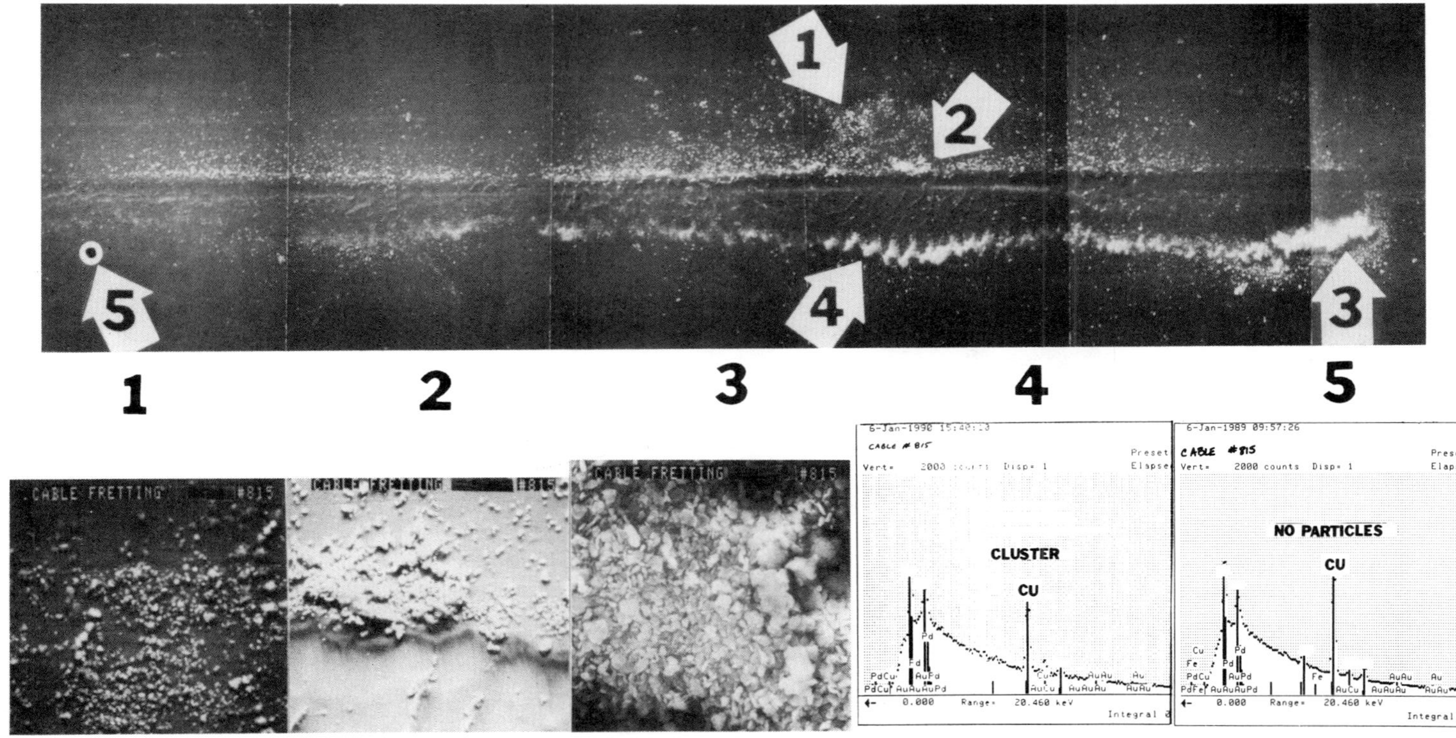

Figure 3. SEM micrograph (top) of the fretting wear scar on the power cable where it was in contact with an Ultem-1000 strain relief, the numbered arrows correspond to the numbered photos (bottom), showing: (Photo 1) loose particles of Kapton, (Photo 2) pitted wear scar, (Photo 3) large agglomerates of particles, adjacent to the wear scar, (Photo 4) EDS plot of particles showing copper present, and (Photo 5) EDS plot showing presence of copper in unworn area of the power cable.

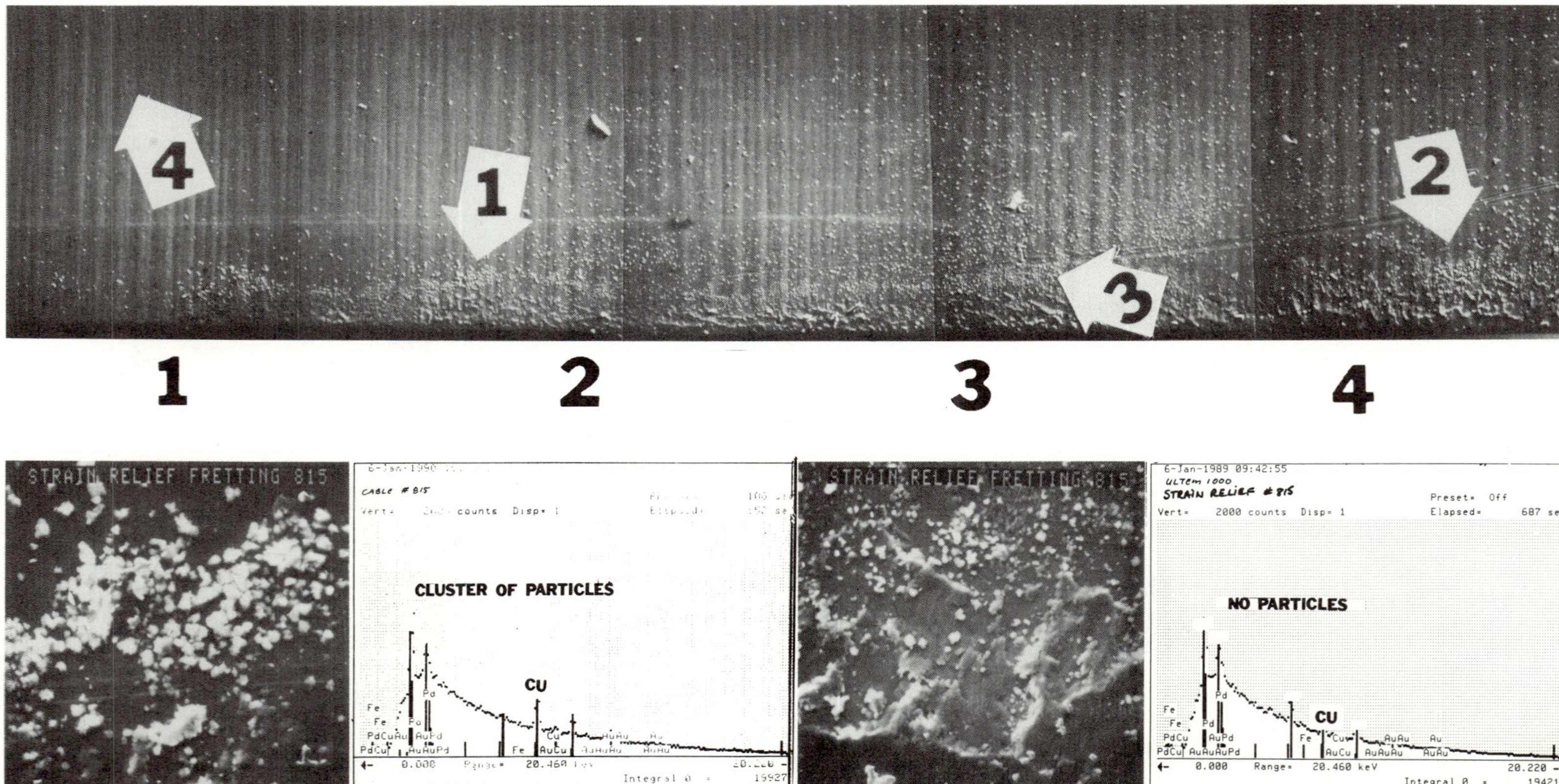

Figure 4. SEM micrograph (top) of the fretting wear debris on the Ultem-1000 strain relief, the numbered arrows correspond to the numbered photos (bottom), showing: (Photo 1) clusters of Kapton particles, (Photo 2) EDS plot of particles showing copper present, (Photo 3) compacted Kapton particles, and (Photo 4) EDS plot of Ultem-1000 showing no presence of copper.

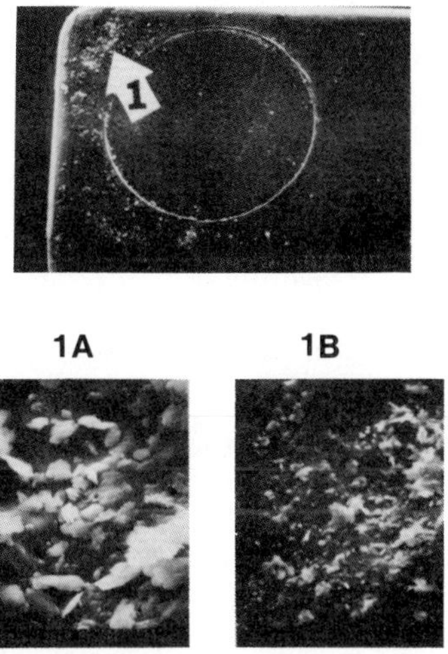

Figure 5. SEM micrograph (top) of the strain relief having a circular ejector pin mark and a raised fretted area in the upper left corner, with detailed photos at higher magnification (bottom), showing: (Photo 1A) the platelet particles of Kapton coming from the power cable, and (Photo 1B) the pitted fretting scar with compacted particles.

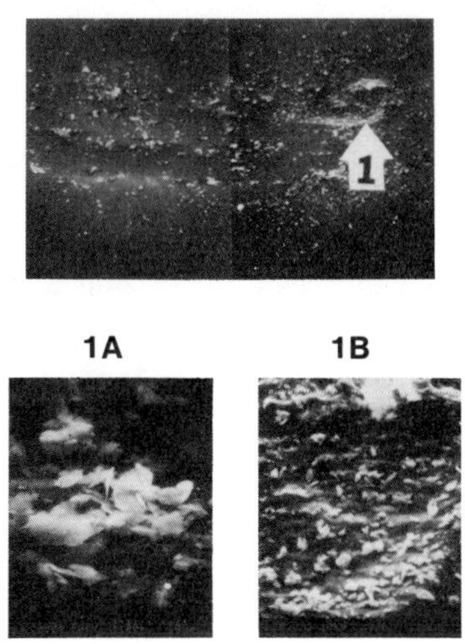

Figure 6. SEM micrograph (top) of the fretting wear scar on the power cable, with detailed photos at higher magnification (bottom), showing: (Photo 1A) the platelet particles of Kapton at 1000X and (Photo 1B) the pitted wear scar.

An SEM micrograph of the power cable fretting wear scar shows one dominant pit surrounded by loose particles, see Figure 6. Photo 1B reveals an area in the wear scar where the particles are produced by the sliding motion of the strain relief. Mechanical ejection of these loose particles allows them to be redeposited around the periphery of the wear scar. They are most likely held in place by electrostatic charges. An agglomeration of these platelet particles is shown in detail in Photo 1A, Figure 6.

Strain reliefs made of carbon-filled polycarbonate were substituted for the Ultem-1000 strain reliefs on 12 actuators, which were then placed on the actuator-access wear-tester. Four actuators, in Table VI, were evaluated for particle count and fretting at 26 to 29 million of seeks (more than three times the number of seeks needed to cause fretting with an Ultem-1000 strain relief). Eight other actuators were accessed from 214 to 242 million seeks. One of these four actuators (89) showed slight fretting at 29 million seeks. The eight actuators that were accessed over 200 million seeks never showed any indications of fretting wear.

An SEM micrograph of the power cable fretting-wear scar of the category 'type-B', from actuator 89, is shown at the top in Figure 7. This wear scar is located on the edge of the power cable where it contacted the flat area of the polycarbonate strain relief. The wear scar is short and very shallow compared to the earlier micrographs of the Ultem-1000 induced-wear scars. The bottom photo shows a higher magnification of the wear particles in the scar.

The polycarbonate strain relief fretting wear debris, at the top of Figure 8, matches up with the power cable wear scar seen earlier in Figure 7. As with the Ultem-1000, the carbon-filled polycarbonate strain relief shows the predominant ridged surface texture created by the injection mold. In the case of the polycarbonate strain relief, the rough surface texture is believed to be the primary cause of the fretting wear. Photos 1 and 2 at the bottom of Figure 8 show compacted Kapton particles and loose Kapton particles, respectively.

Table VI. Actuator Access Wear Test Results for Fretting Power Cable.

Actuator Serial Number	Millions of Seeks	Strain Relief	Visible Fretting
124	29	Polycarbonate (Black)	No
125	29	P	No
88	29	P	No
89	29	P	Slight
830	216	P	No
831	216	P	No
832	214	P	No
833	214	P	No
826	242	P	No
827	242	P	No
814	242	P	No
815	242	P	No

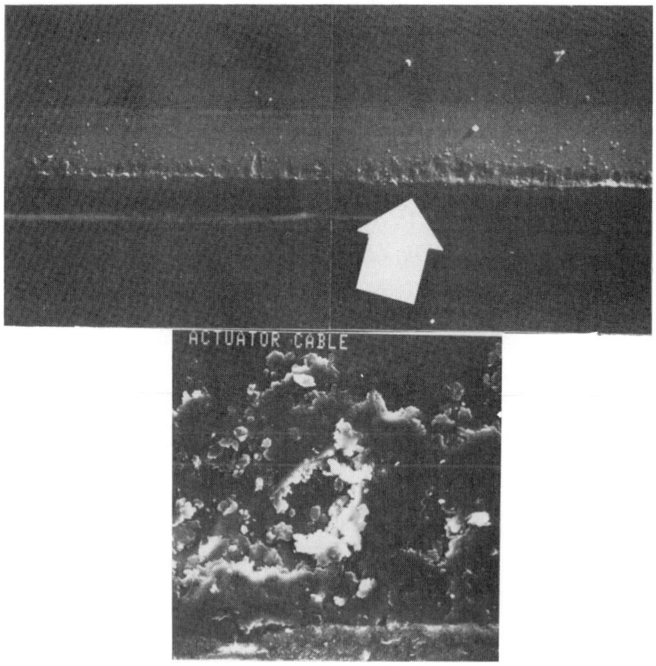

Figure 7. SEM micrograph of the cable fretting scar against polycarbonate strain relief at 50X (top), showing details of the compacted particles of Kapton in the pitted area at 500X (bottom).

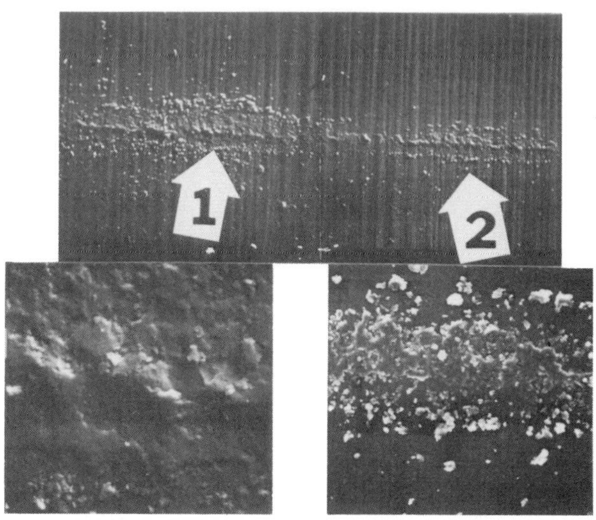

Figure 8. SEM micrograph of the fretting debris on polycarbonate strain relief at 50X (top), the numbered arrows correspond to the higher magnification photos directly below, showing: (Photo 1) compacted Kapton particles, and (Photo 2) loose Kapton particles.

Another life test actuator, 814, also had a loose strain relief caused by deformation from the ejector pin impact. The strain relief had such a loose fit with the former that it began to slide back and forth with every seek. As a consequence of the sliding motion, the strain relief tang was abraded by the aluminum former, see Figure 9, generating a mass of polycarbonate fibers and some fine aluminum particles, see Figure 10.

A third fretting wear problem was discovered when two actuators were observed to have minute wear scars on the power cable where it contacts the far edge of the strain relief land. SEM micrographs revealed the complete absence of a blend along the 0.5 mm radius and the strain relief land, see Figure 11. The sharp edge produces high contact stresses in the power cable which promotes fretting wear, see Figure 12.

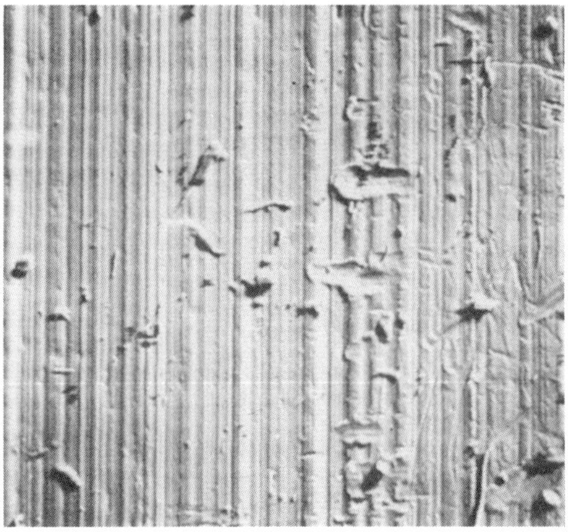

Figure 9. Abrasive wear scar on the aluminum former caused by the sliding of a loose strain relief.

Figure 10. Mass of polycarbonate fibers and aluminum particles from the former abrasion caused by the sliding of a loose strain relief.

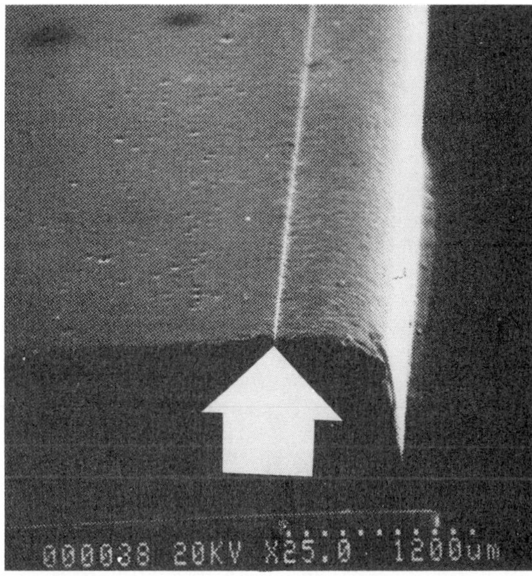

Figure 11. Strain relief land showing the sharp edge where the radius on the end does not blend.

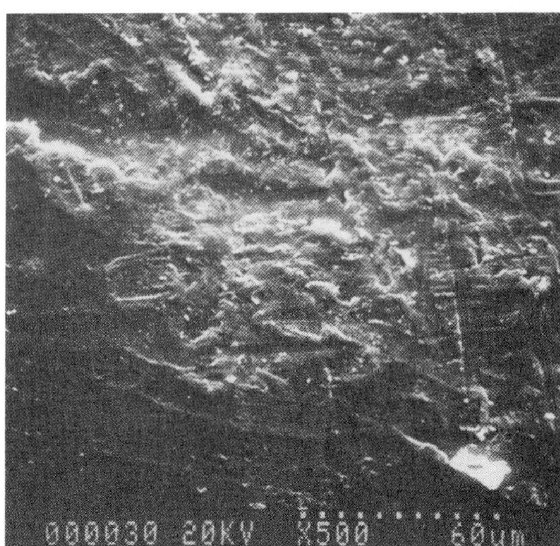

Figure 12. Fretting wear scar on the power cable caused by the lack of a blend radius on the strain relief land.

Six dimensional tolerance analyses were performed to assess the misalignment of the power cable to the strain relief and the snap-fit engagements of the strain relief onto its mating part, the former. The results of these tolerance analyses were summarized earlier in Table II, but are worth repeating and examining in more detail. Three of the tolerance analyses indicate there are dimensional incompatibilities between the parts.

First, the tolerance analysis of the side clearance between the strain relief-tooth and the former shows a statistical clearance of $+0.37 \pm 0.093$ mm. This clearance should be loose, as it is, to allow the tooth to engage the former properly.

Second, the tolerance analysis of the side clearance between the strain relief tang to the former shows a statistical clearance of $+0.42 \pm 0.145$ mm. Ideally, these parts should be designed to have an interference, which would hold the strain relief firmly against the former. Since these parts always have a loose fit, the strain relief is not restrained from rocking side-to-side relative to the former, and will rub the cable.

Third, the snap-fit engagement between the strain relief tooth and the former is shown to be statistically -0.105 ± 0.124 mm. The overlap of the two parts is required to hold them firmly together, otherwise the loose strain relief could fret against either the former or the cable. This analysis shows sometimes the two parts could become loose.

Fourth, the power cable is retained between the strain relief and the former by the snap fit engagement between the two parts. Power cable retention is statistically -0.143 ± 0.069 mm, which shows it is always firmly held in place. Any looseness would cause fretting between the power cable and the strain relief.

Fifth, the edge overlap between the power cable and the strain relief is statistically 2.705 ± 1.304 mm, which indicates that the cable will be abraded by one of the two edges of the strain relief. These 22 contributing tolerances are categorized into three general groups: (1) former to strain relief lateral misalignment, (2) power cable to strain relief angular misalignment, and (3) strain relief to former angular misalignment.

Sixth and finally, a dimensional tolerance analysis was made to determine the alignment of the cable connections at each end. The statistical alignment is 0 ± 0.644 mm, which shows that there is no excessive power cable twist against the strain relief.

CONCLUSIONS

The material incompatibility between the Kapton cable and the Ultem-1000 strain relief was the main contributor to the fretting wear problem. By changing the strain relief material to carbon-filled polycarbonate the fretting wear was basically eliminated. The findings from this work were implemented prior to shipment of the product to any customer.

Five mechanical features collectively represent the secondary causes responsible for the fretting wear of the power cable. The first mechanical cause of fretting is a rough surface on the strain relief in contact with the cable. This ridge-like surface finish of Ra 0.39 μm is caused by machining marks in the injection mold that exacerbates the wear by causing high point-contact stresses along the ridge tops. The second mechanical cause is the improper location of the ejector pins of the injection mold. The impact of these pins deforms the strain relief and thereby prevents it from properly seating in the former. The third mechanical cause is the lack of a blend radius on the end of the strain relief land. The fourth mechanical contributor to fretting is caused by improper dimensions and tolerances of the strain relief that result in misalignments severe enough to force the cable to contact the sharp edge of the strain relief. The fifth, and final, mechanical cause of fretting is the sloppy snap-fit of the strain relief and former which allows the strain relief to move against the cable.

Greater attention should be paid to the design details, and extensive life-testing of model parts must be done to prevent plastic parts from becoming reliability problems. Simple plastic parts can have problems that are not obvious to the experts during a design review and can only be determined through proper testing.

REFERENCES

1. D. Klaffke, M. Woydt, and K. H. Habig, in "Engineering Materials for Advanced Friction and Wear Applications," F. A. Smidt and P. J. Blau, editors, ASM International Conference Proceedings, pp. 71-78, American Society of Metals, Metals Park, Ohio, 1988.
2. Y. Berthier, L. Vincent, and M. Godet, presented at ASME Tribology Conference, paper no. 87-TRIB-35, 1987.
3. P. J. Blau, "Friction and Wear Transitions of Materials," pp. 120-128, Noyes Publications, New Jersey, 1989.

SCANNING UV LASER REMOVAL OF CONTAMINANTS FROM SEMICONDUCTOR AND OPTICAL SURFACES

T.J. Magee and C.S. Leung

STI/XMR Technology Group, XMR Inc.
5403 Betsy Ross Drive
Santa Clara, California 95054

The cleaning of semiconductor and optical surfaces by a scanning UV laser cleaning process is effective in removing molecular films, chemical and metallic as well as particulate contaminants. The cleaning process utilized a pulsed excimer laser, operating at 308 nm and a nominal pulse width of 80 nsec, with a patented optical beam homogenizer to provide a "top hat" profile to ensure precise, controlled removal of contaminants. The technology has been demonstrated to be rapid and effective in removing a wide range of contaminants. This paper discusses various aspects of employing a scanning UV laser system to remove contaminants from surfaces of various substrates. Very low laser energy densities (0.05 to 0.5 J/cm^2) are effective at UV wavelengths of 193 to 308 nm in removing surface contaminants without the use of toxic or hazardous chemicals and without imparting substantial chemical or physical change at the surface.

INTRODUCTION

The increasing requirements for chemically clean, particulate-free surfaces in the semiconductor industry has provided the impetus for the development of improved procedures for control of semiconductor surfaces, both prior to device line insertion and as an in-situ process step. Over the past five years the optics industry has also expressed a concern for control of film-type contaminants on surfaces within both ground based manufacturing facilities and space environments. Since much of the technology proposed for this area is in an early stage of development, successful deployment for removal of all film-type contaminants and particulates under a wide range of environmental conditions has not yet been demonstrated.

In this paper we will present an overview of scanning UV laser cleaning technology developed over the last ten years and deployed for removal of film contaminants and particulates on semiconductors, thin film metal coatings, ceramics, optical surfaces, and polymers.

Particles on Surfaces 3, Edited by K.L. Mittal
Plenum Press, New York, 1991

HISTORICAL DEVELOPMENT

In the early 1980's considerable attention was devoted to the use of IR and other lasers for removing organic film contaminants from the surfaces of semiconductors and other materials.[1-4] Removal of particulates was accomplished by vaporization, often producing damage at the surface. In-situ Auger electron spectroscopy within a high vacuum chamber showed that effective removal of hydrocarbon contamination was not accomplished until the surface was exposed to laser energy densities that produced melting of the near surface layer.[1-4] This was subsequently deemed unacceptable for the majority of semiconductor device applications.

In the period 1981-1982, research at XMR resulted in the discovery of a process step that proved effective in reducing film contamination and particulates on a surface prior to ion implantation or laser doping. It was found that when a surface was exposed to lower UV laser energy densities, considerably below the threshold level for laser induced melt, sweeping of the surface with the laser beam prior to doping produced higher electrical activation and more consistent control of sheet resistivities. Subsequent evaluation using Auger electron spectroscopy scanning electron microscopy, and transmission electron microscopy showed that hydrocarbons and particulates were effectively removed at low energy density exposures without producing resolvable damage at the surface.

Since this time samples subjected to UV laser cleaning have included Si, Ge, CdTe, GaAs, InSb, InP, HgCdTe, fused silica, sapphire, Al, Be, reflective optical coatings, SiO_2 layers on Si, metal silicide thin films, Al on Si, bulk metal surfaces, thin metallic films on metal substrates, and polymers.[5-9] The laser cleaning procedures have been used as a separate process step or in conjuction with discrete processes in sub-modular systems.[5,10]

PARTICULATES/FILM CONTAMINANTS ON SURFACES AND REMOVAL BY UV LASER CLEANING

Particulates are bound to semiconductor surfaces by capillary forces or electrostatic interaction between charges existing on the surface and the particulate. The particles are generally tightly bound to surfaces and for submicron sized particulates where the charge per unit area is high, physical removal by wetting or chemical cleaning is difficult.[11]

For impurities in the form of hydrocarbon residues or metallics bound directly to surfaces, the bonding can be attributed to :(1) dipole-dipole interaction; (2) direct overlap of d-orbitals of metallics with hybridization orbitals; and (3) binding of d-orbitals to delocalized orbitals of adsorbed organic molecules on surface sites.

For simply bound states (physisorption) where charge interaction is the dominant mechanism, some of the bound impurities can be removed by thermal desoprtion or heating. However, not all of the impurities are removed and some hydrocarbons may actually be converted to alternate species. For more complexly bound impurities (chemisorption) removal cannot be accomplished by conventional procedures.

Removal of hydrocarbon trace residues using ozone or UV lamp exposures has been shown to be generally effective,[12] although long term exposures are usually not desirable and microparticulates are retained on surfaces after exposures. Many of the identifiable surface complexes and hydrocarbons are strongly absorbing in the ultraviolet wavelength range of commercially available UV lasers. Molecular dissociation or bond breaking is achieved at relatively low fluences in the 193 nm to 308 nm wavelength range. In additon, since the latent heat of vaporization is typically in the range, 40 - 100 kJ/mole, sufficient energy density is available for volatilization of the minute quantities present on surface. Even for high molecular weight organic structures, energy is available to break bonds, giving smaller fragments that can be photothermally vaporized.

Particulates are removed from surfaces by a photo-acoustic effect.[5] Absorption of energy within a shallow region of a substrate produces a stress wave that is reflected from the back to the front surface. Measurements of the magnitude of the induced stress wave have been reported using "Sandia-type" quartz gauges[13-15] for metal[16-18] and thin film metal or SiO_2 coated semiconductors.[19] At high laser energy densities, momentum transfer from the ejected plasma creates the stress wave within the material.[18] At the extremely small energy densities required for laser cleaning, the rapid expansion and contraction of the surface produces a low level stress wave responsible for the removal of particulates.[5,8,16,19] The acoustic mismatch between the particulate and substrate results in forward momentum transfer to the particle. Since the electrostatic and capillary binding energies of the particle are small compared to the energy imparted by the acoustic wave, particles are ejected at relatively high velocities from the surface.

Table I summarizes the laser removal mechanisms for transparent and absorbing particles on surfaces. Although vaporization of particles appears the most intuitive mechanism, it is the least desirable because of damage induced in the underlying material. The photoacoustic ejection by short pulse width lasers is the most efficient method and produces no damage in the substrate. For extremely fine particles or complexes bound at the surface, a bond breaking or dissociation is the prevalent mechanism and may be used effectively for adsorbed "particles".

Table I. Removal of Particulate Contaminants by Laser Cleaning.

Removal Mechanism	Transparent Particles	Absorbing Particles
Vaporization	■Absorption in substrate, heat transfer to particle exceeding vaporization threshold of particle ■Potential damage or vaporization of substrate ■CW or pulsed lasers	■Sufficient absorption, heat transfer to vaporize contaminant, heat transfer to substrate ■Potential damage or vaporization of substrate ■CW or pulsed lasers
Opto-Acoustic Wave	■Pulsed laser only ■Depends on rapid thermal expansion and contraction of substrate material to generate stress wave ■Particle ejection caused by momentum transfer at acoustic mismatch boundary of particle and substrate	■Pulsed laser only ■Depends on rapid thermal expansion of material within particle to generate stress wave ■Particle ejection caused by momentum transfer at acoustic mismatch boundary of particle and substrate
Simple Bond Breaking (Electrostatic)	■Pulsed UV laser ■Photon energy exceeds bond energy	■N/A ■N/A

SCANNING UV LASER CLEANING SYSTEM REQUIREMENTS

To perform UV laser surface cleaning, a multimode excimer laser is used.[20,21] Frequency conversion applied to solid state lasers to produce a UV output is generally considered undesirable because of the interference effects produced at the surface by the single mode structure of the beam. For reproducibility in an actual process application, the "shot-to-shot" repeatability must be maintained and beam discharge variations minimized. In addition, the beam intensity profile produced by the laser at the target must be shaped or "homogenized" to be effective in a cleaning process. Focussed Gaussian beams were found to be unacceptable, producing non-uniform particulate and film contaminant removal across the beam spot at the target and thermalization or conversion of hydrocarbon contaminants at the low intensity trailing edges of the Gaussian beam profile. This phenomenon has also been noted in recent attempts to use CO_2 laser cleaning on thin film metallic mirror surfaces.[22] Thermal conversion of hydrocarbon contaminants at the lateral edges of the Gaussian beam resulted in ineffective cleaning and products that were difficult to remove in subsequent steps.

To provide control of the beam for process applications, a patented beam homogenizer[23] was developed in 1981. Figure 1 shows the beam intensity profile emerging from the laser before and after beam homogenization. Although no significant "hot" or "cold" spots exist in the intensity profile from the laser, the varying intensity is not adequate for cleaning. After homogenization, the intensity profile is flat within ±5% across the central portion of the beam with pronounced near vertical slopes at the edges.

SCANNING UV LASER CLEANING APPLICATIONS

To provide cleaning of a variety of surfaces, a XeCl UV gas discharge laser having a pulse width of 80 nsec. and operating at a wavelength of 308 nm was directed through the beam homogenizer onto the target surface. The output of the laser was measured and at the focal point of the homogenizer the beam spot dimensions were adjusted using X and Y controls on the homogenizer to produce irradiation areas of 0.06 to 1.0 cm^2. To obtain real time monitoring of microparticulate removal, the beam was directed through a UV microscope system and the surface monitored in visible light with an attached video camera. For these experiments, the spot size was 75 x 75 μm. In all cases, relative displacement and scanning of the beam was accomplished using X-Y drive stages. A lateral overlap factor of <5% was chosen for all applications.

Surfaces were exposed to laser irradiation either after wet chemical cleaning or deliberate contamination with hydrocarbon residues, fine carbon (0.14 μm diameter), SiO_2, Fe_2O_3, $CaSO_4$ or MgO particles. In separate tests, dioctyl phthalate or water/tricholoroethane layers were also placed on surfaces for removal efficiency tests.

Figure 2 shows micrographs obtained at a contaminated/laser cleaned boundary after laser cleaning at a laser pulse repetition rate of 5 Hz and energy densities of 0.3 and 0.2 J/cm^2 for beryllium and silicon, respectively. The "as-received" beryllium surface was contaminated with SiO_2 particles and solvent residues prior to laser cleaning in air. The silicon wafer was subjected to a conventional RCA clean procedure, and then contaminated with residues from a trichloroethane rinse. Fine carbon particles (0.14 μm diameter) in suspension were swept across the surface and the entire wafer allowed to dry. The wafer was then placed in a vacuum chamber and evacuated to a pressure of ≈10^{-6} Torr for laser cleaning. Subsequent evaluation in an SEM of the surfaces after cleaning showed that all resolvable particles had been removed by the laser cleaning process. In addition, we detected no damage or physical ablation of the substrate within the laser cleaned regions.

To examine the effect of multiple pulses on cleaning of thicker organic residues on silicon surfaces, we subjected a selected area to a single pulse at an energy density of 0.2 J/cm^2. In Figure 3 we show that a single pulse produces volatilization and fragmentation of the contaminant region (≤5 μm thick). A second pulse then removed all traces of contaminants and fragmentation from the irradiated region.

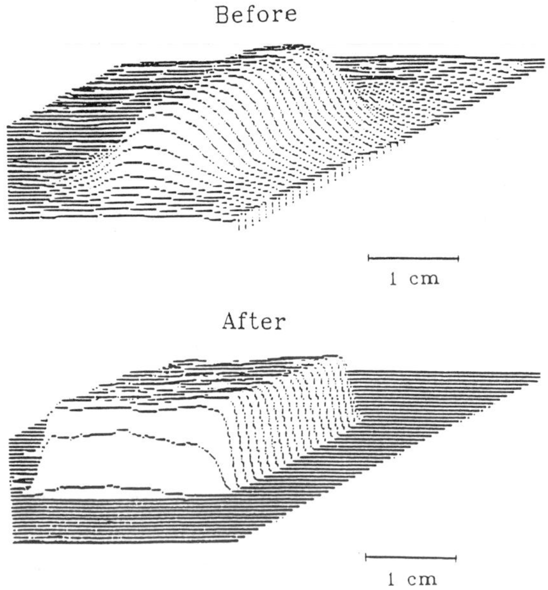

Figure 1. Laser intensity profiles before and after homogenization.

Organic plasticizers (dioctyl phthalate) have been of concern in surface contamination studies for a number of years. There is an extremely high surface affinity for collection and the contaminated regions often proved difficult to clean by conventional techniques. To test the efficiency of the scanning UV laser cleaning technique we deposited DOP layers containing carbon and silicon dioxide particulates on an Al-coated (5000 Å thick) silicon wafer and allowed the surface to age at room temperature for two weeks. The resulting layer was irregular in thickness, typically ≤40 μm. The wafer was then cleaned in air at a laser energy density of 0.23 J/cm^2 using a pulse repetition rate of 5 Hz. Figure 4 shows the removal of the DOP/particulate layer at the boundary of the laser cleaned region. The effective removal of the contaminant is attributed to dissociation, fragmentation, and volatilization of segments of the long chain molecule. Comparative tests conducted with a pulsed CO_2 laser, using conditions identified by other investigators[22] for removing organic films, show no effective or efficient removal of the 40 μm DOP layer, thereby discounting thermalization as the dominant mechanism.

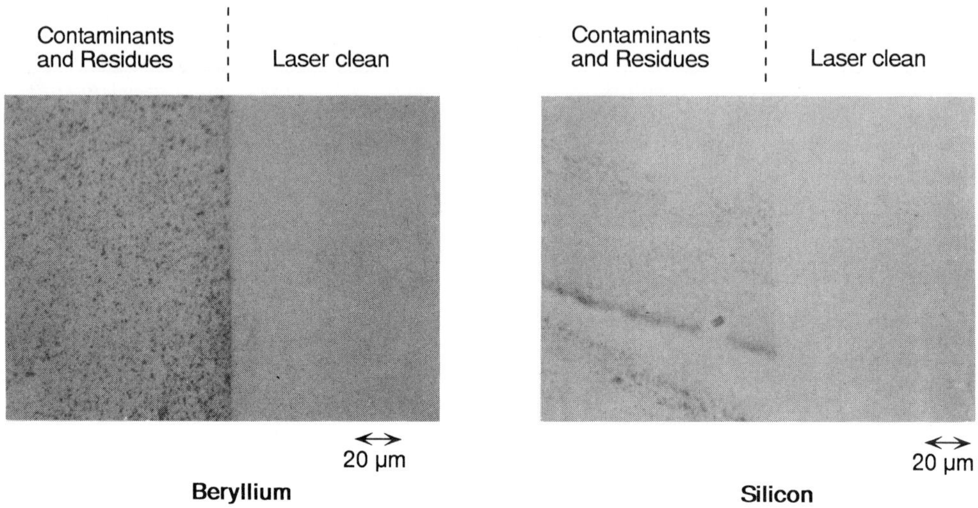

Figure 2. UV excimer laser removal of contaminants.

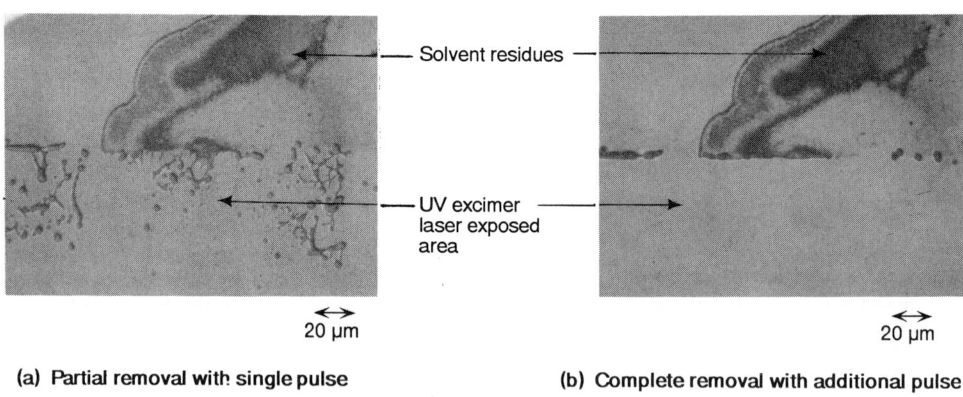

Figure 3. UV excimer laser removal of contaminants (organic residues) from silicon surface.

The use of scanning UV laser cleaning for removing contaminants on mirrors at 35 and 300K in a vacuum environment has been recently reported.[8,9] Removal of heavily contaminated regions on quartz and Al/quartz surfaces at room temperature is shown in Figure 5. Similarly, when a "superfinished" Au/Ni/Al mirror surface was contaminated with solvent residues and fine carbon particles, the UV laser cleaning effectively removes the contaminants (Figure 6). To evaluate the effectiveness of the cleaning procedure we used BRDF (Bidirectional Reflectance Distribution Function) measurements[24] to assess optical scattering of the surface at 10.6 μm. Figure 7 shows the spectra obtained, as a function of angle on the "as-received," contaminated, and after laser cleaning of the Au/Ni/Al surface.

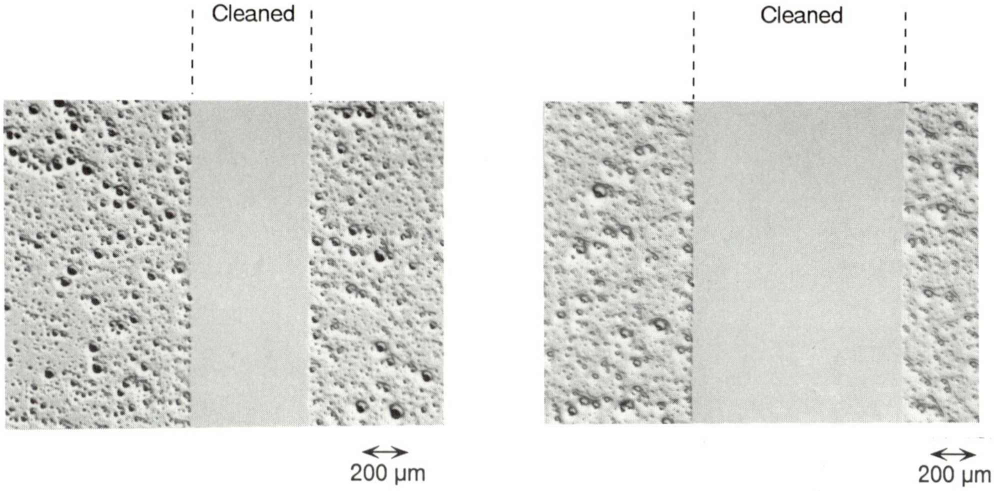

Figure 4. UV excimer laser removal of Model contaminant (dioctyl phthalate) on two areas of Al-coated silicon substrate.

After contaminating the surface, the scatter intensity increased significantly above the as-received value. Interestingly, after UV laser cleaning, scatter is reduced below that of the as-received surface. This test was then repeated three times with essentially identical results. To further evaluate the nature of the laser cleaned surface, the sample was allowed to age in a clean room (class 1000) at 300K for approximately three months. Measurements of the BRDF spectra showed that the scattering was again equivalent to the as-received value. After UV laser cleaning, the BRDF spectra were identical to the "laser cleaned" spectra of Figure 7. Although more detailed analysis is required, one can speculate that impurities within the electroplated Au layer form complexes at the surface, producing an equilibrium or passivated surface in normal air exposures. After UV laser cleaning, the complexes are removed and scattering at 10.6 μm reduced.

Figure 5. UV excimer laser removal of contaminants from optical flats.

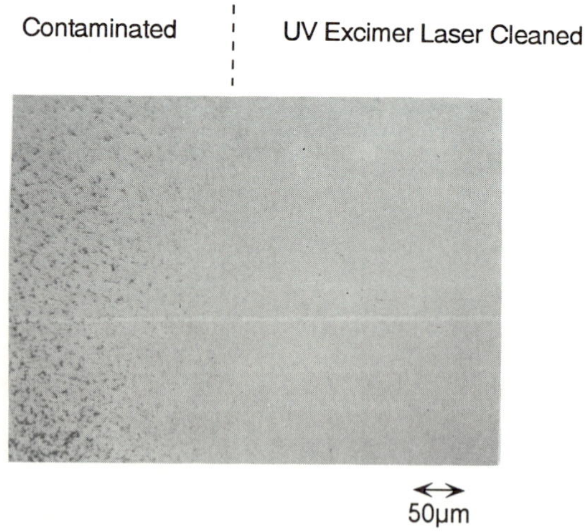

Figure 6. UV laser cleaning of superfinished Au/Ni/Al mirror.

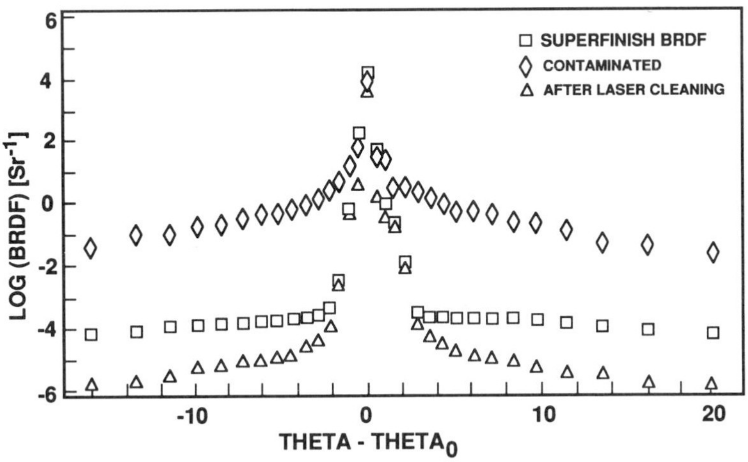

Figure 7. BRDF spectra obtained on "as-received", after contamination and after UV laser cleaning of Au/Ni/Al mirror surface.

CONCLUSIONS

In this paper we have provided overview of scanning UV laser cleaning technology developed over the past ten years. It has been shown that organic residues and particulates can be effectively removed from a variety of semiconductor and optical surfaces. Removal of microparticultes is accomplished by ejection as a result of the low level stress wave induced by the laser irradiation. Removal of organic residue traces occurs either through photodissociation or volatilization in direct heat transfer from the substrate. Although the technology is reasonably mature for discrete scanning laser cleaning applications, further development is required for potential implementation in integrated or cluster tool systems.

REFERENCES

1. D. M. Zehner, C, W. White and G. W. Ownby, Appl. Phys. Lett. 36(1), 56(1980).

2. G. W. Ownby, C. W. White and P. M. Zehner, Method of using laser irradiation for the production of atomically clean crystalline silicon and germanium surfaces, U.S. Patent No. 4,292,093, 1981.

3. A. McKinley, A. W. Parko, G. J. Hughes, J. Fryer and R. H. Williams, J. Phys. (D): Appl. Phys. 13, L193(1980).

4. D. C. Rodway, A. G. Cullis and H. C. Webber, Appl. Surface Sci. 6, 76(1980).

5. T. J. Magee and R. J. Pressley, Applications of UV lasers in semiconductor material processing, Proc. Conf. on Laser Science, Japanese Inst. Physics and Chemistry (1985).

6. T. J. Magee and C. S. Leung, DARPA/XMR Technical Reports, Contract No. N00014-86-C-0674, (1986-1988).

7. D. Worth amd K. V. Reddy, Ultraviolet laser treating of surfaces, U.S. Patent No, 4,803,021, 1989.

8. T. J. Magee and R. A. Osiecki, XMR UV Laser cleaning technology, Proc,. MSX Contamination Workshop, Appl. Phys. Laboratory, John Hopkins University, Laurel, Maryland, p. 499, 1990.

9. R. A. Osiecki and T. J. Magee, Ultraviolet cleaning of mirrored surfaces, Proc. SPIE International Symposium on Optical and Optoeletronic Applied Science and Engineering, San Diego, 1990.

10. T. J. Magee, J. F. Osborne, P. Gildea and C. S. Leung, Laser planarization of nonrefractory metal during integrated circuit fabrication, U.S. Patent No. 4,758,533, July 19, 1988.

11(a) K. L. Mittal, editor, "Particles on Surfaces 1: Detection, Adhesion and Removal," Plenum Press, New York, 1988.

(b) K. L. Mittal, editor, "Particles on Surfaces 2: Detection, Adhesion and Removal," Plenum Press, New York, 1989.

12. J. R. Vig. J. Vac. Sci. Technol. A3, 1027(1985).

13. R. A. Graham, F. W. Neilson and W. B. Benedick, J. Appl, Phys. 36, 1775(1965).

14. W. J. Halpin, O. E. Jones and R. A. Graham, ASTM Special Technical Publ. No. 336(1963).

15. G. A. Jones and W. J. Halpin, Rev. Sci. Instrum. 39, 258(1968).

16. T. J. Magee, R. A. Armistead and P. Krehl, J. Physics (D): Appl. Phys. 8, 498 (1975).

17. T. J. Magee. P. Krehl, K. Hirschberg and J. Terry, Rev. Sci. Instrum. 45(7), 907 (1974).

18. P. Krehl, F. Schwirzke and A. W. Cooper, J. Appl. Phys. 46, 4400(1975).

19. J. F. Jacobson, "Examination of Laser Produced Pressure Pulses in a Gallium Arsenide Solar Cell," M.S. Thesis, Naval Postgraduate School, Monterey, California (1976).

20. D.J. Clark and T. S. Fahlen, Gas transport laser system, U.S. Patent No. 4,611,327, September 9, 1986.

21. T. S. Fahlen and B. Mass, Electrical excitation circuit for gas lasers, U.S. Patent No. 4,549,091, October 22,1985.

22. L. G. Piper, M. B. Frisk, V. Pierce and B. D. Green, Laser cleaning of cryogenic optics, Proc. SPIE International Symposium on Optical and Optoelectronic Applied Science and Engineering, San Diego, 1990.

23. T. S. Fahlen, S.B. Hutchinson and T. McNulty, Optical beam integration system, U.S. Patent No. 4,733,944, March 29, 1988.

24. J. E. Harvey, Proc. SPIE 107, 41 (1977).

ABOUT THE CONTRIBUTORS

SHIGEYUKI AKIYAMA has been with Horiba Co. Ltd. since 1967 where he has been engaged in the research and development relating to water analysis. He graduated from Tokushima University in 1963.

LAURENCE ARCHIBALD is currently Manager of the Advanced Materials Division of Pera International, U.K., where he is involved in surface engineering, composite materials and environmental engineering. He obtained a Ph.D. in Electrochemistry at the University of Nottingham. He has been involved in all aspects of surface finishing for over 20 years and is a Fellow of the Institute of Metal Finishing. He is the author of numerous papers on subjects such as electroplating, anodising, corrosion, surface cleaning, metalworking fluids and RFI shielding.

JACK B. BARENGOLTZ is a senior physicist in the Contamination Control and Management Group in the Space Materials Science and Engineering Section at the Jet Propulsion Laboratory, California Institute of Technology. He received Ph.D. in Physics from the University of Illinois in 1970.Since joining JPL, he has conducted research and performed analysis of contamination for space and energy programs. His specialties are particulate, molecular, and microbiological contamination as it applies to spaceflight systems. He also has 14 years of experience in space radiation environments, shielding and effects; micrometeoroid impact modeling; radio-microbiology; and electrostatic discharges in materials.

J. SAMUEL BATCHELDER is with the IBM Research Division, T.J. Watson Research Center, Yorktown Hts., New York.

EDWARD J. BAWOLEK has been since 1989 a student and research assistant pursuing the Ph.D. degree in Electrical Engineering at Arizona State University in Tempe, AZ. Before joining the Ph.D. program, he had worked at Union Carbide, GTE and Intel. At Intel he managed yield improvement projects and development process monitors using automated defect detection equipment. His research interests include wafer processing and contamination detection technology.

DIANNA S. BLAIR has been with Sandia National Laboratories in Albuquerque, NM since 1985. Since 1988 she has become involved in the areas of infrared spectroscopy, microscopy/FT-IR, gas chromatography/FT-IR and thermogravimetric analysis/FT-IR as applied to materials characterization. Her research interests include the use of quantitative spectroscopy with multivariate statistical analyses for the study of materials degradation. She received an M.S. in Chemical Engineering in 1985 from the University of New Mexico.

JULIO R. BLANCO is an Assistant Professor of Physics at California State University, Northridge with 15 years experience in theoretical and experimental optics. He received a Ph.D. in Physics from The Pennsylvania State University in 1985. He has published over 15 articles in optical spectroscopy measurements, thin film technology, low scatter surfaces studies and effective medium theory.

AHMED A. BUSNAINA is an Associate Professor of Mechanical Engineering and Director of the Center for Particulate Control in Process Equipment at Clarkson University, Potsdam, New York. He is the author of over 96 publications and presentations on computational fluid mechanics, microcontamination and environmental flows and combustion. He is also the author of three software packages for fluid flow simulation on personal computers and mainframes.

BRENT D. CARLSON is an Associate Process Engineer with FSI International. During his 7 years at FSI he has worked with several types of semiconductor processing equipment including Megasonic cleaning and drying technologies. He is currently with the spray processing applications group.

ROBERT J. CHAMPETIER is a Senior Staff Scientist with 30 years of experience in optical, thermal and contamination technologies pertaining to space systems He received his A.B. in Physics from the University of California, Los Angeles in 1959. He has contributed to many unpublished government reports in experimental and analytical research for surveillance infrared sensors, low scatter optics, X-ray telescopes, nuclear and laser hardening, and metrology. He has 5 US patents and has published several articles.

ROBYN S. COLEMAN is currently a Senior Applications Engineer at KLA Instruments Corporation responsible for managing the Applications Lab and Applications Training Courses for process engineers. She has also been involved in the development of test vehicles for characterizing defect detection systems and the testing and evaluation of new products at KLA. Before she worked at Eaton Semiconductor, and Solitec, Inc. She received her B.S. degree in Chemical Engineering from MIT in 1984.

D. FEIJOO is in the School of Engineering, Duke University, Durham, North Carolina.

CINDY FOURNIER is a Defect Reduction Technician in the Yield Group at Intel Fab 6 in Chandler, Arizona and has been with Intel for 8 1/2 years. She has worked in the Yield Group for 2 years and her primary responsibility is to reduce polysilicon and polysilicon related defects on semiconductors devices. Also, she is responsible for administrating a factory wide Particle Control Program. She also worked for Siemens Inc. in Scottsdale, Arizona for 2 years and at General Semiconductors Inc. in Tempe, Arizona for 3 1/2 years.

ALICE FUKUSHIMA is currently a Senior Laboratory Specialist working in Gas Chromatography and Probe/Mass Spectrometry for the Materials Laboratory in IBM San Jose. She has an A.S. in Chemical Technology from Merritt College, Oakland, CA.

RYOSUKE FUKUSHIMA has been with Horiba Co. Ltd. since 1972. From 1972 to 1988 he was engaged in research and development relating to water analysis. Currently, he is involved in the Product Development and Product Planning Department of Horiba. He graduated from Kyoto Technical University in 1972.

ULRICH GÖSELE is a faculty member in The School of Engineering, Duke University, Durham, North Carolina.

DONALD C. GRANT is a Research Manager at FSI International where he directs research in new product development. He holds a BS degree in Chemical Engineering from Case Western Reserve University and has obtained an M.S. degree in the Particle Technology Laboratory at the University of Minnesota. He has more than 15 years experience in the analysis and purification of fluids and more than 50 publications and presentations in this area.

VAUGHN P. GROSS is a Staff Engineer/Scientist with IBM. He is presently in the Memory Reliability Site Assurance (Quality) Dept. at The Essex Junction, VT wafer/module fabrication facility and has been involved with IBM's "ESD" program since 1978. He has authored/coauthored several papers and has presented in conferences. He is Chairman of the IBM General Technology Division Defect Density ESD/ESA/EMI Task Group.

LILLIAN E. GULBRANDSEN is an Analytical Specialist at FSI International. She has a BA in Physics from St. Olaf College. She has 20 years experience in surface analysis including Scanning Auger Microscopy, X-ray Photoelectron Spectroscopy, and Secondary Ion Mass Spectrometry. She currently runs a laboratory with SIMS, Auger, SEM and FTIR capabilities.

E. DAN HIRLEMAN is Professor in the Department of Mechanical and Aerospace Engineering at Arizona State University in Tempe, AZ, where he has been since 1977. He received his Ph.D. degree in Mechanical Engineering in 1977 from Purdue University. He was a Purdue Rhodes Scholar Nominee and received National Science Foundation and Howard Hughes Doctoral Fellowships. He has 63 technical publications and 6 patents awarded/pending in laser diagnostics and related fields. He received the ASU College of Engineering Award for Significant Accomplishment in Research in 1980, was named Mechanical Engineering Professor of the Year in 1982, and is cited in Who's Who in Frontier Science and Technology and American Men and Women of Science.

PHILIP C.D. HOBBS is a Research Staff Member at the IBM Thomas J. Watson Research Center in Yorktown Hts., NY. He received his Ph.D. degree in Applied Physics from Stanford University in 1987. His current research interests include ultra-low noise optical measurements, analogue and digital signal processing, optical and atomic force microscopy, and contamination monitoring and control technology.

DIDIER HOUI is Researcher at The Institut de Mecanique des Fluides, URA-CNRS in Toulouse, France. He is Doctor of the Toulouse Polytechnic National Institute. He was recipient of the Filtration Gold Medal in 1986.

MITSUSHI ITANO has been with Daikin Industries Ltd. since 1987 and has been engaged in the development of high-quality fluoride chemicals. He received his M.S. degree in Environmental Engineering from Osaka University. Currently, he is a visiting researcher of the Faculty of Engineering, Tohoku University, where he is carrying out research on advanced wet processing for ULSI manufacturing.

ISMAIL KASHKOUSH is a Ph.D. student in the Department of Mechanical and Aeronautical Engineering at Clarkson University. He received his MS degree in Mechanical Engineering from Cairo University, Egypt in 1986.

ICHIRO KAWANABE has been with Hashimoto Chemical Co. Ltd. since 1980 and has been engaged in the development of high-quality fluoride chemicals. He received B.S. degree in Chemical Engineering from Kagoshima University. Currently, he is a visiting researcher of the Faculty of Engineering, Tohoku University, where he is carrying out research on gas-phase etching of oxide films and related process technologies.

FREDERICK W. KERN, JR. is a Staff Engineer at IBM Corporation, General Technology Division, in Essex Junction, VT., where he is responsible for process equipment design for high density memory devices. From 1974 to 1978 he worked at the Joseph C. Wilson Center for Technology of Xerox Corporation. He received a B.S. degree in Chemical Engineering from the University of Rochester, and has authored papers on equipment aerodynamics, particle transport mechanisms in cleanrooms, process equipment design and evaluation. He serves on the advisory board for the Center for Advanced Materials Processing at Clarkson University in Potsdam. Currently, he is on temporary assignment, as a visiting scholar, at the Solid State Electronics Laboratory, Department of Electronics at Tohoku University in Sendai, Japan.

JOACHIM KNOTH has been with the GKSS Research Centre since 1966. In 1967 he received his Diploma in Physics from the University of Hamburg. Total Reflection X-ray Fluorescence Spectrometry has been his research interest since 1975. As a developer of TXRF-instrument for wafer control he received an I + R 100 award from the "Research & Development" magazine in 1989.

ROBERT F. KUNESH is a Staff Engineer at IBM's General Technology Division, Essex Junction, VT. He received his MS in Materials Science from the University of Vermont. He is responsible for the design of semiconductor manufacturing equipment used to produce high-density memory devices.

V. LEHMANN is with Siemens Research Laboratories, Munich, Germany.

BRIAN LESLIE is with Tencor Instruments, Mountain View, California.

CHARLES S. LEUNG is currently with XMR Inc./Amoco Technology Company in Santa Clara, California, where he is employed as a Senior Scientist. He received a Ph.D. degree in Physical Chemistry from the University of California, Davis in 1977. His principal duties involve research and development activities in the advanced material and process characterization area and the development of new semiconductor processing techniques. He is the author or co-author of 26 journal articles and holds two patents.

DANIEL L. LLOYD is Vice President and Director of Engineering for the Cold Jet Corporation. He has extensive experience in the fields of Mechanical Engineering, Aerospace Engineering, Applied Physics, Facilities Engineering and Engineering Management. Prior to joining the Cold Jet Company, he was with The Procter & Gamble Company. He has 25 years of experience in engineering and new product development areas.

THOMAS J. MAGEE is currently with XMR Inc./Amoco Technology Company in Santa Clara, California, where he serves as Director of Semiconductor Research. He received a Ph.D. degree in Solid State Physics in 1970 and has been involved in semiconductor process research and equipment development since that time. He has authored or co-authored over 150 papers and reports and holds four patents.

CHRISTOPHER F. McCONNELL is the President and co-founder of CFM Technologies, Inc. of Lionville, PA. Before founding CFM, he worked in production and marketing for The Dow Chemical Company. He holds an MS in Chemical Engineering from Purdue University, and an MBA from Harvard University Graduate School of Business Administration. He has published several papers in the field of particle removal and holds 12 patents related to semiconductor wet processing.

JOHN McDOWELL is Senior Designer at IBM in San Jose, which he joined in 1970. He received his A.A. from Pennsylvania State University. He is presently working in the Advanced Component Development Department and has coauthored two papers on disk drive contamination.

BETTY MILLER is a Defect Reduction Engineer in the Yield Group at Intel Fab 6 in Chandler, Arizona, and her primary responsibility is to reduce polysilicon and polysilicon related defects on semiconductors devices. She has also worked as a Process Engineer at the Intel Mask Shop in Santa Clara, California, and prior to that, she was with Johns Manville in Denver, Colorado. She received her B.S. degree in Chemical Engineering at the University of Arizona in 1986.

K. MITANI is with the R&D Center of Shin-Etsu Handotai in Gunma, Japan.

KASHMIRI LAL MITTAL * is presently employed at the IBM US Technical Education in Thornwood, N.Y. He received his M.Sc. (First Class First) in 1966 from Indian Institute of Technology, New Delhi, and Ph.D. in Colloid Chemistry in 1970 from the University of Southern California. He has organized and chaired a number of very successful international symposia and in addition to this volume, he has edited 35 more books as follows: Adsorption at Interfaces, and Colloidal Dispersions and Micellar Behavior (1975); Micellization, Solubilization, and Microemulsions, Volumes 1 & 2 (1977); Adhesion Measurement of Thin Films, Thick Films and Bulk Coatings (1978); Surface Contamination: Genesis, Detection, and Control, Volumes 1 & 2(1979); Solution Chemistry of Surfactants, Volumes 1 & 2 (1979); Solution Behavior of Surfactants: Theoretical and Applied Aspects, Volumes 1 & 2 (1982); Adhesion Aspects of Polymeric Coatings, (1983); Physicochemical Aspects of Polymer Surfaces, Volumes 1 & 2 (1983); Surfactants in Solution, Volumes 1, 2 & 3 (1984), Adhesive Joints: Formation, Characteristics, and Testing (1984), Polyimides: Synthesis, Characterization and Applications, Volumes 1 & 2 (1984); Surfactants in Solution, Volumes 4, 5 & 6 (1986); Surface and Colloid Science in Computer Technology (1987); Particles on Surfaces 1: Detection, Adhesion and Removal, (1988); Particles in Gases and Liquids 1: Detection, Characterization and Control (1989); Particles on Surfaces 2: Detection, Adhesion and Removal (1989); Surfactants in Solution, Volumes 7-10 (1989); Metallized Plastics 1: Fundamental and Applied Aspects (1989); Polymers in Information Storage Technology (1989); Particles in Gases and Liquids 2: Detection, Characterization and Control (1990); and Acid-Base Interactions: Relevance to Adhesion Science and Technology (1991). Also he is Editor of the Series, Treatise on Clean Surface Technology, the premier volume appeared in 1987. In addition to these books he has published about 60 papers in the areas of surface and colloid chemistry, adhesion, polymers, etc. He has given many invited talks on the multifarious facets of surface science, particularly adhesion, on the invitation of various societies and organizations in many countries all over the world, and is always a sought-after speaker. He is a Fellow of the American Institute of Chemists and Indian Chemical Society, is listed

* As the editor of this volume.

in American Men and Women of Science, Who's Who in the East, Men of Achievement and many other reference works. He is or has been a member of the Editorial Boards of a number of scientific and technical journals, and is the Editor of the Journal of Adhesion Science and Technology, which made its debut in 1987. In 1990 he received the Charles B. Dudley Award of the American Society for Testing and Materials.

MASAYUKI MIYASHITA is with Hashimoto Chemical Co. Ltd., where he is engaged in the development of high-quality fluoride chemicals. He received B.S. degree in chemistry from Ritumeikan University in Kyoto, Japan in 1984. At present he is a visiting researcher of the Faculty of Engineering, Tohoku University, where he is conducting research on gas-phase etching of oxide films and related process technologies.

KENNETH D. MURRAY joined IBM Process Engineering as a Technician in 1985. Since then, his work has focused on IC manufacturing line support and contamination control efforts in tools/processes at the General Technology Division's wafer fabrication facility in Essex Junction, VT. He is author/coauthor of technical disclosures, papers, patent applications and has presented at conferences.

R. NAGARAJAN is an Advisory Engineer at IBM San Jose, where he is responsible for contamination control on disk drive parts. He received his Ph.D. degree in Chemical Engineering from Yale University in 1986. He serves on the Editorial Advisory Board of the Journal of the IES, is Technical Advisor to IES Working Group on Revisions to MIL-STD-1246B and was recipient of the John Martin Outstanding Younger Member Award from the Institute of Environmental Sciences (IES) in April 1990. He has authored several publications and has chaired various technical sessions.

ARMAND P. NEUKERMANS is Vice President for R&D at Tencor Instruments. He holds degrees in Electrical and Mechanical Engineering from Louvain University in Belgium (1962), and the Ph.D. in Applied Physics from Stanford University. He has worked for General Electric, IBM, and Xerox in the area of magnetic recording and xerographic physics. From 1973 to 1985 he was with Hewlett-Packard, most recently as Manager of the Applied Physics Department. He is the author of 23 publications and holds 21 patents.

TADAHIRO OHMI is presently a Professor in the Department of Electronics, Faculty of Engineering, Tohoku University, Sendai, Japan. He received his Ph.D. degree in Electrical Engineering from the Tokyo Institute of Technology in 1966. He is internationally renowned and is engaged in many and varied research activities. His current interests include ultra high-speed ULSIs, Current Overshoot Transistors, LSI, HBT LSI and SOI on metal substrates, Base Store Image Sensors (BASIS), high speed flat panel displays, and advanced semiconductor process technologies, i.e., ultra clean technologies. He has 260 original papers and 190 patent applications to his credit, and has been an invited speaker in many international conferences. He received the Ichimura Award in 1979, the Tashima Award in 1987 and the Inoue Harushiga Award in 1989. He serves as the general chairman of the International Symposium on Power Semiconductor Devices and the Institute of Basic Semiconductor Technology Development (the Ultra Clean Society).

JIRI PECEN is an Electrical Engineer at Tencor Instruments, which he joined in 1979. He received his M.S. in Electronics from the Czech Institute of Technology, Prague. Until 1979 he worked for the Ore Research Institute in Prague on the development of new instruments for underground geophysical prospecting. He is co-author of several publications and patents.

WALTER PRATER is currently an Advisory Engineer at IBM in San Jose, which he joined in 1978. He received an M.S. degree in Mechanical Engineering from San Jose State University. He is registered in California as both a Mechanical Engineer and a Manufacturing Engineer. He has authored six papers on disk drive design and contamination.

REED W. ROSENBERG is a Senior Development Engineer at AIRCO Electronic Gasses, in San Marcos, California, and is responsible for analytical and corrosive gasses development. He has bachelors degree in Chemistry from the University of Arizona. From 1984 until 1988, he worked for the J.C. Schumacher Co. as an Analytical Chemist, responsible for analytical development. At present, he is a visiting researcher of the Faculty of Engineering at Tohoku University, Sendai, Japan.

PHILIPPE SCHMITZ is a Researcher at the Institut de Mecanique des Fluides, URA-CNRS, in Toulouse, France. He is Doctor of the Toulouse Polytechnic National Institute (INPT).

PAUL J. SCHUBRING is a Ph.D. candidate in the Department of Theoretical and Applied Mechanics, Cornell University. He received BSME degree from Rensselaer Polytechnic Institute.

HEINRICH SCHWENKE has been since 1975 group leader for research on Total Reflection X-ray Fluorescence Spectroscopy at GKSS. He received his Diploma in Physics from the University of Hamburg, Germany in 1965 and subsequently joined the GKSS Research Centre. His interests have been in the application of the above technique to trace element analyses mainly in the environmental field, and surface analyses. He holds several patents.

MICHAEL SOBELMAN is a software engineer with five years experience in technical software development. He received an M.S. in Physics from California State University, Northridge in 1990.

R. STENGL is with Siemens Research Laboratories, Munich, Germany.

GARVIN J. STONE is a Senior Associate Engineer working in particle analysis in the Development Materials Laboratory of IBM in San Jose. He has a BA in Aquatic Biology from the University of California, Santa Barbara, and has published several papers on contamination and automated SEM/EDX image analysis.

WILLIAM A. SYVERSON is a Staff Engineer at IBM General Technology Division, Essex Jct., VT. He received BSME and BSNS from University of Wisconsin and MBA (cum laude) from Bryant College, Smithfield, RI. He has published a number of technical reports.

T.Y. TAN is in the School of Engineering, Duke University, Durham, North Carolina. He is also affiliated with the Microelectronics Center of North Carolina, Research Triangle Park.

CHRIS TEUTSCH is a Defect Reduction Manager at Intel's Fab6 in Chandler, Arizona. Previously, he worked with Intel's corporate engineering group as a photolithography and defect reduction staff engineer. He has 10 years of experience in the semiconductor industry including 4 years spent at GTE Laboratories in Tempe, Arizona. He received his B.S. degree in Chemical Engineering, Summa Cum Laude, at Arizona State University in 1980 and his MBA degree at Arizona State University in 1984.

MONICA VARGAS is with IBM in San Jose which she joined in 1979. She has worked in the Materials Laboratory for the last 10 years, specializing in Failure Analysis of Disk Drive Components using SEM/EDX techniques. She has coauthored three papers on wear studies, superconductivity and photolithography.

NICOLAOS VATISTAS has been a research associate at the Department of Chemical Engineering of Pisa, Italy since 1981. He obtained a doctorate in Chemical Engineering from University of Pisa in 1977. He has worked on some theoretical aspects of chemical reactor mixing, on new types of electrodes in the field of electrochemical engineering and on particulate and crystallisation fouling. He is author or co-author of over 20 publications and one patent.

HWA-CHI WANG is a Senior Scientist at the Chicago Research Center of American Air Liquide, where he is responsible for particle and surface contamination research. He was a member of the research staff at the Air & Industrial Hygiene Laboratory in Berkeley, CA before joining American Air Liquide in 1987. He has authored or co-authored more than 40 technical papers dealing with contamination issues with a special focus on particle-surface interactions. He is also actively involved in the SEMI Standard Committee and co-chairs a Task Force Group on cylinder sampling techniques. He received his Ph.D. in 1984 from the University of Illinois at Urbana-Champaign.

KENNETH J. WARD began working at Sandia National Laboratories in Albuquerque, NM in 1984. The same year, he received a Ph.D. in Inorganic Chemistry from the University of Washington in Seattle. He is currently developing combined instrumentation methods for analytical chemistry applications, specifically microscopy/FT-IR and gas chromatography/FT-IR. This work particularly emphasizes quantitative spectroscopy for complex mixture characterization using multivariate statistical analyses. He also applies multivariate statistics and FT-IR to medical and process analyses problems to enhance the precision of in-situ determinations.

ULRICH WEISBROD joined the research staff of GKSS in 1987. He holds a Diploma in Physics from the TH Darmstadt (1983) and also a Doctorate (1988) from the University of Frankfurt/Main. He has been primarily inovlved with calculations for the quantitation of TXRF-data.

INDEX